Studies in Systems, Decision and Control

Volume 175

Series editor

Janusz Kacprzyk, Polish Academy of Sciences, Warsaw, Poland
e-mail: kacprzyk@ibspan.waw.pl

The series "Studies in Systems, Decision and Control" (SSDC) covers both new developments and advances, as well as the state of the art, in the various areas of broadly perceived systems, decision making and control–quickly, up to date and with a high quality. The intent is to cover the theory, applications, and perspectives on the state of the art and future developments relevant to systems, decision making, control, complex processes and related areas, as embedded in the fields of engineering, computer science, physics, economics, social and life sciences, as well as the paradigms and methodologies behind them. The series contains monographs, textbooks, lecture notes and edited volumes in systems, decision making and control spanning the areas of Cyber-Physical Systems, Autonomous Systems, Sensor Networks, Control Systems, Energy Systems, Automotive Systems, Biological Systems, Vehicular Networking and Connected Vehicles, Aerospace Systems, Automation, Manufacturing, Smart Grids, Nonlinear Systems, Power Systems, Robotics, Social Systems, Economic Systems and other. Of particular value to both the contributors and the readership are the short publication timeframe and the world-wide distribution and exposure which enable both a wide and rapid dissemination of research output.

More information about this series at http://www.springer.com/series/13304

Nabil Derbel • Jawhar Ghommam • Quanmin Zhu
Editors

New Developments and Advances in Robot Control

Springer

Editors
Nabil Derbel
Sfax Engineering School
University of Sfax
Sfax, Tunisia

Quanmin Zhu
University of the West of England
Bristol, UK

Jawhar Ghommam
Department of Electrical
and Computer Engineering
College of Engineering
Sultan Quaboos University
Muscat, Oman

ISSN 2198-4182 ISSN 2198-4190 (electronic)
Studies in Systems, Decision and Control
ISBN 978-981-13-2211-2 ISBN 978-981-13-2212-9 (eBook)
https://doi.org/10.1007/978-981-13-2212-9

Library of Congress Control Number: 2018964939

This Springer imprint is published by the registered company Springer Nature Singapore Pte Ltd.
The registered company address is: 152 Beach Road, #21-01/04 Gateway East, Singapore 189721, Singapore

Preface

The book *New developments and Advances in Robot Control* is intended to encourage technology transfer in the field of autonomous systems and robotics. The rapid development of new mechatronic systems has been strong propellers for the advancement of control systems theory and application, as well the fields of sensors, industrial processes, and computer methods. Nonetheless, distinct robotic systems with variety of profiles have still to be developed and controlled in a way that good performances are guaranteed. This book provides an update on selected topics in the field of robotics. It also offers an opportunity for researchers to present an extended exposition of their new works in all aspects of robotic systems for a wider and rapid dissemination by opening up new challenges to understanding and developing novel control strategies for such robotic systems. Hence, this book is fairly balanced between mathematical theory and practical applications of control engineering aspects for robotic systems ranging from robot manipulators to aerial vehicles. This makes the book suitable for final year undergraduates and postgraduates, lecturers, and practitioners in the areas of robotic systems and control engineering. It is then hoped that the compiled references by each authored chapter would provide insightful and valuable information about related work in the field and make up a new starting up for further research studies.

The book consists of 16 contributed chapters by scientific experts who are specialized in various topics addressed in this book, and it focuses on multidisciplinary applications in the field of robotics. This volume has been organized into the following chapters, which are regrouped into two parts.

- The first part of the book includes nine chapters on controlling robotic manipulators. The first chapter introduces a novel resource allocation framework for cloud robotics and explores the trends of robotic technology in Industry 4.0. The second chapter proposes a novel approach based on 4×2 visual approach to control a robotic manipulator. In Chap. 3, the parallel robotic machine is extensively studied and controlled with different control approaches. Real-time experimentation has been conducted to demonstrate the effectiveness of the proposed approaches. Chapter 4 addresses the modeling problem of a cable-driven parallel robot. Chapters 5 and 6 employ the backstepping techniques along with advanced robust control schemes to control the 7-DOF Annat robot manipulator. In Chap. 7, the nonlinear observer is employed to tackle the problem of fault detection and isolation for robotic manipulator. Chapter 8 proposes a novel robust technique for controlling a robotic manipulator using the Crone control formalism. Finally, Chap. 9 applies the sliding mode control technique to control the upper-extremity exoskeleton robot for rehabilitation purposes.

- The second part, containing seven chapters, covers the control of different types of vehicles and discusses in separate chapters different adopted control strategies. Chapter 10 applies Particle Swarm to design an optimal PID 2-DOF controller for a hybrid vehicle. Chapter 11 describes a digital stabilizing controller for two-wheeled robot, where the matrix fraction description theory is fully employed along with the digital PID controller for the tracking and stabilization of the two-wheeled robot. Chapter 12 proposes a path-planning algorithm based on optimized fuzzy logic controller to steer a mobile robot in its environment. Along the same spirit of path-planning, Chap. 13 offers a strategy for the navigation of mobile robots in a cluttered environment, using artificial potential function based on fuzzy logic design. Chapter 14 extends the navigation of single mobile robot to multiples agents moving in a coordinated fashion as they try to cover a spatial region in an optimal way. New sets of flocking-based algorithms are proposed to illustrate the efficacy of the strategy in covering wide area in an optimal way. Finally, the last two chapters expose the reader to new challenge in the motion control of unmanned aerial vehicle, in particular, Chap. 15 solves the tracking control problem of quadrotor-type UAV. A combination of the backstepping and the RISE techniques is employed to guarantee robust three-dimensional tracking algorithm for quadrotor-UAV. Chapter 16 addresses the problem of designing a model-based estimation algorithm for the unobservable flapping angles while a near-hover flight is considered.

We would like to acknowledge the efforts of many reviewers who devoted some of their time in providing feedback to the chapter's contributors; thus, all chapters were meticulously peer reviewed. We would also acknowledge the efforts of all authors who contributed to this book. They are among the researchers who keep enlarging the envelope on the state of the art in the field of robotics. Lastly, we would like to sincerely express our deepest gratitude to the Springer editorial staff for their continuous support, assistance, and significant improvement in the manuscript. Without their help, the book would not be published as scheduled.

Sfax, Tunisia — Nabil Derbel
Muscat, Oman — Jawhar Ghommam
Bristol, UK — Quanmin Zhu
June 2018

Contents

Chapter 1
Cloud Robotic: Opening a New Road to the Industry 4.0

Manal Aissam, Mohammed Benbrahim, and Mohammed Nabil Kabbaj

Abstract Cloud Robotics (CR) is a rising field of robotics rooted in cloud computing, cloud storage, and other Internet technologies centered around the benefits of converged infrastructure and shared services. It allows robots to benefit from the powerful computational, storage, and communications resources of modern data centers. In addition, it removes overheads for maintenance and updates, and reduces dependence on custom middleware. This chapter reviews the concept of cloud-enabled robotics with some of the most applications related to it. Also, it explores the trends of robotic technology in Industry 4.0. Since the advent of Information and Communication Technologies (ICT), economies around the world have ground dramatically as companies can compete on a global scale. The Fourth Industrial Revolution will be based on cyber-physical systems, the Internet of Things and Internet of Services.

Keywords Cloud robotics · Cloud computing · Robotization · Industry 4.0

1.1 Introduction

The principle of Cloud Computing (CC) dates back to the 1990s, where the foundations for the distribution of computing power as well as distributed data storage in networked computer systems were laid. CC is one of the dominant computing paradigm, Since past few years Robotics applications have also started to build around these paradigms. With using cloud infrastructure in conjunction with robotics, the development of the term CR from Kuffner (2010) has led to a wide array of new research interests.

M. Aissam (✉) · M. Benbrahim · M. N. Kabbaj
Faculty of Science Dhar El Marhraz, University of Sidi Mohamed Ben Abdellah, Fez, Morocco
e-mail: manal.aissam@usmba.ac.ma; mohammed.benbrahim@usmba.ac.ma; n.kabbaj@usmba.ac.ma

N. Derbel et al. (eds.), *New Developments and Advances in Robot Control*, Studies in Systems, Decision and Control 175,
https://doi.org/10.1007/978-981-13-2212-9_1

CR is a rapidly evolving field that allows robots to offload computation-intensive and storage-intensive jobs into the cloud. Robots are limited in terms of computational capacity, memory and storage. Cloud provides unlimited computation power, memory, storage and especially collaboration opportunity.

Robotic systems have brought significant economic and social impacts to human lives over the past few decades. For example, industrial robots (especially robot manipulators) have been widely deployed in factories to do tedious, repetitive, or dangerous tasks, such as assembly, painting, packaging, and welding. These preprogrammed robots have been very successful in industrial applications due to their high endurance, speed, and precision in structured factory environments. To enlarge the functional range of these robots or to deploy them in unstructured environments, robotic technologies are integrated with network technologies to foster the emergence of networked robotics.

According to current research linking robots to the Internet, CR and Automation build on emerging research in cloud computing, machine learning, big data, open source software, and major industry initiatives in the "Internet of Things", "Smarter Planet", "Industrial Internet", and "Industry 4.0".

Industry 4.0 is the current trend of automation and data exchange in manufacturing technologies. It includes cyber-physical systems, the Internet of things and Cloud Computing. So the CR will revolutionize the industry 4.0.

This survey is organized around the following essential concepts:

1. Cloud-Enabled Robotics
2. Cloud Computing
3. Industry 4.0

1.2 Literature Review

Networked Robotics has a long history but CR is a relatively recent term. "Robotics on Cloud" is a concept that utilizes data collected from robots and sensor networks by creating cloud storage to save the information and reusing it at the time of need. It uses the help of the Internet to increase a robot's capabilities by reducing on-board computation and providing services on demand. Since robots will be connected to the cloud, they can communicate with each other and exchange useful information among each other. This off board computing is called Cloud Computing (see Sect. 1.3.1). Massive parallel computation and real-time sharing of data resources are the basic services provided by Cloud Computing.

The term "cloud-enabled robotics" was presented by James Kuffner for the first time at the IEEE RAS International Conference on Humanoid Robotics in 2010 (Kuffner 2010). He was first to point out the potential of distributed networks combined with robotics, primarily to enhance the robot agents limited capabilities. CR presented a need of fast, reliable and ubiquitous network connections. The Internet infrastructure has gone through a radical change in the past 10 years in terms of bandwidth and Quality of Service, and is now able to enable to support the cloud reliably.

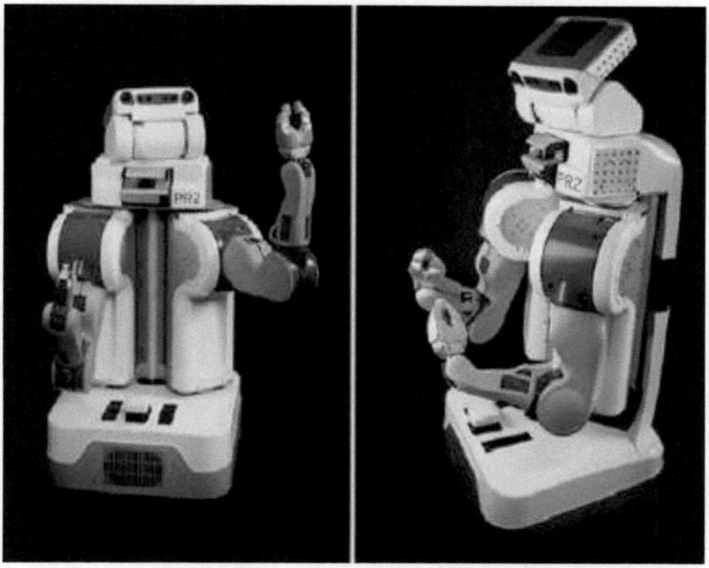

Fig. 1.1 The PR2 research robot

At the 2011 Google I/O developer's conference, Google announced a new initiative called "cloud robotics" in conjunction with robot manufacturer Willow Garage and introduced their theory and foreseen application of CR. Google has developed an open source (free) operating system for robots, with the unsurprising name "ROS" – or Robot Operating System. In other words, Google is trying to create the MS-DOS (or MS Windows) of robotics (Kohler et al. 2011).

The PR2 was the first demonstration about how to make robots smarter and more energy-efficient using these techniques. PR2 is a generic research platform widely used for cloud applications (see Fig. 1.1, Kharel et al. 2014). The PR2 has demonstrated various capabilities relying on the Robot Operating System (ROS) (Quigley et al. 2009).

In a human-centered, real world environment, the robot can encounter a lot of computation-intensive tasks. One of the most common problems is the recognition and transfer of an unknown object. This is a basic human action, yet– even nowadays–it is a complex and difficult task to be carried out by a home robot. Willow Garage's PR2 performed this task employing cloud-based technologies in the following sequence (Jordan et al. 2013):

1. **Capture data with robot:** detecting objects with a 2D camera and 3D scanner.
2. **The processing of data on cloud servers:** analysing the 2D image.
3. **The robot applies the processed data:** after proceeding, the robot applies the grasping movement.
4. **Feedback to the server:** Finally, the robot sends the results of the grasping action, so that train the server's database for further uses.

Fig. 1.2 The suckerbot

The process has online and offline elements, and it is based on both models.

The Cloud can also be used to facilitate open challenges and design competitions. For example, the African Robotics Network with support from IEEE Robotics and Automation Society hosted the "$10 Robot" Design Challenge in the summer of 2012. This open competition attracted 28 designs from around the world including a winning entry from Thailand that modified a surplus Sony game controller, adapting its embedded vibration motors to drive wheels and adding lollipops to the thumb switches as inertial counterweights for contact sensing, which can be built from surplus parts for US $8.96.

Suckerbot, designed by Tom Tilley of Thailand, a winner of the $10 Robot Design Challenge (see Fig. 1.2, Kehoe et al. 2015).

Thus, several research groups continue to explore the use of cloud technologies in robotic applications. For example, research groups Scientists in The Netherlands have created a new World Wide Web for robots. It's called RoboEarth (Waibel et al. 2011), and the corporate-sponsored collaborators who developed it describe it simply as "a World Wide Web for Robots." After four years of research, the team has arranged to host its first public demonstration of the technology: in 2014 in The Netherlands, they set four robots loose in a hospital under staged conditions. These robots became "collaboratively working together to help patients in a hospital". These robots use RoboEarth as a knowledge base, communication medium, and computational resource to offload some of their heavy computation.

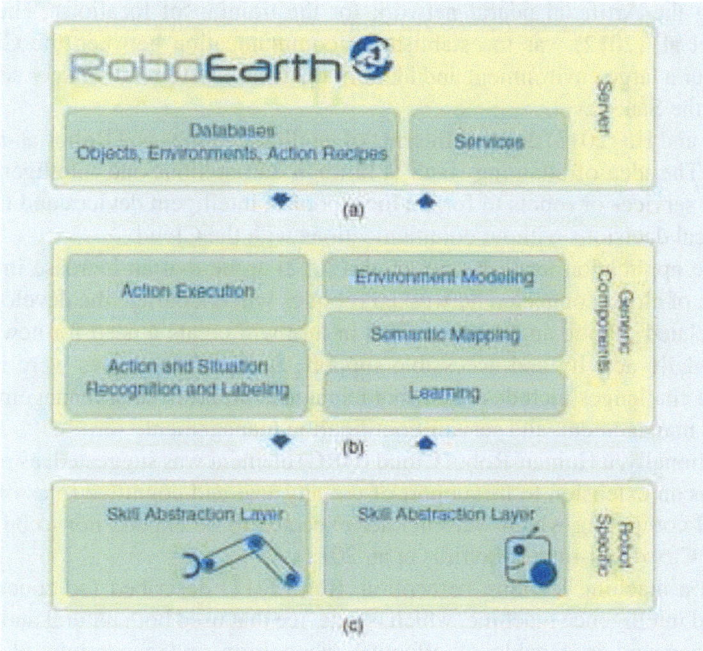

Fig. 1.3 RoboEarth's architecture

RoboEarth is implemented based on a three-layered architecture (see Fig. 1.3, Waibel et al. 2011). The core of this architecture is a server layer that holds the RoboEarth database.

In the robot Cloud centre (Du et al. 2011) designed a framework following the general Cloud computing paradigm, in order to address the current limitations in capacity and versatility of robotic applications.

The work of Mouradian et al. (2014) concentrated on the IaaS aspects of robotic applications as Cloud computing services. It suggested an architecture that enabled cost efficiency through virtualization and dynamic task delegation to robots, including the robots that might belong to other Clouds.

Gherardi et al. (2014) introduced a PaaS approach for configurable product line based-on Cloud robotics applications. With this method, robotics developers eliminated the need for low-level decision making by end users required to configure the complex architecture of distributed systems on the robot and the Cloud.

Mohanarajah et al. (2014) presented the design and implementation of Rapyuta. Rapyuta was an open source cloud robotics platform. It allowed robots to distribute complex and heavy computation overhead over secured customizable computing environments in the Cloud.

Using the Artificial neural network for the training of locations. The idea of Ansari et al. (2012) was to establish the communication between the Cloud and robot over a large environment and identify the location from the images sent by the robot at the SaaS level.

Chen and Hu (2013) debated Internet of intelligent things and Robot-as-a-Service (RaaS). The idea of attaining RaaS is through autonomous and intelligent mobile physical services or robots to form a local pool of intelligent devices and that could make local decisions without communications with the Cloud.

At the application level, Kamei et al. (2012) argue that an increase in industry adoption of cloud connected networked robotics will encourage the development of Cloud-related robotic applications. This in turn will create a need for new services such as daily activity and accessible support. Some of the issues were identified as future challenges include multirobot management, multi-area management, user attribute management, and service coordination management.

Additionally, a Human-Robot Cloud (HRC) element was suggested, as part of the Cloud, as an extension to its support of the physical and cognitive roles of humans in Cloud computing which were neither expected to be experts nor to be engaged with the Cloud full-time (Mavridis et al. 2012).

From a machine learning perception, Ren (2011) described the concept of an advanced intelligence machine, which is a device that used both natural and artificial intelligence and is capable of affective recognition and generation of affective speech and behavior.

Nakagawa et al. (2012) proposed a distributed service framework using Robot Service Network Protocol (RSNP) to integrate various devices including robots with internet services.

A robot developer should study many languages for development of firmware, to solve this problem, Jang and Kim (2013) developed a script language-based template for the source code generation and exchange.

The applications that will emerge for cloud robots are of many kinds; some are emerging now, others are at an early stage of development.

1.3 From Cloud Computing to Cloud Robotics

1.3.1 Cloud Computing

According to the National Institute of Standards and Technology (NIST), Cloud computing is a model for enabling-ubiquitous, convenient, on-demand network access to a shared pool of configurable computing resources (e.g., networks, servers, storage, applications, and services) that can be rapidly provisioned and released with minimal management effort or service provider interaction (Fig. 1.4).

This cloud model is composed of five essential characteristics, three service models, and four deployment models (Fig. 1.5).

Fig. 1.4 Cloud computing

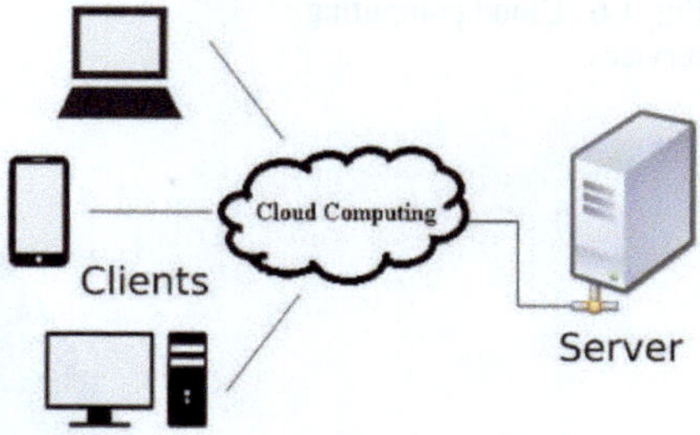

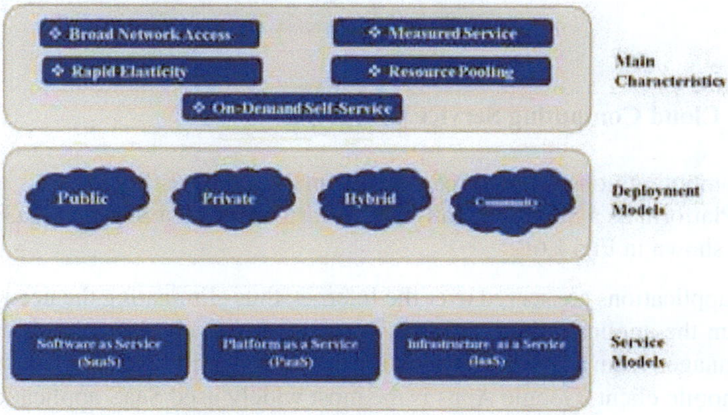

Fig. 1.5 Cloud computing infrastructure

1.3.1.1 Cloud Computing Deployment Models

Cloud computing services can be private, public or hybrid. Private cloud services are delivered from a business' data center to internal users. This model offers versatility and convenience, while preserving the management, control and security common to local data centers. Internal users may or may not be billed for services through IT charge back. In the public cloud model, a third-party provider delivers the cloud service over the Internet. Public cloud services are sold on demand, typically by the minute or hour. Customers only pay for the CPU cycles, storage or bandwidth they consume. Leading public cloud providers include Amazon Web Services (AWS), Microsoft Azure, IBM Soft Layer and Google Compute Engine. Hybrid cloud is a combination of public cloud services and on-premises private cloud with orchestration and automation between the two. Companies can run mission-critical workloads or sensitive applications on the private cloud while using the public cloud for bursting workloads that must scale on demand. The goal of hybrid cloud is to create a unified, automated, scalable environment that takes advantage of all that a public cloud infrastructure can provide while still maintaining control over mission-critical data.

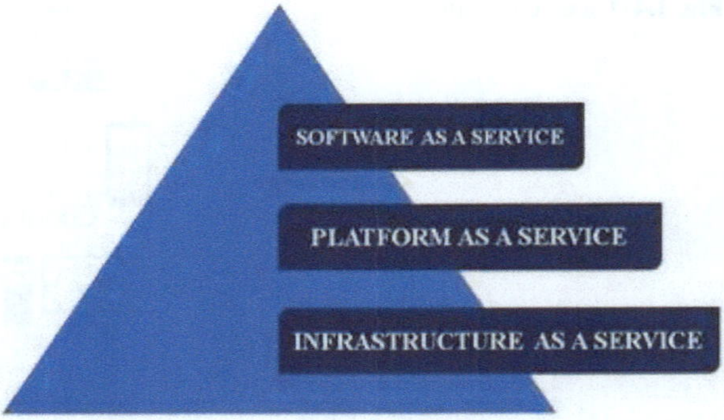

Fig. 1.6 Cloud computing services

1.3.1.2 Cloud Computing Service Categories

Cloud computing consists of three fundamental models as Software as a Service (SaaS), Platform as a Service (PaaS) and Infrastructure as a Service (IaaS) (Koken 2015) as shown in Fig. 1.6:

- SaaS applications are served over the Internet, thus eliminating the need to install and run the application on the users system (Mathur and Nishchal 2010). They are managed from a centralized location and accessed remotely by a web browser or a mobile client. Google Apps is the most widely used SaaS application suit.
- PaaS refers to a computing platform served by cloud infrastructure. PaaS offers developers to get a hold of all the systems and environments required for the life cycle of software, be it developing, testing, deploying and hosting of web applications. Some examples are Amazon Web Services (AWS) and Microsoft's Azure (Rimal et al. 2009).
- IaaS provides the required infrastructure as a service. The client doesn't need purchase the required servers, data center or the network resources. The essence of IaaS model is a pay-as-you-go financial model. Amazon and Microsoft are also IaaS providers.

1.3.1.3 Cloud Architecture

When talking about a cloud computing system, it's helpful to divide it into two sections: the front end and the back end. They connect to each other through a network, usually the Internet. The front end is the side the computer user, or client, sees. The back end is the "Cloud" section of the system. The front end includes the client's computer (or computer network) and the application required to access the cloud computing system. Not all cloud computing systems have the same user interface. Services like Web-based e-mail programs leverage existing Web browsers like Internet Explorer or Firefox. Other systems have unique applications that provide network access to clients (see Fig. 1.7).

Fig. 1.7 Cloud computing architecture

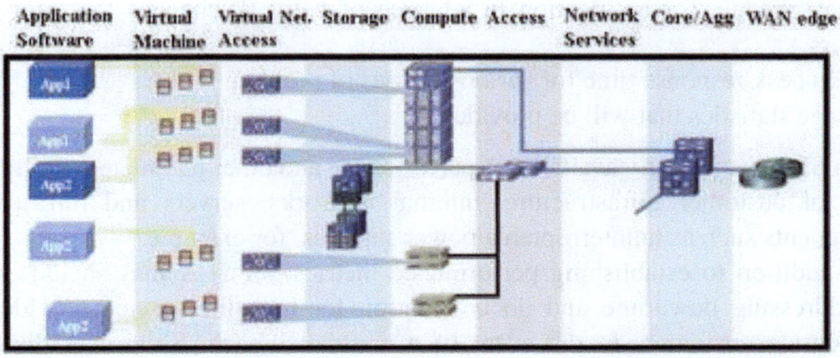

Fig. 1.8 Layers of a cloud architecture

A central server administers the system, monitoring traffic and client demands to ensure everything runs smoothly. It follows a set of rules called protocols and uses a special kind of software called middleware. Middleware allows networked computers to communicate with each other (see Fig. 1.8). A Cloud architecture can be divided into different layers (Bakshi 2011):

- *WAN edge:* are edge routers act as DC/WAN;
- *Core/Agg:* are switches that provide OSI (Open Systems Interconnection) layer 2 between compute nodes and access points;
- *Network Services:* Provides intelligence to the network via capabilities such as security firewalls;
- *Access:* This is where the network can be virtualized via VLANs and new policies and an access control list can be applied;
- *Compute:* runs the hypervisor to support virtual machines;
- *Storage:* consists of storage elements that are Storage Disk arrays, Storage Tape, Fiber Channel and Ether switches;
- *Virtual Network Access:* It acts as a virtual access layer for MVs.

1.3.1.4 Service Level Agreement

A Service-Level Agreement (SLA) is a contract between a service provider and its internal or external customers that documents what services the provider will furnish and defines the performance standards the provider is obligated to meet. SLAs establish customer expectations with regard to the service provider's performance and quality in a number of ways. Some metrics that SLAs may specify include:

- Availability and uptime, the percentage of the time services will be available.
- Specific performance benchmarks to which actual performance will be periodically compared.
- Application response time.
- The schedule for notification in advance of network changes that may affect users.
- Help desk response time for various classes of problems.
- Usage statistics that will be provided.

An SLA may specify availability, performance and other parameters for different types of customer infrastructure, internal networks, servers and infrastructure components such as uninterruptable power supplies, for example.

In addition to establishing performance metrics, an SLA may include a plan for addressing downtime and documentation for how the service provider will compensate customers in the event of a contract breach. Service credits are a typical remedy. Here, the service provider issues credits to the customer based on an SLA-specified calculation. Service providers, for example, might provide credits commensurate with the amount of time it exceeded the SLA's performance guarantee.

The SLA will also include a section detailing exclusions, that is, situations in which an SLA's guarantees and penalties for failing to meet them don't apply. The list might include events such as natural disasters or terrorist acts. This section is sometimes referred to as a force majeure clause, which aims to excuse the service provider from events beyond its control.

1.3.2 Cloud Robotics

1.3.2.1 Definition

Cloud Robotics (CR) is an emerging field within robotics, currently covering various application domains and robot network paradigms. CR was born from the merger of cloud technologies and service robotics, which was preceded by a change in paradigm in both domains. Cloud technology-based computing–or simply Cloud Computing–is one of the most dynamically growing areas of Info-Communication Technologies (ICT) (Jordan et al. 2013).

CR allows robots to take advantage of the rapid increase in data transfer rates to offload tasks without hard real time requirements. This is of particular interest for mobile robots, where on-board computation entails additional power requirements which may reduce operating duration and constrain robot mobility as well as increase costs.

1.3.2.2 Robots Classification

Robots have some constraints in terms of computational capacity, memory and storage. CR helps them to overcome these challenges. Opportunity to use cloud allows cost effective robots to be produced. Robots can be classified as traditional robots and cloud-enabled robots. Cloud technologies not only empower robots but also it allows them to network each other regardless of distance. Cloud-enabled robots are divided into two categories as standalone robots and networked robots. Classification of robots is shown in Fig. 1.9 (Koken 2015).

Robots can do a wide variety of works such as grasping, identifying objects, SLAM (Simultaneous Localization and Mapping) (Durrant-Whyte and Bailey 2006), monitoring, networking and some other actuating works. Robots can grasp formerly known objects easily. They can also grasp novel objects with the help of cloud. In Hu et al. (2012), a study about grasp planning in the presence of shape uncertainty and how cloud computing can facilitate parallel Monte Carlo sampling is presented. Standalone robots can benefit from cloud in terms of computation power, storage capacity and memory. However, networked robots can make networks, share their information through cloud and can perform collaborative works. CR infrastructure with standalone robots and networked robots is presented in Fig. 1.10.

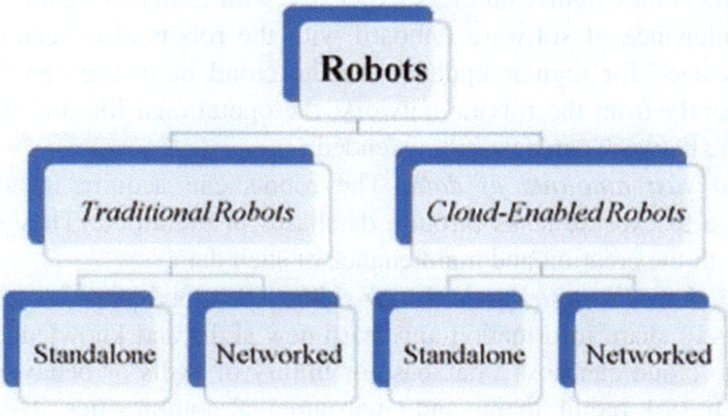

Fig. 1.9 Classification of robots

Fig. 1.10 Standalone and networked robots

1.3.2.3 Benefits and Risks

For us humans, with our non-upgradeable, offline brains, the possibility of acquiring new skills by connecting our heads to a computer network is still science fiction. It is a reality for robots.

Cloud Robotics can allow the robot to access vast amounts of processing power, data and offload compute-intensive tasks like image processing and voice recognition and even download new skills instantly, Matrix-style.

Future robotic applications will benefit from cloud robotics, which provides the following advantages over traditional networked robots (Hu et al. 2012).

- ***Ability to offload computation-intensive tasks to the cloud.*** The robots only have to keep necessary sensors, actuators, and basic processing power to enable real-time actions (e.g., real-time control). The battery life is extended, and the robotic platform becomes lighter and less expensive with easier to maintain hardware. The maintenance of software onboard with the robots also becomes simpler, with less need for regular updates. As the cloud hardware can be upgraded independently from the robotic network, the operational life and usefulness of the robotic network can be easily extended.
- ***Access to vast amounts of data.*** The robots can acquire information and knowledge to execute tasks through databases in the cloud. They do not have to deal with the creation and maintenance of such data.
- ***Access to shared knowledge and new skills.*** The cloud provides a medium for the robots to share information and learn new skills and knowledge from each other. The cloud can host a database or library of skills or behaviors that map to different task requirements and environmental complexities. The RoboEarth project (Waibel et al. 2011) is trying to turn this into a reality.

Any old robot full of dust, stored in the closets of the research labs of the world can be reused again by using the hardware and software infrastructure of the cloud.

CR has certainly benefited many enterprises by reducing costs and allowing them to concentrate on their core business competence rather than IT and infrastructure issues. But, there are still distinct disadvantages of CR; especially relating to security. Here are the top CR risks we identified:

- ***Environmental security:*** The concentration of computing resources and users in a cloud computing environment also represents a concentration of security threats. Because of their size and significance, cloud environments are often targeted by virtual machines and bot malware, brute force attacks, and other attacks.
- ***Data privacy and security:*** Hosting confidential data with cloud service providers involves the transfer of a considerable amount of an organization's control over data security to the provider. For example, every cloud contains a huge information from the clients include personal data. If a household robot is hacked, users could have risk of their personal privacy and security, like house layout, life snapshot, home-view, etc. It may be accessed and leaked to the world around by criminals. Another problem is once a robot is hacked and controlled by someone else, which may put the user in danger.
- ***Ethical problems:*** Some ethics of robotics, especially for Cloud based robotics must be considered. Since a robot is connected via networks, it has risk to be accessed by other people. If a robot is out of control and carries out illegal activities, which should be responsible for it.

1.3.3 Applications by Domain

Unquestionably, CR is a fascinating research topic, yet it has to identify its target applications and finds its particular domains. It is believed that CR is a way to introduce cognition into the domain of robotics, which is envisioned and strongly supported by the European Union through the FP calls.

The CR is used in several fields, The most important applications are in the following fields:

- **Industrial robots:** As highlighted by the Germany Industry 4.0 (MacDougall 2014) Plan "Industry is on the threshold of the fourth industrial revolution. Driven by the Internet, the real and virtual worlds are growing closer and closer together to form the Internet of Things. Industrial production of the future will be characterized by the strong individualization of products under the conditions of highly flexible (large series) production, the extensive integration of customers and business partners in business and value-added processes, and the linking of production and high-quality services leading to so-called hybrid products." In manufacturing, such cloud based robot systems could learn to handle tasks such as threading wires or cables, or aligning gaskets from professional knowledge base. A group of robots can share information for some collaborative tasks. Even more, a consumer is able to order customized product to manufacturing robots directly with online order system. Another potential paradigm is shopping-delivery robot system- once an order is placed; a warehouse robot dispatches the item to an autonomous car or autonomous drone to deliver it to its recipient.

- **Cloud medical robots:** a medical cloud (also called a healthcare cluster) consists of various services such as a disease archive, electronic medical records, a patient health management system, practice services, analytics services, clinic solutions, expert systems, etc. Health care system is a very complicated system and involved in a variety of related technologies (Ma et al. 2015). A robot can connect to the cloud to provide clinical service to patients, as well as deliver assistance to doctors (e.g. a co-surgery robot). Moreover, it also provides a collaboration service by sharing information between doctors and care givers about clinical treatment. Assistive robots: A domestic robot can be employed for healthcare and life monitoring for elderly people. The system collects the health status of users and exchange information with cloud expert system or doctors to facilitate elderly people's life, especially for those with chronic diseases. For example, the robots are able to provide support to prevent the elderly from falling down, emergency healthy support such as heart disease, blooding disease. Care givers of elderly people can also get notification when in emergency from the robot through network.
- **Autonomous mobile robots:** self-driving cars are cloud robots (Boeglin 2015). The cars use the network to access Google's enormous database of maps and satellite and environment model (like Street view) and combines it with streaming data from GPS, cameras, and 3D sensors to monitor its own position within centimeters, and with past and current traffic patterns to avoid collisions. Each car can learn something about environments, roads, or driving, or conditions, and it sends the information to the Google cloud, where it can be used to improve the performance of other cars.

1.3.4 Case Study Example

To apply CR functions in a real word scenario, the team of Sherpa project (Marconi et al. 2012) has developed a robotics platform to support search and rescue activities in hostile environments such as the alpine project. In this approach, to rescue survivor victims after avalanches, a human operator cooperates with a heterogeneous robotic system. The robotic team is principally composed of unmanned aerial vehicles (UAVs) with different characteristics and equipped with different types of sensors in order to retrieve information from the rescue scene and assist the rescuer during a mission. The Sherpa team is composed of:

- **A human rescuer:** who is a professional of the rescuing and surveillance mission.
- **Small scale rotary-wing Unmanned Aerial Vehicles:** share with rescue members visuals information of victims detected with cameras.

- **A ground rover:** is used as a transportation module for the rescuer equipment and as a hardware station with computational capabilities.
- **A fixed-wing UAV and an unmanned helicopter:** are used for constructing a 3D map of the rescuing area, as communication hub between the platforms in presence of critical terrain morphologies, for patrolling large areas not necessarily confined in the neighborhood of the rescuer.

The unmanned aerial vehicles (UAVs), have been widely used in several applications such as industrial building inspection and surveillance (Cacace et al. 2015), photogrammetry (Colomina and Molina 2014), remote sensing (Mitchell et al. 2012) and many others. In the same philosophy as Sherpa, similar works (Cacace et al. 2016) demonstrate the ability of a human rescuer to orchestrate a heterogeneous multi-robot system.

1.4 Cloud Robotics and Industry 4.0

1.4.1 Industry 4.0

The Industrial Revolution was a time in the eighteenth century when many important inventions were made. Many of these inventions made work easier and cheaper. As these inventions created new manufacturing and industry, many people also moved away from farms into cities. It was a time of very rapid change in the world.

Over the course of history, mankind has perfected its industry by not only relying on technical evolution but also by reinventing it as new resources have created new technical means. Therefore, industry has benefited from qualitative advancements which have sometimes been so ingrained in a certain time period and have had such an overwhelming impact that we have dubbed them "revolutions".Here is a quick look back in time at these first three industrial revolutions to define the contours of a fourth revolution which is taking shape right before our very eyes (Fig. 1.11).

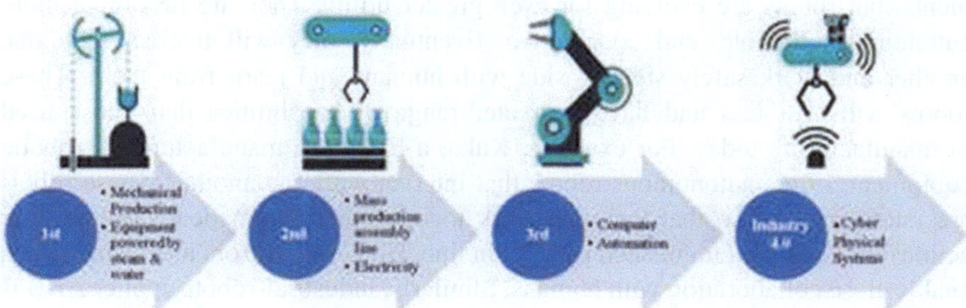

Fig. 1.11 Industrial revolutions timeline

Today we stand on the cups of a fourth industrial revolution; one which promises to marry the worlds of production and network connectivity in an "Internet of Things" which makes "Industry 4.0" a reality.

The term "Industry 4.0" originates from a project in the high-tech strategy of the German government, which promotes the computerization of manufacturing. Exactly, this term was revived in 2011 at the Hannover Fair. In October 2012 the Working Group on Industry 4.0 presented a set of Industry 4.0 implementation recommendations to the German federal government.

Therefore, some have compared Industry 4.0 with the Fourth Industrial Revolution. However, the latter refers to a systemic transformation that includes impact on civil society, governance structures, and human identity in addition to solely economic/manufacturing ramifications. The first industrial revolution mobilized the mechanization of production using water and steam power; the second industrial revolution then introduced mass production with the help of electric power, followed by the digital revolution and the use of electronics and IT to further automate production. The term "fourth industrial revolution" has been applied to significant technological developments several times over the last 75 years, and is up for academic debate. Industry 4.0, on the other hand, focuses on manufacturing specifically in the current context, and thus is separate from the fourth industrial revolution in terms of scope.

Industry 4.0 emphasizes the idea of consistent digitization and linking of all productive units in an economy. There are several technological areas that underpin Industry 4.0, which are horizontal and vertical system integration, the internet of things, cybersecurity, the cloud, big data analytics, simulation, additive manufacturing (3d printing), augmented reality, and robot (Rüssmann et al. 2015). The figure below shows the technologies related to Industry 4.0 (Fig. 1.12).

1.4.2 Robot in Industry 4.0

Manufacturers in many industries have long used robots to tackle complex assignments, but robots are evolving for even greater utility. They are becoming more autonomous, flexible, and cooperative. Eventually, they will interact with one another and work safely side by side with humans and learn from them. These robots will cost less and have a greater range of capabilities than those used in manufacturing today. For example, Kuka, a European manufacturer of robotic equipment, offers autonomous robots that interact with one another. These robots are interconnected so that they can work together and automatically adjust their actions to fit the next unfinished product in line. High-end sensors and control units enable close collaboration with humans. Similarly, industrial-robot supplier ABB is launching a two-armed robot called YuMi that is specifically designed to assemble products (such as consumer electronics) alongside humans. Two padded arms and computer vision allow for safe interaction and parts recognition.

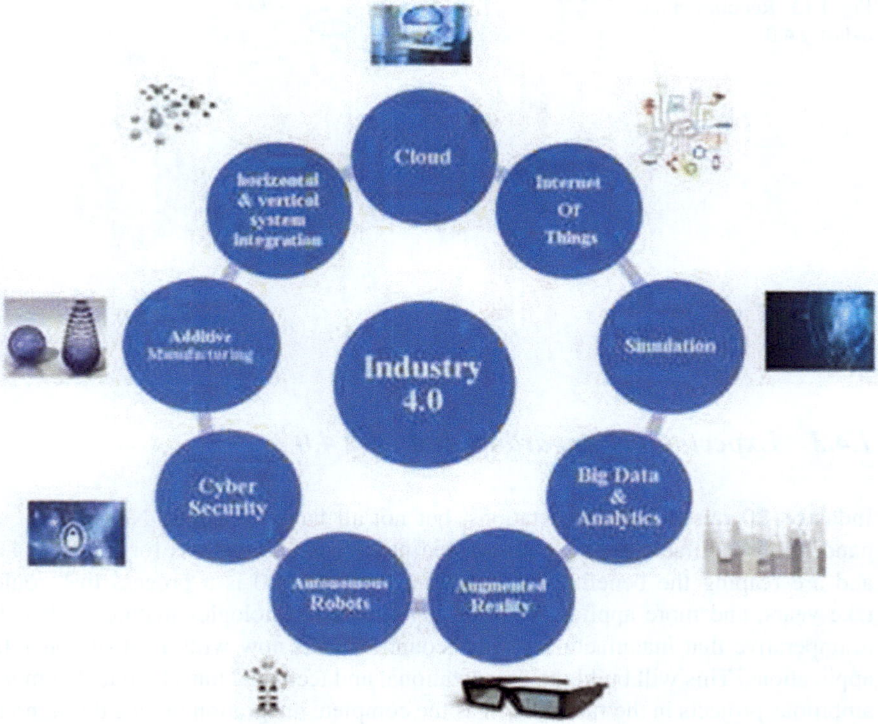

Fig. 1.12 Technologies related to industry 4.0

Industrial robots, which are one of the key drivers in Industry 4.0, have progressed greatly since the last 10 years of the twentieth century. They are becoming more productive, flexible, versatile, safer, and collaborative and are thus creating an unprecedented level of value in the entire ecosystem. Smart factories, which will be at the heart of Industry 4.0, will embark information and communication technology for an evolution in the supply chain and production line that brings a much higher level of both automation and digitization. It means machines using self-optimization, self-configuration and even artificial intelligence to complete complex tasks in order to deliver vastly superior cost efficiencies and better quality goods or services (Fig. 1.13).

With all this evolution, various companies have focused on the development of robots, such as Fanuc, Universal Robots, Gomtec, . . . (Bahrin et al. 2016).

Also, we can notice that various organizations and key players have substantial research departments which support product development and automation of manufacturing processes, while also marketing their know-how to other industrial enterprises or providing platforms for 'smart' services for third parties. This can further stimulate the implementation of Industry 4.0. And in this case, there are many organizations that are interested in the robotic industry of Industry 4.0 (Bahrin et al. 2016).

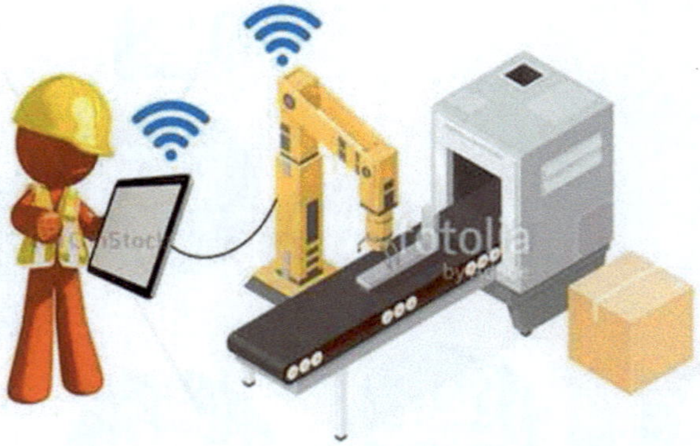

Fig. 1.13 Robotics in the industry 4.0

1.4.3 Expectations Regarding Industry 4.0

Industry 4.0 raised high expectations, but not all have been met. Nonetheless, a handful of manufacturers have discovered numerous applications for Industry 4.0 and are reaping the benefits. Implementing Industry 4.0 is a process that could take years, and more applications will develop as technologies mature further. It is imperative that manufacturers in all countries start now with a set of concrete applications. This will build the organizational and technical muscle to tackle more ambitious projects in the future, such as the complete integration of data throughout the product life cycle. Here are some expectations regarding Industry 4.0:

- **More Flexibility and Adaptability:** Fixed structures will be replaced by adaptable networks and Local Intelligence will help to handle complexity.
- **More Modularity and Autonomy:** Cyber Physical Products will be introduced in all levels of Automation also Granularity will adjust according to market needs.
- **Highest Productivity:** Resource optimization will be done in the network based swarm optimization, Plug & Produce features will bring down development cost significantly and Management will do more optimization than fire fighting.
- **New Business Models:** "app-store" and "cloud" will provide the new knowledge management, (Big Data) shall become a new driver and Open Source Approaches shall provide new opportunities.

1.5 Conclusion

Overall, the exciting field of cloud robotics has only just begun to unravel paving the way to new and bold experiments within automation and control. Having the capacity to exploit such large computing resources and allocate them to robots is an advancement. Using a cloud infrastructure to handle an ever growing and increasingly complicated set of algorithms, utilized in vision acquisition systems and formation control, is a more logical approach. In this chapter, we have also shown an approach to introduce robotic services in the cloud and in Industry 4.0, linking a real-life factory with virtual reality, will play an increasingly important role in global manufacturing. Human-robot collaboration will have a breakthrough in this period.

References

Ansari, F., Pal, J., Shukla, J., Nandi, G., & Chakraborty, P. (2012). A cloud based robot localization technique. In *Internaional Conference on Contemporary Computing*, Noida, India (pp. 347–357).

Bahrin, M. A. K., Othman, F., Azli, N. H. N., & Talib, M. F. (2016). Industry 4.0: A review on industrial automation and robotic. *Jurnal Teknologi*, *78*(6–13), 137–143.

Bakshi, K. (2011). Considerations for cloud data centers: Framework, architecture and adoption. In *IEEE Aerospace Conference*, Big Sky, MT, USA (pp. 1–7).

Boeglin, J. (2015). The costs of self-driving cars: Reconciling freedom and privacy with tort liability in autonomous vehicle regulation. *Yale Journal of Law and Technology, 17*, 172–203.

Cacace, J., Finzi, A., Lippiello, V., Loianno, G., & Sanzone, D. (2015). Aerial service vehicles for industrial inspection: Task decomposition and plan execution. *Applied Intelligence, 42*, 49–62.

Cacace, J., Finzi, A., Lippiello, V., Furci, M., Mimmo, N., & Marconi, L. (2016). A control architecture for multiple drones operated via multimodal interaction in search & rescue mission. In *IEEE International Symposium on Safety, Security, and Rescue Robotics*, Lausanne, Switzerland (pp. 233–239).

Chen, Y., & Hu, H. (2013). Internet of intelligent things and robot as a service. *Simulation Modelling Practice and Theory, 34*, 159–171.

Colomina, I., & Molina, P. (2014). Unmanned aerial systems for photogrammetry and remote sensing: A review. *ISPRS Journal of Photogrammetry and Remote Sensing, 92*, 79–97.

Du, Z., Yang, W., Chen, Y., Sun, X., Wang, X., & Xu, C. (2011). Design of a robot cloud center. In *Tenth International Symposium on Autonomous Decentralized Systems (ISADS)*, Tokyo/Hiroshima, Japan (pp. 269–275).

Durrant-Whyte, H., & Bailey, T. (2006). Simultaneous localization and mapping: Part I. *IEEE Robotics and Automation Magazine, 13*, 99–110.

Gherardi, L., Hunziker, D., & Mohanarajah, G. (2014). A software product line approach for configuring cloud robotics applications. In *IEEE 7th International Conference on Cloud Computing (CLOUD)*, Anchorage, AK, USA (pp. 745–752).

Hu, G., Tay, W., & Wen, Y. (2012). Cloud robotics: Architecture, challenges and applications. *IEEE Network, 26*, 21–28.

Jang, W.-S., & Kim, R. (2013). Template design of automatic source code generation based on script language used in cloud robot compiling environment. *CST, 27*, 184–185.

Jordan, S., Haidegger, T., Kovs, L., Felde, I., & Rudas, I. (2013). The rising prospects of cloud robotic applications. In *IEEE 9th International Conference on Computational Cybernetics*, Tihany, Hungary (pp. 327–332).

Kamei, K., Nishio, S., Hagita, N., & Sato, M. (2012). Cloud networked robotics. *IEEE Network, 26*, 28–34.

Kehoe, B., Patil, S., Abbeel, P., & Goldberg, K. (2015). A survey of research on cloud robotics and automation. *IEEE Transactions on Automation Science and Engineering, 12*, 398–409.

Kharel, A., Bhutia, D., Rai, S., & Ningombam, D. (2014). Cloud robotics using ROS. *International Journal of Computer Applications* (pp. 18–21).

Kohler, D., Hickman, R., Conley, K., & Gerkey, B. (2011). Cloud robotics. In *Google I/O 2011 Developer Conference*, Mountain View, CA, USA.

Koken, B. (2015). Cloud robotics platforms. *Interdisciplinary Description of Complex Systems, 13*, 26–33.

Kuffner, J. J. (2010). Cloud-enabled robots. In *IEEE-RAS International Conference on Humanoid Robots*, Nashville, TN, USA (pp. 1–9).

Ma, Y., Zhang, Y., Wan, J., Zhang, D., & Pan, N. (2015). *Robot and cloud-assisted multi-modal healthcare system*. New York: Springer.

MacDougall, W. (2014). *Industrie 4.0: Smart manufacturing for the future*. Berlin: Germany Trade and Invest.

Marconi, L., Melchiorri, C., Beetz, M., Pangercic, D., Siegwart, R., Leutenegger, S., Carloni, R., Stramigioli, S., Bruyninckx, H., Doherty, P., Kleiner, A., Lippiello, V., Finzi, A., Siciliano, B., Sala, A., & Tomatis, N. (2012). The sherpa project: Smart collaboration between humans and ground-aerial robots for improving rescuing activities in alpine environments. In *IEEE International Symposium on Safety, Security, and Rescue Robotics*, College Station, TN, USA.

Mathur, P., & Nishchal, N. (2010). Cloud computing: New challenge to the entire computer industry. In *First International Conference on Parallel, Distributed and Grid Computing*, Solan, India (pp. 223–228).

Mavridis, N., Bourlai, T., & Ognibene, D. (2012). The human-robot cloud: Situated collective intelligence on demand. In *IEEE International Conference on Cyber Technology in Automation Control, and Intelligent Systems (CYBER)*, Bangkok, Thailand (pp. 360–365).

Mitchell, J. J., Glenn, N. F., Anderson, M. O., Hruska, R. C., Halford, A., Baun, C., & Nydegger, N. (2012). Unmanned aerial vehicle (UAV) hyperspectral remote sensing for dryland vegetation monitoring. In *4th Workshop on Hyperspectral Image and Signal Processing*, Shanghai, China.

Mohanarajah, G., Hunziker, D., D'Andrea, R., & Waibel, M. (2014). Rapyuta: A cloud robotics platform. *IEEE Transactions on Automation Science and Engineering, 12*, 481–493.

Mouradian, C., Errounda, F., Belqasmi, F., & Glitho, R. (2014). An infrastructure for robotic applications as cloud computing services. In *IEEE World Forum on Internet of Things*, Seoul, South Korea (pp. 377–382).

Nakagawa, S., Igarashi, N., Tsuchiya, Y., Narita, M., & Kato, Y. (2012). An implementation of a distributed service framework for cloud-based robot services. In *38th Annual Conference on IEEE Industrial Electronics Society*, Montreal, QC, Canada (pp. 4148–4153).

Quigley, M., Gerkey, B., Conley, K., Faust, J., Foote, T., Leibs, J., Berger, E., Wheeler, R., & Ng, A. (2009). ROS: An open-source robot operating system. *Proceedings of ICRA Workshop on Open Source Software*, Kobe, Japan.

Ren, F. (2011). Robotics cloud and robotics school. In *7th International Conference on Natural Language Processing and Knowledge Engineering (NLP-KE)*, Tokushima, Japan (pp. 1–8).

Rimal, B., Eunmi, C., & Lumb, I. (2009). A taxonomy and survey of cloud computing systems. *Fifth International Joint Conference on INC, IMS and IDC*, Seoul, South Korea (pp. 44–51).

Rüssmann, M., Lorenz, M., Gerbert, P., Waldner, M., Justus, J., Engel, P., & Harnisch, M. (2015). *Industry 4.0: The future of productivity and growth in manufacturing industries* (Vol. 9). Boston Consulting Group.

Waibel, M., Beetz, M., Civera, J., D'Andrea, R., Elfring, J., Galvez-Lopez, D., Haussermann, K., Janssen, R., Montiel, J., Perzylo, A., Schiessle, B., Tenorth, M., Zweigle, O., & Molengraft, R. V. D. (2011). Roboearth: A world wide web for robots. *IEEE Robotics and Automation Magazine, 18*, 69–82.

Chapter 2
4 × 2D Visual Servoing Approach for Advanced Robot Applications

Mohamad Bdiwi, Jozef Suchý, and Matthias Putz

Abstract Actually, vision information can be used as sensory input in open-loop as well as in the closed-loop. However, the visual servoing approach is performed only inside the closed-loop robot control, because in the open-loop the vision sensor will represent the initial extraction of the features to generate directly the robot motion sequence and these features and motion could be off-line generated. On the contrast, closed-loop robot system uses the vision as real time sensor and it consists of two phases: tracking and control. Tracking provides a continuous estimation and update of features during the robot/object motion. Based on this information, a real time control loop will be generated. The main contribution of this work is a proposed visual servoing approach which will benefit from the images which are obtained by Kinect camera (RGB-D camera). The proposed visual servoing approach is called 4 × 2D visual servoing which combines the correspondent color and depth images to build two new images. Using these 4 images the control error signals will be calculated in order to track the objects. Firstly, this chapter will present all types of visual servoing then it will introduce the 4 × 2D visual servoing approach and the visible side coordinate system, after that it will illustrate the concept of the proposed approach and how the error signal will be calculated. In addition to that, this approach proposes a coordinate system which is called visible side coordinate system.

Keywords Visual servoing · Robot vision · RGB-D camera systems

M. Bdiwi (✉) · M. Putz
Fraunhofer IWU, Chemnitz, Germany
e-mail: mohamad.bdiwi@iwu.fraunhofer.de

J. Suchý
Technical University of Chemnitz, Chemnitz, Germany

N. Derbel et al. (eds.), *New Developments and Advances in Robot Control*, Studies in Systems, Decision and Control 175,
https://doi.org/10.1007/978-981-13-2212-9_2

2.1 Introduction

The first who used vision feedback was Y. Shirai (Shirai and Inoue 1973), he has described how a visual feedback loop can be used to correct the position of a robot to increase task accuracy. Visual servoing term has been first introduced by Hill and Park (1979). Since that time considerable efforts have been devoted to the visual control of robot manipulators, e.g. Weiss et al. (1985, 1987). Numerous papers, e.g. Hutchinson et al. (1996), Hashimoto (2003) and Corke and Hutchinson (2000), have discussed the principles of the arts of visual servoing for robot manipulators in details. Figure 2.1 illustrates the main categories of the visual servoing approaches depending on different issues. These issues are: (1) Controller architecture, (2) Camera configuration, (3) Observing end-effector, (4) Control law error.

Based on whether the control is applied to the joints directly or as a position command to a robot controller, visual servoing is divided into two types; (1) Direct servoing, (2) Dynamic look-and-move. Camera configuration includes two categories which are: (1) Camera type and (2) Camera location. Regarding the observing the end-effector, the visual servo control can be also classified whether if the end-effector is observed or not: (1) End-effector close loop and (2) End-effector open loop. Chesi and Hashimoto (2002) has presented different types of visual servoing and it has compared the stability of the visual servoing in respect to the camera configuration eye-in-hand and eye-to-hand, where Flandin et al. (2000) has

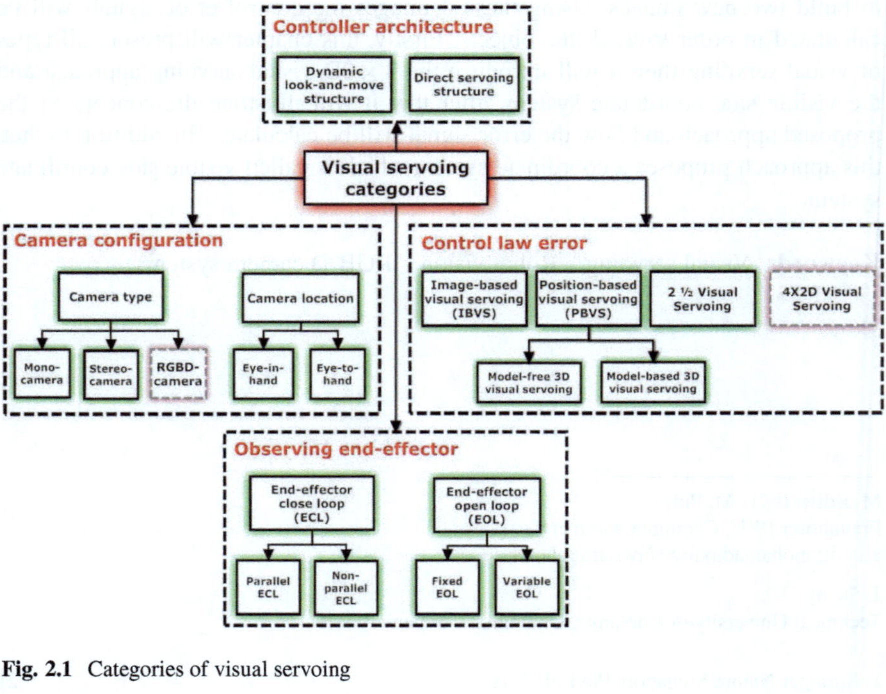

Fig. 2.1 Categories of visual servoing

presented a cooperation approach which integrates a fixed camera with one mounted on the robot. Next section will illustrate the different types of visual servoing based on control law error in details.

2.2 Related Work and Motivation

The main types of visual servoing which depending on the error used to compute the control law are: position-based visual servo system, image-based visual servo system and 2D$\frac{1}{2}$visual servoing. In the position based control system, the error will be computed in the Cartesian space. In other words, the features are extracted from the image and they are used to estimate the pose of the target object with respect to the camera coordinate system. In general, position-based visual servoing techniques can be classified into two groups: model-based and model-free visual servoing.

Model-based 3D visual servoing: In this case, 3D model of the target object is available. The desired camera pose with respect to the object frame could be estimated from the error between the 3D model of the target object and the current image features. Figure 2.2 presents scheme of model-based visual servoing. The extracted features of the object will be compared with its model in order to estimate the object's pose respect to the camera coordinate system.

Model-free 3D visual servoing: In this mode, the 3D structure of the object or the environment is completely unknown. In other words, there is no 3D model of the target object (Model-free). Model-free 3D visual servoing is based on teaching-by-showing step. In the teaching phase the robot will be driven to the desired position in order to store the corresponding reference image features, after that when the object or the robot have been moved, the control error will drive the robot to the desired position from the current position using the displacement between the reference image features and current image features.

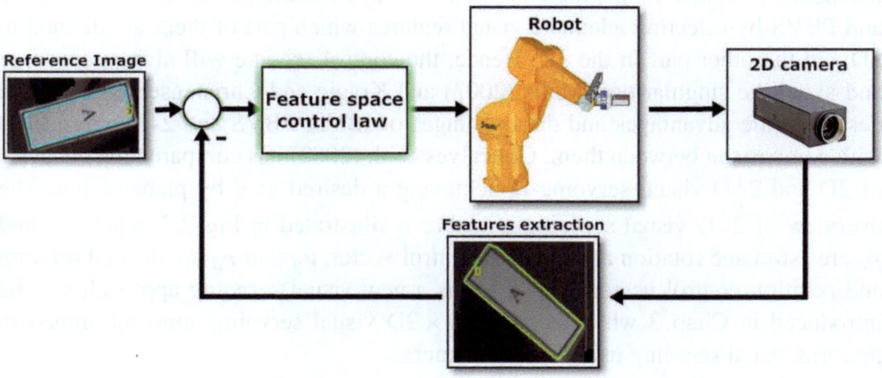

Fig. 2.2 Image based visual servoing

3D information could be reconstructed by fusing many 2D features, such as brightness, texture, color, size, and transparentness of the object. Furthermore photographic measurement or stereo camera could be used to reconstruct the 3D information. Photographic measurement uses the object shape and size to estimate the object position and orientation.

The image based visual servoing is a special case from model-free control because it does not need the knowledge of the 3D model of the target object and the control error will be formulated directly in 2D image space, so it will be called image error function. In the image-based visual servoing, the interaction matrix (image Jacobian matrix) will be used. This matrix describes how image feature parameters will change with respect to the change of the object or manipulator pose. This approach works with single camera and there is no need to the stereo computations.

In general, both methods position-based and image-based visual servoing have some drawbacks, e.g. in the position-based approach there is no performed control in the image space, which could imply that the target object may get out of the camera view during the servoing. Furthermore, a model of the target or teaching phase is needed to estimate the pose of the camera with respect to the target. In the image-based approach the depth estimation is needed to compute the interaction matrix and to design the control law. In addition to that, the matching between the desired and the current position will be insured only when the current position is in a neighborhood of the desired position.

A new approach was suggested by Malis et al. (1999) in order to reduce these drawbacks. This approach is called $2\frac{1}{2}$D visual servoing.

$2\frac{1}{2}$D visual servoing (Malis et al. 1999) is based on the estimation of the partial camera displacement from the current to the desired camera poses at each iteration of the control law. The extracted data from the partial camera displacement allow designing a decoupled control law which controls the six camera degree of freedom. In other words, $2\frac{1}{2}$D visual servoing can be used to decouple the translational and the rotational control loop. Since the position and orientation errors is controlled by two independent control loops, there is possibility to combine the advantages of IBVS and PBVS by selecting adequate visual features which part of them are defined in 2D and the other part in the 3D. Hence, the control scheme will always converge and avoid the singularities. Malis (2002) and Kragic and Christensen (2002) have described the advantages and disadvantages of IBVS, PBVS and $2\frac{1}{2}$D approaches with comparison between them. Goncalves et al. (2002) has compared the behavior of 2D and $2\frac{1}{2}$D visual servoing in achieving a desired goal by planar robot. The overview of $2\frac{1}{2}$D visual servoing structure is illustrated in Fig. 2.3, where u_e and p_e are estimated rotation and position control vector, u_d and p_d are desired rotation and position control vector. In this thesis a new visual servoing approach will be introduced in Chap. 3 which is called 4×2D visual servoing approach aimed to perform visual servoing using RGBD camera.

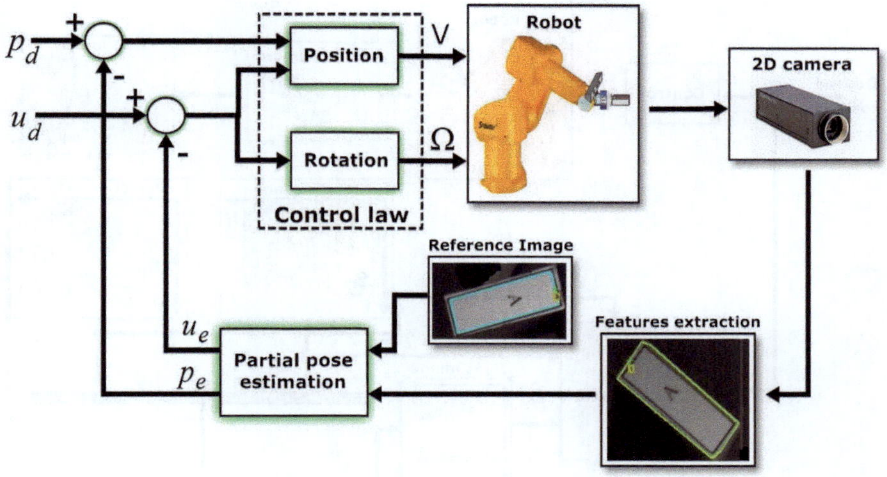

Fig. 2.3 $2\frac{1}{2}$D visual servoing

2.3 Concept of 4 × 2D Visual Servoing Approach

All the previous visual servoing approaches are proposed with consideration of using 2D camera, e.g.: (1) A single 2D camera will be used in case of IBVS. (2) In the case of PBVS stereo 2D camera system or single 2D camera with multiple viewpoints or multiple focusing factors could be used. In the last years the RGB-D cameras (such as Kinect camera) are spread widely everywhere. RGB-D cameras deliver depth and color images simultaneously and give us the opportunity to use this information in the visual servoing tasks. To exploit all the capabilities and advantages of the RGB-D cameras, this work will suggest a new approach of visual servoing. This approach will be called 4 × 2D visual servoing. 4 × 2D visual servoing combines the correspondence of color and depth images to build (u, w) and (w, v) images. By using these 4 images [RGB, depth, (u, w) and (w, v)] the system can calculate the control error signals at every subspace directly from the obtained images as is shown in Fig. 2.4. In PBVS approach many methods are used to estimate the 3D pose of the object from 2D images such as stereo vision, photometric stereo, shape from shadow etc. These approaches have many problems and disadvantages, for example the stereo vision has two main problems (1) Correspondence problem: determining which pixel on the left image corresponds to which pixel on the right image. (2) Reconstruction problem: define a number of correspondence pairs to find location and 3D structure of the target object. Furthermore, the using of IBVS approach in the 3D visual servoing applications is limited.

Otherwise 4 × 2D visual servoing approach is a simple approach which allows to use IBVS in 3D space and to estimate the pose of the target object when PBVS is needed. Next sections will discuss some suggested concepts and notations to

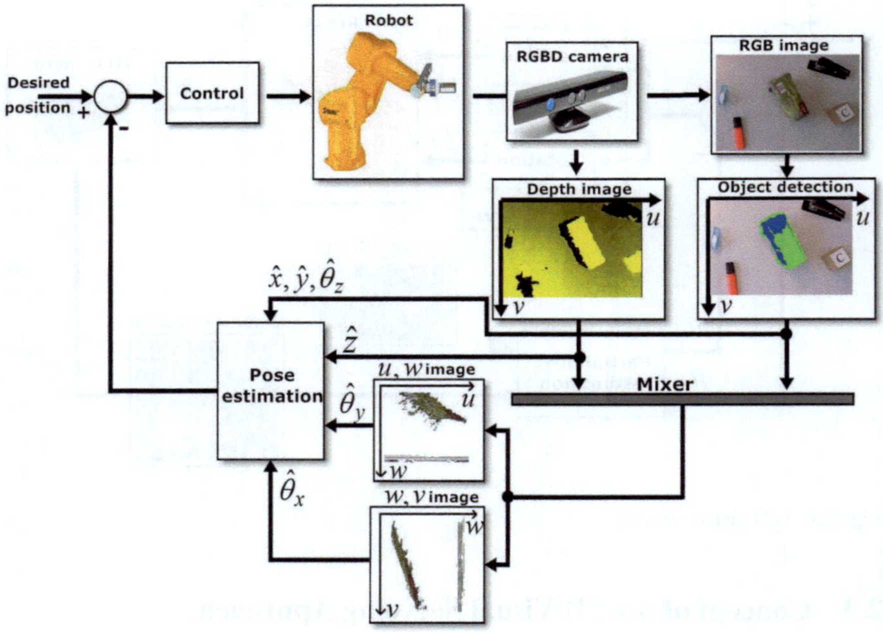

Fig. 2.4 4 × 2D visual servoing approach

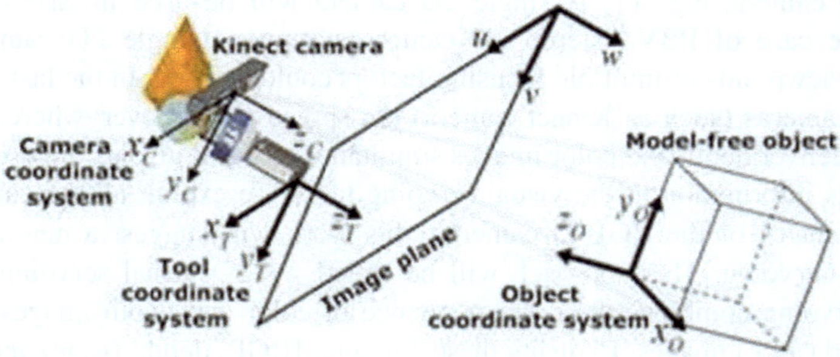

Fig. 2.5 Coordinate systems

perform the 4 × 2D visual servoing approach, such as mixing of RGB/depth images, visible side coordinate system, projections in the 3D image space etc.

Figure 2.5 shows three different coordinate systems which are camera (C), tool (T) and object (O) coordinate systems. The (T) coordinate frame is attached to the tool of the robot which could be 2 or 3 fingers gripper. There are two other coordinate systems, which are world coordinate system (W) and a new suggested coordinate system (S). The (S) coordinate frame is attached to the visible side of the target object which will be illustrated in the next section.

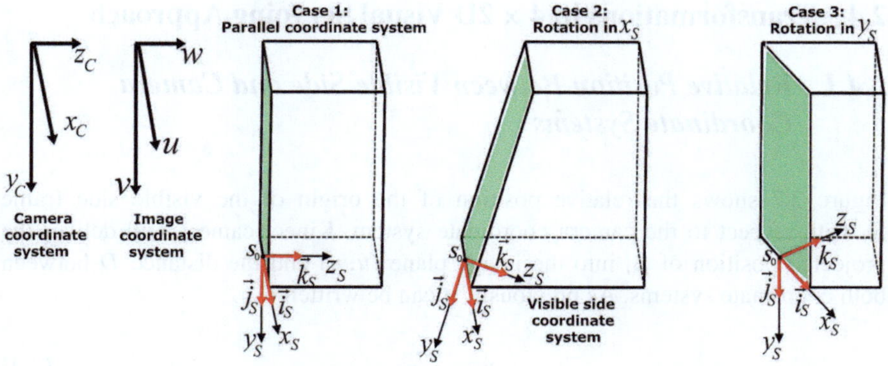

Fig. 2.6 Examples of visible side coordinate system

Figure 2.6 shows some examples with different poses of the visible side planes according to the camera coordinate system. In the first case the visible side (S) coordinate system is parallel to the camera (C) coordinate system. In the second case there is the rotation about x_s, whereas in the third case the rotation is about y_s direction.

The transformation from S coordinate system to the image plane coordinate system will be performed depending on perspective projection matrix as follows: In the first case of Fig. 2.6, where the S coordinate system is parallel to the camera coordinate system:

$$S \times \begin{bmatrix} u \\ v \\ 1 \end{bmatrix} = \begin{bmatrix} \lambda_x & S_c & u_0 & 0 \\ 0 & \lambda_y & v_0 & 0 \\ 0 & 0 & 1 & 0 \end{bmatrix} \times \begin{bmatrix} 1 & 0 & 0 & t_x \\ 0 & 1 & 0 & t_y \\ 0 & 0 & 1 & t_z \\ 0 & 0 & 0 & 1 \end{bmatrix} \times \begin{bmatrix} x_s \\ y_s \\ 0 \\ 1 \end{bmatrix} \tag{2.1}$$

In the second case of Fig. 2.6, where the rotation is about x_s direction:

$$S \times \begin{bmatrix} u \\ v \\ 1 \end{bmatrix} = \begin{bmatrix} \lambda_x & S_c & u_0 & 0 \\ 0 & \lambda_y & v_0 & 0 \\ 0 & 0 & 1 & 0 \end{bmatrix} \times \begin{bmatrix} 1 & 0 & 0 & t_x \\ 0 & \cos\theta_x & -\sin\theta_x & t_y \\ 0 & \sin\theta_x & \cos\theta_x & t_z \\ 0 & 0 & 0 & 1 \end{bmatrix} \times \begin{bmatrix} x_s \\ y_s \\ 0 \\ 1 \end{bmatrix} \tag{2.2}$$

In the third case of Fig. 2.6, where rotation is about direction:

$$S \times \begin{bmatrix} u \\ v \\ 1 \end{bmatrix} = \begin{bmatrix} \lambda_x & S_c & u_0 & 0 \\ 0 & \lambda_y & v_0 & 0 \\ 0 & 0 & 1 & 0 \end{bmatrix} \times \begin{bmatrix} \cos\theta_x & 0 & \sin\theta_x & t_x \\ 0 & 1 & 0 & t_y \\ -\sin\theta_x & 0 & \cos\theta_x & t_z \\ 0 & 0 & 0 & 1 \end{bmatrix} \times \begin{bmatrix} x_s \\ y_s \\ 0 \\ 1 \end{bmatrix} \tag{2.3}$$

2.4 Transformations in 4 × 2D Visual Servoing Approach

2.4.1 Relative Position Between Visible Side and Camera Coordinate Systems

Figure 2.7 shows the relative position of the origin of the visible side frame S_0 with respect to the camera coordinate system. Kinect camera will deliver the projected position of S_0 into the image plane (u, v) and the distance D between both coordinate systems. As previously, it can be written:

$$u = \lambda \frac{x_c}{z_c} \tag{2.4}$$

$$v = \lambda \frac{y_c}{z_c} \tag{2.5}$$

where (x_c, y_c, z_c) coordinate of point S_0 relative to camera coordinate system, u and v can be calculated from the image plane and λ is camera focal length.

From the depth image the distance D between the origin S_0 and the camera coordinate system, it can be written:

$$D^2 = x_c^2 + y_c^2 + z_c^2 \tag{2.6}$$

In Eqs. (2.4), (2.5) and (2.6) four values are known u, v, D, λ and three others are unknown x_c, y_c and z_c. If we could calculate these three values, we can define the relative position between the visible side and the camera coordinate systems. The previous equations could be rewritten as follows:

$$x_c = z_c \ \frac{u}{\lambda} \tag{2.7}$$

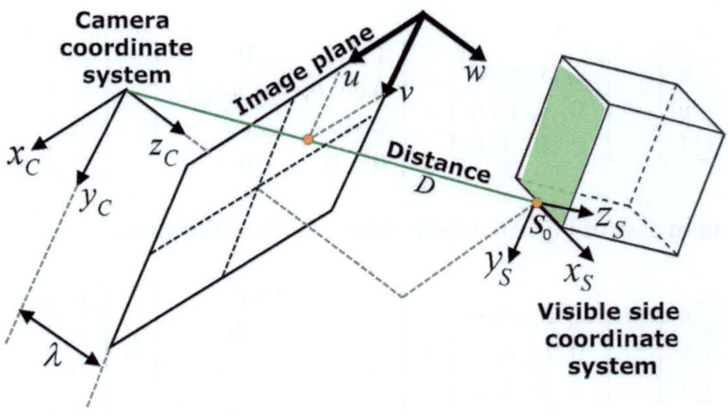

Fig. 2.7 Visible side and camera coordinate system

$$y_c = z_c \; \frac{v}{\lambda} \tag{2.8}$$

$$z_c^2 = D^2 - x_c^2 - y_c^2 \tag{2.9}$$

By substituting (2.7) and (2.8) in (2.9):

$$z_c^2 = D^2 - \left(z_c \; \frac{u}{\lambda}\right)^2 - \left(z_c \; \frac{v}{\lambda}\right)^2 \tag{2.10}$$

This gives:

$$z_c = \pm \frac{\lambda \; D}{\sqrt{\lambda^2 + u^2 + v^2}} \tag{2.11}$$

By substituting (2.11) in (2.7), x_c can be calculated:

$$x_c = \pm \frac{u \; D}{\sqrt{\lambda^2 + u^2 + v^2}} \tag{2.12}$$

By compensation (2.11) in (2.8), y_c can be calculated:

$$y_c = \pm \frac{v \; D}{\sqrt{\lambda^2 + u^2 + v^2}} \tag{2.13}$$

From (2.11), (2.12) and (2.13), the relative position of the origin S_0 with respect to the camera coordinate system (x_c, y_c, z_c) can be calculated.

2.5 Relative Orientation Between Visible Side and Camera Coordinate Systems

This section presents how to calculate the orientation of the object with respect to the camera coordinate system. According to the approach suggested in this work, the orientation of the object will be calculated from three 2D images (u, v), (u, w), and (w, v) with the help of image moments. First of all we will explain how to calculate the 2D orientation of an object with the help of image moments after that we will generalize this concept to be used in the three images (3D spaces of images).

2.5.1 Object's Orientation in Image Plane

The orientation of the object in 2D image plane is defined as the angle between the horizontal axis and the axis, where the object can be rotated with minimal inertia

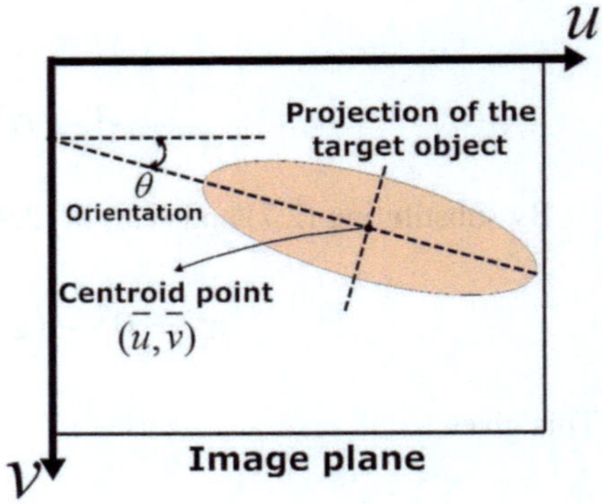

Fig. 2.8 Orientation of the target object

(i.e. the direction of the major semi-axis). In this direction the object has its longest extension. Figure 2.8 presents the projection of an object in the (u, v) image plane.

Let's assume that the target object found in the current image $I(u, v)$ which is shown in Fig. 2.8 is represented by the indicate function $S(u, v).S(u, v)$, when pixel belongs to the target object, otherwise $S(u, v) = 0$. There are two types of moments: non-central moments m_{pq} and central moments μ_{pq}:

$$m_{pq} = \sum_{v=0}^{V-1}\sum_{u=0}^{U-1} u^p v^q S(u, v) \tag{2.14}$$

where the order of the moment is $(p + q)$, so when $p = 0$ and $q = 0$ the m_{00} moment can be calculated as follows:

$$m_{pq} = \sum_{v=0}^{V-1}\sum_{u=0}^{U-1} u^0 v^0 S(u, v) = \sum_{v=0}^{V-1}\sum_{u=0}^{U-1} S(u, v) \tag{2.15}$$

where m_{00} is the total number of pixels belonging to the object. In the same way the centroid point of the object can be calculated:

$$\overline{u} = \frac{m_{10}}{m_{00}} \tag{2.16}$$

$$\overline{v} = \frac{m_{01}}{m_{00}} \tag{2.17}$$

The central moments of the object $S(u, v)$ can be calculated:

$$\mu_{pq} = \sum_{v=0}^{V-1}\sum_{u=0}^{U-1} (u - \overline{u})^p (v - \overline{v})^q S(u, v) \tag{2.18}$$

The main advantage of central moments is that they are invariant to translations of the object. The disadvantage of the central moments is their dependency on the size of the object (scale). Usually, the area of the object could be used as a scaling factor. Hence, by dividing the central moments with power of the area of the object, we get the central normalized moments ν_{pq}:

$$\nu_{pq} = \frac{\mu_{pq}}{m_{00}^{\frac{p+q}{2}+1}} \tag{2.19}$$

The main advantage of this kind of moments is their invariancy to the scale and translations. Using the central moments, the orientation of the 2D object can be calculated as follows:

$$\theta = \frac{1}{2} \arctan \frac{2\mu_{11}}{\mu_{20} - \mu_{02}} \tag{2.20}$$

2.5.2 *Object's Orientation in the 3D Image Space*

As shown previously, in our experiment we have supposed that the camera coordinate system is parallel to the tool coordinate system, which means the object's visible side seen by the camera will take the main role to specify the strategies of performing the grasping or contacting tasks. Therefore, this side will be projected on the two other planes (u, w) and (w, v). The projections of the visible side to the two other planes will be calculated by combining the RGB and depth images. These projections will give us the opportunity to achieve the contacting or grasping tasks with easier way to extract the vision error between the actual and target position of the object. Three different examples will be presented in Figs. 2.9, 2.10 and 2.11 to get better overview of visible side's projection.

As shown in the previous figures, there are three 2D objects which are the projections of the visible side of the object in the 3D image space. Hence, there are three angles $\Delta\theta_1, \Delta\theta_2$ and $\Delta\theta_3$ that can be calculated easily from three image planes. $\Delta\theta_1$ is the rotation angle between the projection of visible side coordinate

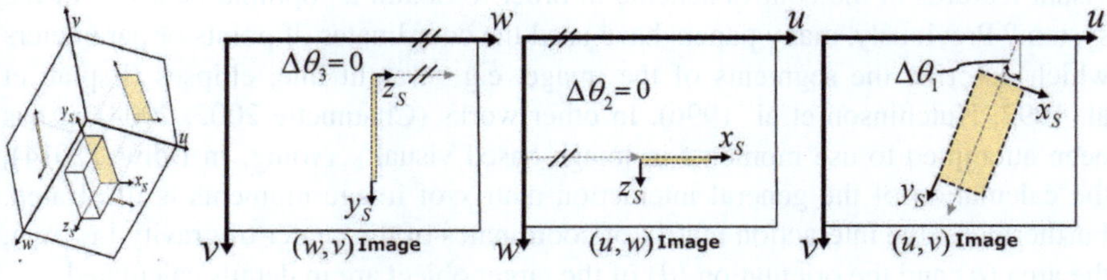

Fig. 2.9 Projections of the visible side (Example 1: Rotation about)

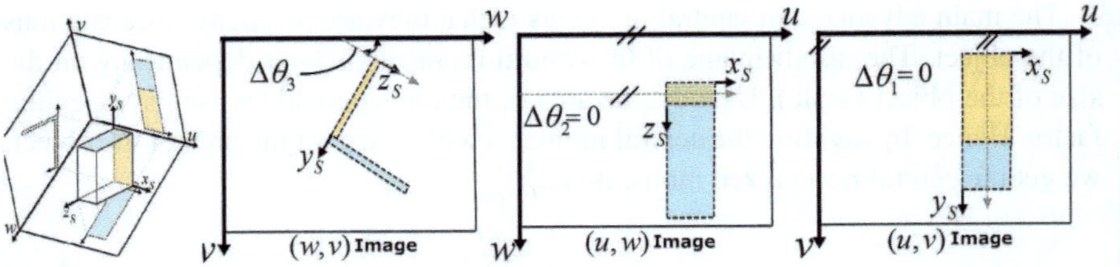

Fig. 2.10 Projections of the visible side (Example 2: Rotation about)

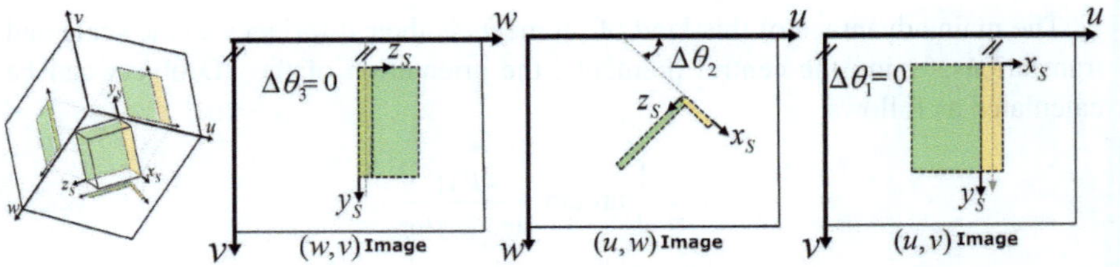

Fig. 2.11 Projections of the visible side (Example 3: Rotation about)

system (x_s, y_s) and the image plane (u, v). $\Delta\theta_2$ is the rotation angle between the projection of visible side coordinate system (x_s, z_s) and the image plane (u, w). $\Delta\theta_3$ is the rotation angle between the projection of visible side coordinate system (z_s, y_s) and the image plane (w, v). As discussed previously, using the central moments of the image (see (2.20)) and depending on the definition of the 2D object's orientation (the angle between the horizontal axes and the direction of the major semi-axis), the orientation of the 2D objects in three image planes can be easily calculated.

2.6 Control Error of 4 x 2D Visual Servoing Approach

The first step of the visual servoing is defining in any image a particular set of features that describes the goal which should be reached. In the image-based visual servoing approach, the determination of the visual features is still considered as open problem. In other words, how does the system choose the most appropriated visual features in the control scheme in order to obtain an optimal behavior of the system? Previously, many papers have used the coordinates of points or parameters which describe the segments of the image, e.g. straight line, ellipses (Espiau et al. 1992; Hutchinson et al. 1996). In other works (Chaumette 2002, 2004), it has been attempted to use moments in image-based visual servoing. In Bdiwi (2014), the calculation of the general interaction matrix of image moments is illustrated. Furthermore, the interaction matrix of coordinates of the center of gravity (x_g, y_g), the area (a) and the orientation (θ) of the target object are in details calculated.

Actually, the image moments have been widely used in pattern recognition and visual servoing for a long time (Prokop and Reeves 1992). However, the main problem of using image moments in visual servoing loop was that the analytical form of the interaction matrix related to the image moments was unavailable. In Bien et al. (1993) the image moments implemented to control only four degrees of freedom of a robot (x, y, z, θ_z) were: the area, the centroid and the main orientation of an object in the image. These four visual features (coordinates x_g and y_g of the center of gravity, area a and orientation θ) can be easily obtained from moments of order less than 3 which are closely related to the velocities of the camera v_x and w_y, v_y and w_x, v_z and w_z respectively. However, using the previous image moments leads to the situation that the robot will execute trajectories which are desirable in the image and which can be indirectly and seemingly contorted in Cartesian space. This potential problem appears when redundant image points coordinates are used, e.g. coupled features.

The work Corke and Hutchinson (2001) has overcome this problem by decoupling the z-axis rotational and translational components of the control from the remaining degrees of freedom, as follows:

$$\dot{\vec{f}} = J_{xy}\dot{\vec{r}}_{xy} + J_z\dot{\vec{r}}_z \tag{2.21}$$

where $\dot{\vec{r}}_{xy} = [v_x\ v_y\ w_x\ w_y]$, $\dot{\vec{r}} = [v_z\ w_z]$, J_{xy}, J_z are respectively components or columns 1, 2, 4, 5 and 3, 6 of Jacobian matrix J and $\dot{\vec{f}}$ is the feature point coordinate error. Hence,

$$\dot{\vec{r}}_{xy} = J_{xy}^{+}\left\{\dot{\vec{f}} - J_z\dot{\vec{r}}_z\right\} \tag{2.22}$$

The z-axis velocity will be based directly on two features (orientation and area of the object).

$$v_z = \gamma_{v_z}(a^\star - a)\ ,\quad w_z = \gamma_{w_z}(\theta_z^\star - \theta_z) \tag{2.23}$$

where γ_{v_z}, γ_{w_z} are the scalar gain coefficient. Another approach Chaumette (2002) has used two supplementary visual features in order to decouple w_y from v_x and w_x from v_y. These features are based on classical skewness moments $s_x = \frac{\mu_{30}}{\mu_{20}^{\frac{3}{2}}}$ and $s_y = \frac{\mu_{03}}{\mu_{03}^{\frac{3}{2}}}$. For the symmetrical objects:

$$\begin{cases} s_x = \frac{1}{K}\sqrt{a}(s_1t_1 + s_2t_2) \\ s_y = \frac{1}{K}\sqrt{a}(s_2t_1 + s_1t_2) \end{cases} \tag{2.24}$$

where:

$$\begin{cases} s_1 = \mu_{03} - 3\mu_{21} & , t_1 = (\mu_{20} - \mu_{02})^2 - 4\mu_{12}^2 \\ s_2 = \mu_{30} - 3\mu_{12} & , \quad t_2 = 4\mu_{11}(\mu_{20} - \mu_{02}) \\ K = \Delta(\mu_{20} + \mu_{02})^{\frac{3}{2}} \end{cases} \tag{2.25}$$

For non symmetrical objects:

$$\begin{cases} p_x = \dfrac{\Delta}{(\mu_{20} + \mu_{02})^2} \\ p_y = a(s_1^2 + s_2^2)(\mu_{20} + \mu_{02})^3 \end{cases} \tag{2.26}$$

Hence, the vector of visual features used for tracking symmetrical objects is as follows:

$$s = (x_g, y_g, a, s_x, s_y, \theta_z) \tag{2.27}$$

whereas the vector of visual features used for tracking the nonsymmetrical objects is:

$$s = (x_g, y_g, a, p_x, p_y, \theta_z) \tag{2.28}$$

For the proposed features and for the configurations where the object is parallel to the image plane at the desired position, the interaction matrix has shown nice decoupling feature and the results are satisfactory even if the object is not parallel to the image plane at the beginning of the positioning task. In this approach, the maximum order of the implemented moments is the third order. However, whenever the used moments have less order then the system will be more robust and it will have more numerical stability. Because of that, one advantage of the proposed approach in this chapter that the control loop can use the calculated orientations from (u, w) and (w, v) images instead of s_x and s_y which can be easily calculated from the second order moments.

Figures 2.12, 2.13 and 2.14 illustrate some experimental results during the rotation of the camera about different axes. The target object and its projections are of green color. Figure 2.12 shows the proportional changing of the orientation in (u, v) image during the rotation of the camera about axis z_c. As shown, there is nice decoupling between the rotation about z_c and the orientation in (w, v) and (u, w) images, they have constant values. In the same way, the experiment will be repeated during the rotation of the camera about axis y_c, as shown in Fig. 2.13. Only the orientation (u, w) in image will be changed, whereas orientations in (u, v) and (w, v) images are constant and they are decoupled from the rotating about y_c. In Fig. 2.14, the experiment will be performed during the rotation of the camera in x_c and only the orientation in (w, v) image will be changed, whereas the orientations in (u, v) and (u, w) images are constant and they are decoupled from the rotating in x_c.

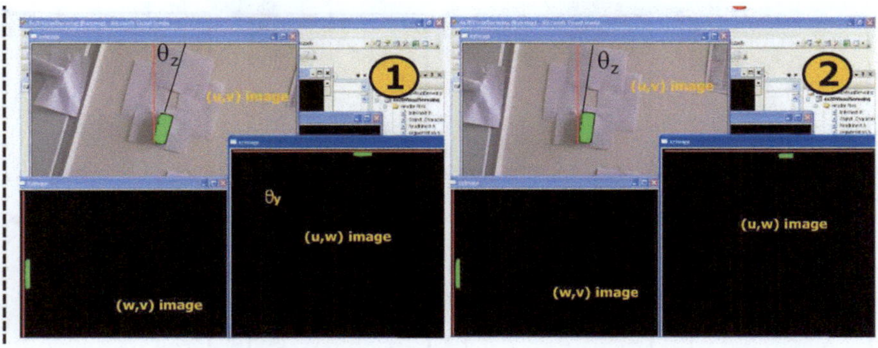

Fig. 2.12 Relation between rotation about z_c and orientation in (u, v) image

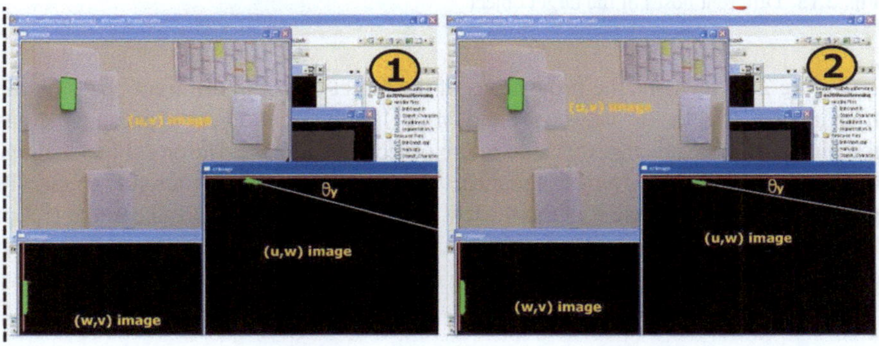

Fig. 2.13 Relation between rotation about y_c and orientation in (u, w) image

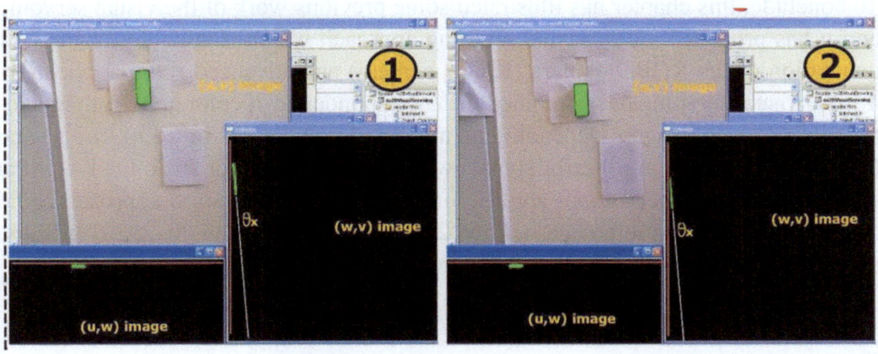

Fig. 2.14 Relation between rotation about x_c and orientation in (w, v) image

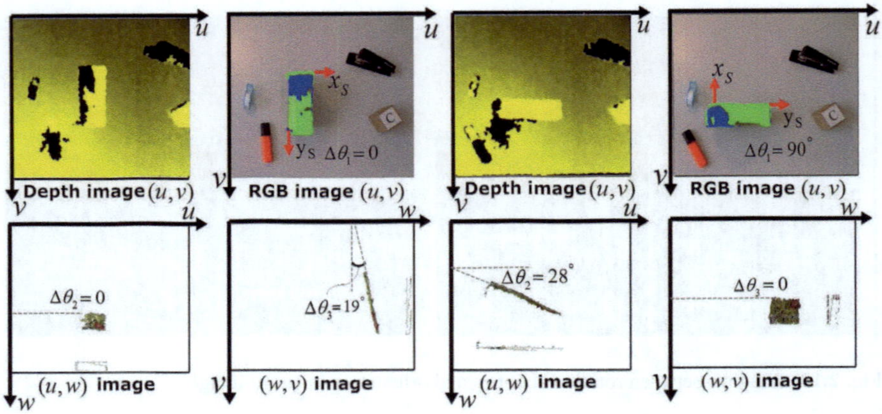

Fig. 2.15 Different poses of the target object

Figure 2.15 presents the projections of the target object in different poses on the four images: (depth, (u, v), (u, w) and (w, v)). The target object is green, whereas the blue part indicates that this side of the object has higher position (object is diagonally positioned). This part of the target object is the position where the robot will grasp the object, because it is the nearest part of the object from the camera (robot tool). It is clear that one can directly calculate the inclination angle of the object either in the first case from (w, v) image or in the second case from (u, w) image.

2.7 Conclusions

To conclude, this chapter has illustrated some previous work of the visual servoing and it has proposed a new visual servoing approach which is suitable for the new generation of the RGBD camera (Kinect camera) and advanced robot application. $4 \times 2D$ visual servoing has combined the correspondence of color and depth images to build two new images. Depending on measured orientations in (u, v), (u, w) and (w, v) images, the control error signals will be calculated in order to track any flat surface. Some advanced robot applications based on the proposed approach could be found in following references (Bdiwi 2014; Bdiwi and Suchý 2012, 2014; Bdiwi et al. 2013; Kolker et al. 2013).

References

Bdiwi, M. (2014). *Development of integration algorithms for vision/force robot control with automatic decision system*. Dissertation, Technical University of Chemnitz, Germany.

Bdiwi, M., & Suchý, J. (2012). Library automation using different structures of vision–force robot control and automatic decision system. In *IEEE/RSJ International Conference on Intelligent Robots and Systems (IROS)*, Portugal.

Bdiwi, M., & Suchý, J. (2014). Integration of vision/force robot control using automatic decision system for performing different successive tasks. In *Processing on 8th German Conference on Robotics*.

Bdiwi, M., Kolker, A., Suchý J., & Winkler A. (2013). Real time visual and force servoing of human hand for physical human–robot interaction: Handing–over unknown objects. In *Workshop on Human Robot Interaction for Assistance and Industrial Robots in IEEE International Conference on Robotics and Automation (ICRA)*, Karlsruhe, Germany.

Bien, Z., Jang, W., & Park, J. (1993). Characterization and use of feature–Jacobian matrix for visual servoing. In K. Hashimoto (Ed.), *In visual servoing* (pp. 317–363). Singapore: World Scientific.

Chaumette, F. (2002). A first step toward visual servoing using image moments. In *IEEE/RSJ International Conference on Intelligent Robots and Systems* (pp. 378–383).

Chaumette, F. (2004). Image moments: A general and useful set of features for visual servoing. *IEEE Transactions on Robotics, 20*(4), 713–723.

Chesi, G., & Hashimoto, K. (2002). Static–eye against hand eye visual servoing. In *Proceedings of 41st IEEE Conference on Decision and Control* (pp. 2854–2859).

Corke, P., & Hutchinson, S. (2000). Real–time vision, tracking and control. In *Proceedings of IEEE International Conference on Robotics and Automation* (pp. 622–629).

Corke, P., & Hutchinson, S. (2001). A new partitioned approach to image–based visual servoing control. *IEEE Transactions on Robotics and Automation, 17*(4), 507–515.

Espiau, B., Chaumette, F., & Rives, P. (1992). A new approach to visual servoing in robotics. *IEEE Transactions on Robotics and Automation, 8*(3), 313–326.

Flandin, G., Chaumette, F., & Marchand, E. (2000). Eye-in-hand/Eye-to-hand cooperation for visual servoing. In *IEEE International Conference on Robotics and Automation* (pp. 2741–2746).

Goncalves, P., Ferreira, P., & Pinto, P. (2002). Comparing visual servoing architectures for a planar robot. In *Proceedings of the 10th Mediterranean Conference on Control and Automation*, Portugal.

Hashimoto, K. (2003). A review on vision–based control of robot manipulators. *Advanced Robotics, 17*, 969–991.

Hill, J., & Park, W. (1979). Real time control of a robot with a mobile camera. In *Proceedings of the 9th SRI International*, USA (pp. 233–246).

Hutchinson, S., Hager, G., & Corke, P. (1996). A tutorial on visual servo control. *IEEE Transactions on Robotics and Automation, 1*, 651–670.

Kolker, A., Winkler, A., Bdiwi, M., & Suchý, J. (2013). Robot visual servoing using the example of the inverted pendulum. In *10th IEEE International Multi–Conference on Systems, Signals & Devices (SSD)*, Hammamet, Tunesien.

Kragic, D., & Christensen, H. (2002). *Survey on visual servoing for manipulation*. Computational Vision and Active Perception Laboratory, Sweden.

Malis, E. (2002). Survey of vision–based robot control. In *European Naval Ship Design, Captain Computer IV Forum*, France (pp. 1–16).

Malis, E., Chaumette, F., & Boudet, S. (1999). $2\frac{1}{2}$D visual servoing. *IEEE International Transactions on Robotics and Automation, 15*, 238–250.

Prokop, R., & Reeves, A. (1992). A survey of moment–based techniques for unoccluded object representation and recognition. *Proceedings Computer Vision, Graphics, Image Processing Conference, 54*, 438–460.

Shirai, Y., & Inoue, H. (1973). Guiding a robot by visual feedback in assembling tasks. *Pattern Recognition Pergamon Press, 5*, 99–108. Great Britain.

Weiss, L., Sanderson, A., & Neuman, C. (1985). Dynamic visual servo control of robots: An adaptive image–based approach. In *Proceedings of IEEE International Conference on Robotics and Automation* (pp. 662–668).

Weiss, L., Sanderson, A., & Neuman, C. (1987). Dynamic sensor based control of robots with visual feedback. *IEEE Journal of Robotics and Automation, 5*, 404–417.

Chapter 3
A Redundant Parallel Robotic Machining Tool: Design, Control and Real-Time Experiments

Hussein Saied, Ahmed Chemori, Micael Michelin, Maher El-Rafei, Clovis Francis, and Francois Pierrot

Abstract In this chapter, we present a machining device, named ARROW robot, designed with the architecture of a redundant parallel manipulator capable of executing five degrees-of-freedom in a large workspace. Machine-tools based on parallel robot development are considered a key technology of machining industries due to their favourable features such as high rigidity, good precision, high payload-to-weight ratio and high swiftness. The mechanism of ARROW robot isolates its workspace from any type of inside singularities allowing it to be more flexible and dynamic. An improved PID with computed feedforward controller is implemented on ARROW robot to perform real-time experiments of a machining task. The control system deals with antagonistic internal forces caused by redundancy through a regularization method, and achieves a stability conservation in case of actuators saturation. The results are evaluated using the root mean square error criteria over all the tracking trajectory confirming the high accuracy and good performance of ARROW robot in machining operations.

H. Saied (✉)
LIRMM, University of Montpellier, CNRS, Montpellier, France

CRSI, Lebanese University, Beirut, Lebanon
e-mail: saied@lirmm.fr

A. Chemori · F. Pierrot
LIRMM, University of Montpellier, CNRS, Montpellier, France
e-mail: chemori@lirmm.fr; pierrot@lirmm.fr

M. Michelin
Tecnalia France, Centre Spatial Universitaire, Montpellier, France
e-mail: micael.michelin@tecnalia.com

M. El-Rafei · C. Francis
CRSI, Lebanese University, Beirut, Lebanon
e-mail: maher.elrafei@ul.edu.lb; cfrancis@ul.edu.lb

N. Derbel et al. (eds.), *New Developments and Advances in Robot Control*, Studies in Systems, Decision and Control 175,
https://doi.org/10.1007/978-981-13-2212-9_3

Keywords Machine tool · Redundant parallel robot · Kinematics · Dynamics · Singularity analysis · Motion planning · Control · PID · Feedforward · Anti-windup

3.1 Introduction

Machining is the process of a controlled material-removal that makes a desired deformation in the shape and size of raw materials. The theme of the material-removal classifies machining in different subtractive manufacturing classes (Kerbrat et al. 2011). It is considered an essential part in the manufacturing of many products based on metal, wood, plastic, ceramic, and composites. The conventional term, "machining", refers to a set of processes such as turning, boring, drilling, milling, broaching, sawing, shaping, planing, reaming, tapping . . . etc (Davim 2008). Nowadays, most of the recently mentioned operations are carried out by Computer Numercial Control (CNC) in which the computers are used to control the movement and act of shaping.

Three principal operations of machining are classified as turning, drilling and milling. All other operations are considered as miscellaneous categories described in what follows:

- Turning operation is performed by rotating the workpiece and facing it over the cutting tool as shown in Fig. 3.1a.
- Milling operation is the operation in which the cutting tool rotates bringing the cutting edges to bear against the workpiece, see Fig. 3.1b.
- Drilling operation is the most common machining process. It is the generation of cylindrical holes by bringing the rotating cutter with cutting edges into the workpiece as illustrated in Fig. 3.1c.

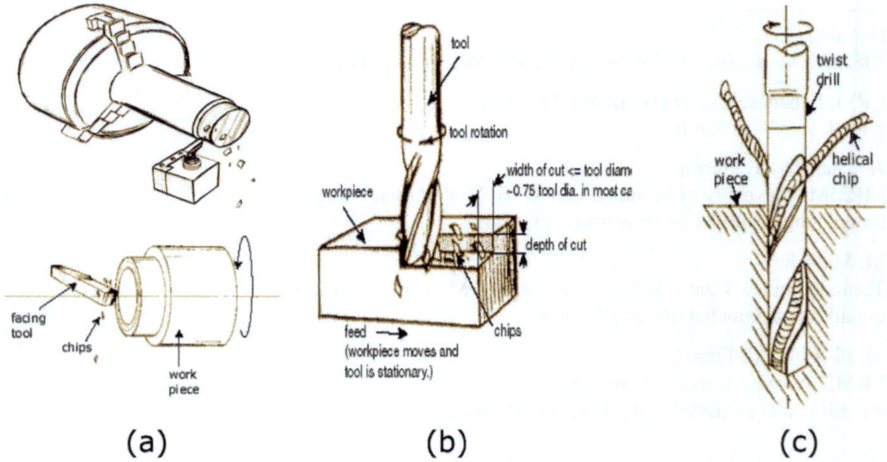

Fig. 3.1 Principal machining operations. **(a)** Turning. **(b)** Milling. **(c)** Drilling (eFunda 2018)

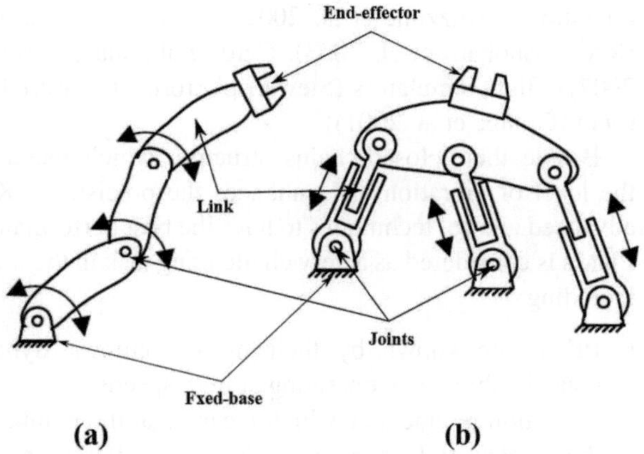

Fig. 3.2 Illustration of Kinematic manipulators. (**a**) Serial manipulators. (**b**) Parallel manipulators (Kucuk 2012)

- Miscellaneous operations are secondary swarf producing operations performed on the workpiece after the principal machining processes using the typical machining tools. For example: boring, broaching, burnishing . . . etc.

The successful design of machine-tools is based on several requirements such as: high accuracy resulting in a best surface integrity, acceptable level of vibration, high speeds and feeds increasing the productivity, high static stiffness of the different machine-tool elements such as structure, joints, and spindles, avoidance of unacceptable natural frequencies that may cause resonance of the machine-tool, and finally adopting modern control techniques (Youssef and El-Hofy 2008).

Traditional machine-tools are based on serial manipulators designed as a series of links connected via actuated joints from the base to the end-effector (see Fig. 3.2). These manipulators, known with their large range of motions, lack to the good accuracy and low stiffness. Moreover, they have low dynamic response and valuable level of vibration (Kucuk 2012; Merlet 2006).

Parallel Kinematic Manipulators (PKMs) offering many advantages over their serial counterparts are welcomed candidates to be the future machining robots. According to Merlet (2006), any structure made up of a fixed base and traveling plate linked together with at least two independent kinematic chains is called a PKM (see Fig. 3.2). PKMs provide large accuracy, low vibration, high acceleration capabilities thanks to the light moving parts, and high stiffness due to the closed-chains structure. However, these manipulators are characterized by a limited workspace especially in the rotational motion. In the addition to their complexity in the forward kinematic solutions, and the considerable number of singularities inside and on the borders of their workspace (Kucuk 2012; Merlet 2006).

Due to the satisfying features of PKMs, they have achieved a real success in various applications other than machining. PKMs are considered the leading robots in the automated industrial applications, particularly in the pick-and-place food packaging tasks which require rapid execution maintaining an acceptable precision (Natal et al. 2015). Also they are used in the field of laser cutting operations (Bruzzone et al. 2002), medical applications (Robot-Assisted Surgery (RAS) (Shoham et al. 2003), CardioPulmonary Resuscitation (CPR) (Li and Xu 2007)), flight simulators (Stewart platforms (Stewart 1965)), Haptic devices (delta robot (Grange et al. 2001)).

Beside their closed-chains structure which increases the stiffness, decreases the level of vibration and enhances the precision, PKMs shall be equipped with advanced control techniques to have the best performance of automation. Control of PKMs is considered as a very challenging task in the literature due to a set of factors including:

- PKMs are known by their higly nonlinear dynamics which may increase considerably when operating at high speeds.
- Actuation redundancy which means that the number of actuators is greater than the number of degrees of freedom (Merlet 2006) exists in some parallel robots.
- Uncertainties and variations of the dynamics and operating conditions are often present in PKMs.

Several control techniques have been proposed and studied in the literature starting from the non-model-based non-adaptive and adaptive controllers, ending up with model-based non-adaptive and adaptive controllers. One can mention few examples from the aforementioned classes of control implemented on PKMs respectively: Proportional-Integral-Derivative (PID) (Natal et al. 2015; Ziegler and Nichols 1942), nonlinear PID (Su et al. 2004), Augmented PD (APD) (Zhang et al. 2007), Adaptive Feedforward plus PD (Codourey et al. 1997; Natal et al. 2012), Desired Compensation Adaptive Control Law (DCAL) (Bennehar et al. 2014a), Adaptive controller based on the Robust Integral of the Sign of the Error (RISE) (Bennehar et al. 2014b), Feedforward Compensation in L1 Adaptive Control (Bennehar et al. 2015).

This chapter introduces a five-degree-of-freedom redundantly actuated PKM named ARROW[1] robot characterized by the good compromise between rapidity and precision developed for machining tasks. In Sect. 3.2, a general overview on ARROW PKM structure is stated, in addition to the kinematic and dynamic modeling with a brief presentation of the analyzed singularities. Section 3.3 describes the machining task to be tested, the required trajectory generation, and the proposed control solution. Experimental results are illustrated, discussed and interpreted in Sect. 3.4. Finally, Sect. 3.5 concludes the chapter and provides future work suggestions.

[1]ARROW project: a national french research project financed by the National Research Agency (ANR). Its main objectives can be summarized in the design of **A**ccurate and **R**apid **R**obots with large **O**perational **W**orkspace, from which the acronym **"ARROW"** has been derived. The project embraces three partners: IRCCyN (Institut de Recherche en Communication et Cybernétique de Nantes), LIRMM (Laboratory of Informatics, Robotics and Microelectroncs of Montpellier) and Tecnalia France.

3.2 ARROW PKM: General Overview, Modeling and Singularity Analysis

The ARROW robot is a redundantly actuated PKM has been designed and developed at LIRMM within the ARROW project. This section covers the general description of ARROW PKM, the detailed inverse kinematic model, the differential kinematics and dynamics associated to ARROW robot, and a brief analysis for the singularities related to the operational workspace. Note that the reader can refer to (Shayya et al. 2014), (Shayya 2015) for further information.

3.2.1 General Overview of ARROW PKM

The structure of ARROW PKM consists of two separate modules as illustrated in Fig. 3.3: the parallel module and the turntable. The parallel module involves six linear actuators each pair lays in the same slider track and equidistant from each other. This configuration facilitates the geometric calibration and reduces the overall cost from manufacturing point of view. These six actuators are linked by means of spherical joints to the kinematic chains which are attached to the articulated traveling plate by revolute joints (see Fig. 3.4). The idea behind using revolute joints is to avoid some possible internal collisions between the arms. The two intermediate parallelogram arms are responsible of the rotation of the traveling plate around the vertical axis parallel to the z-axis, while its position and orientation are originated from the settle of all the actuators together. Note that the traveling plate can reach

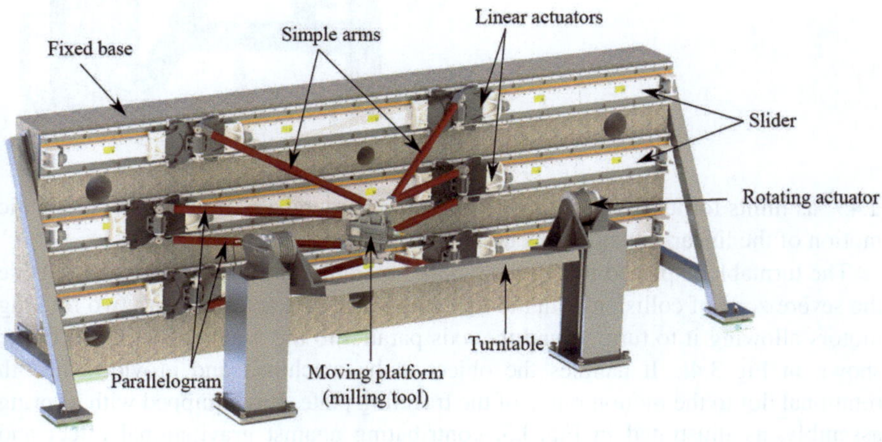

Fig. 3.3 CAD view of the ARROW machining PKM

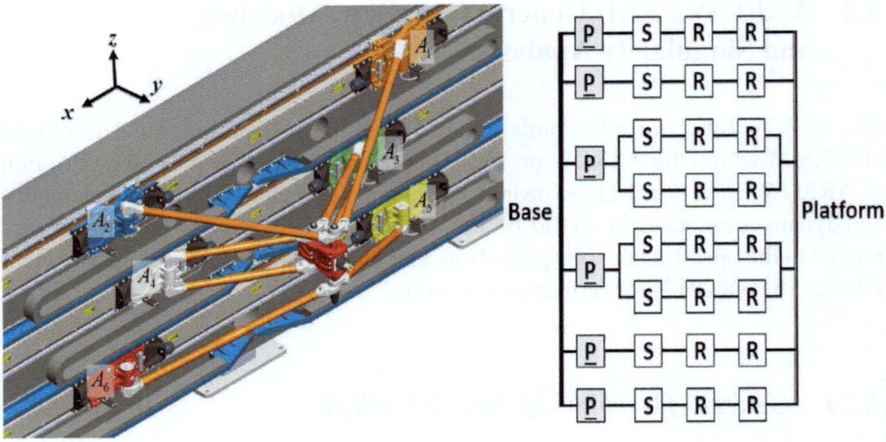

(a) (b)

Fig. 3.4 ARROW PKM. (**a**) 3D CAD view of the parallel module. (**b**) Graph diagram (Shayya 2015)

Fig. 3.5 Close side view to the CAD drawing of ARROW PKM showing the springs used in the turntable (Shayya 2015)

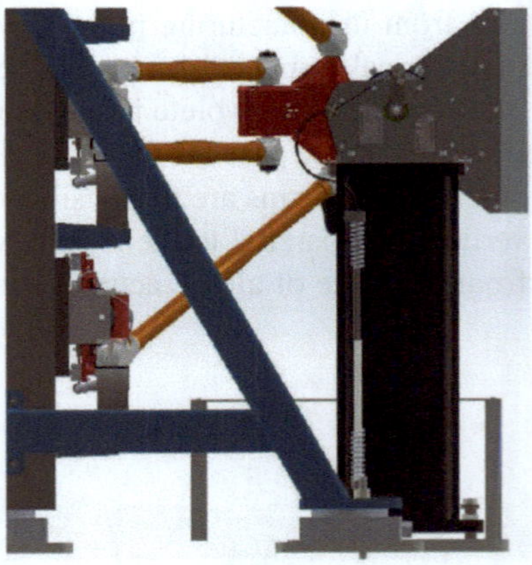

±45° as limits for its allowable range of rotation around z-axis. Thus, the prismatic motion of the linear actuators sets up the traveling plate in 3T-1R[2] dofs.

The turntable is placed facing the parallel module (of Fig. 3.3) and kept outside the severe zone of collision with the traveling plate. It is actuated with two rotating motors allowing it to turn around the axis parallel to the x-axis of the fixed frame shown in Fig. 3.4a. It handles the object to be machined and provides the 5th rotational dof to the motion range of the traveling plate. It is equipped with a spring assembly, as illustrated in Fig. 3.5, contributing against gravitational effect and

[2]"T" corresponds to translational motion and "R" corresponds to rotational motion.

allows maintaining the turntable orientation fixed without any need to the motors torques. This is beneficial regarding energy saving and reduces control efforts.

From a control point of view, the main part to be controlled is the parallel module. It holds the machining tool represented as a traveling plate which is a milling machine in our case. Implementing robust control algorithms to the linear actuators will set up the machine-tool in a precise tracked trajectory for machining purpose.

3.2.2 Modeling of ARROW PKM

This part explains in details the kinematic, differential kinematic and dynamic models of ARROW robot. In view of the reported mutations in Shayya (2015) that have been suggested to implement ARROW PKM avoiding internal arm collisions, the Forward Kinematic Model (FKM) is difficult to establish if not impossible. Thus, only the Inverse Kinematic Model (IKM) will be presented and clarified.

Figure 3.6 shows the parallel module with some notations to be used in modeling. Six arms, simple and parallel, are grouped and numbered from I to VI. The lengths $L_i (i = 1, 2, 5, 6)$ of the simple arms are equal to L_s, while lengths $L_i (i = 3, 4)$ of the parallel arms are equal to L_p.

Two frames are defined for the modeling: Base frame (O, e_x, e_y, e_z) and Platform frame $(P, e_{xp}, e_{yp}, e_{zp})$ (see Figs. 3.7 and 3.9). The following recorded points are coordinated in the base frame:

- $\mathbf{A}_i = (x_i \ y_i \ z_i)^T = (q_i \ y_i \ z_i)^T$ is the point attached to the spherical joint for $i = 1, 2, 5, 6$ and the center of the linear actuator for $i = 3, 4$. q_i is the position of the linear actuator in the x-direction $\forall i = 1, .., 6$ (see Fig. 3.6).
- $\mathbf{B}_i = (x_{bi} \ y_{bi} \ z_{bi})^T \ \forall i = 1, .., 6$ and $\mathbf{D}_i = (x_{di} \ y_{di} \ z_{di})^T$ for $i = 1, 2, 5, 6$ are the points attached to the revolute joints at the level of the platform, and that

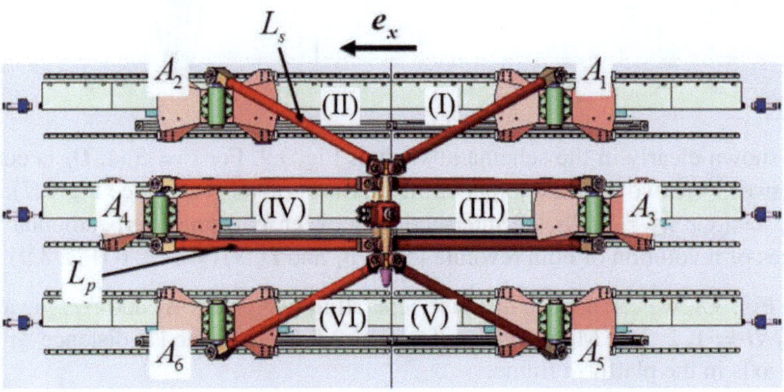

Fig. 3.6 Facing CAD view of the parallel module of ARROW PKM with some notations (Shayya 2015)

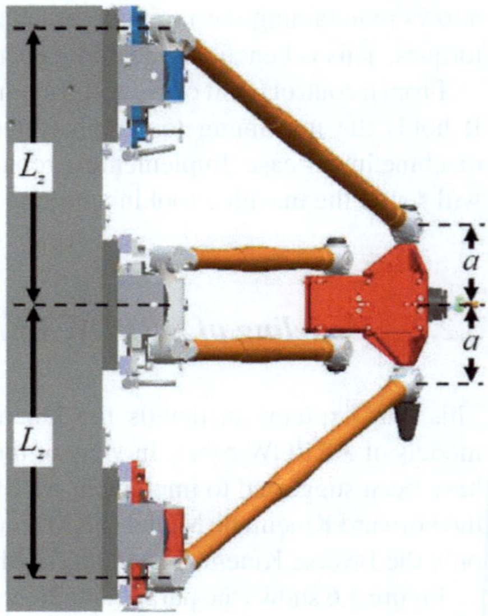

Fig. 3.7 Side CAD view of the traveling plate of ARROW PKM (Shayya 2015)

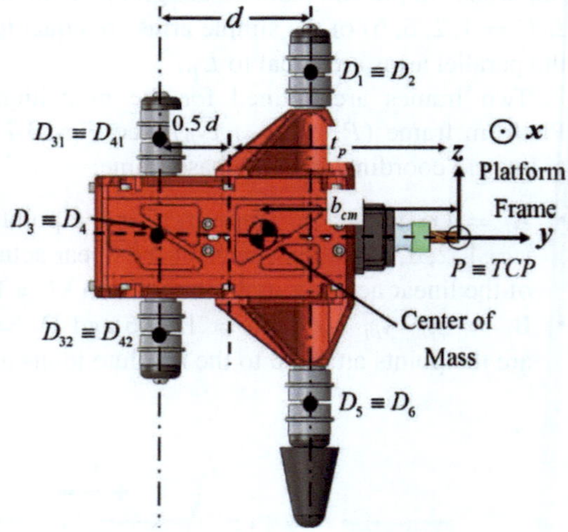

Fig. 3.8 Side CAD view of the parallel module of ARROW PKM (Shayya 2015)

is shown clearly in the schematic view of Fig. 3.9. For $i = 3, 4$, $\mathbf{D}_i$ is centered between the two revolute joints of the platform on its y-axis (see Fig. 3.7).

- $\mathbf{C}_i = (x_{ci}\ y_{ci}\ z_{ci})^T$ is the intersection between two lines perpendicular to the axes of revolution of both revolute joints $\mathbf{B}_i$ and $\mathbf{D}_i$ $\forall i = 1, .., 6$ (Fig. 3.9).

In Fig. 3.8, we can see the distance separating two points $\mathbf{A}_i$ and $\mathbf{A}_{i+2}$ evaluated by L_z $\forall i = 1, 2, 3, 4$. Moreover, D_i, for i = 1,2,5,6, are located at distance "a" from the y-axis in the platform frame.

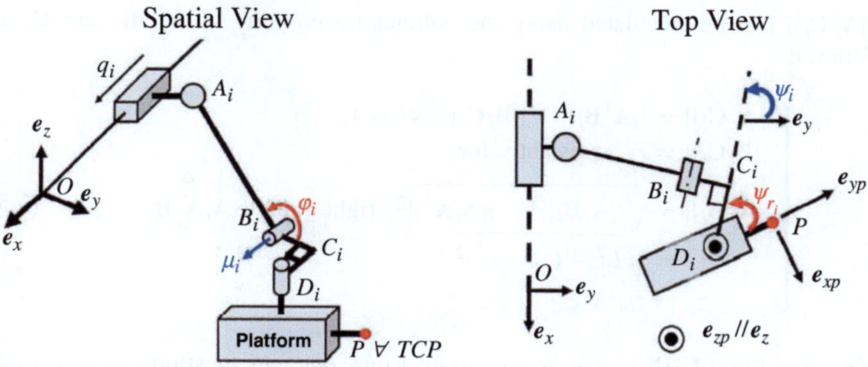

Fig. 3.9 ARROW PKM: schematic view of a kinematic chain with different notations

Two mandatory assembly conditions for the mechanism to function properly are given in the following two equalities:

$$L_z \neq a, \text{ Here } L_z \text{ is chosen greater than } a. \tag{3.1}$$

$$q_i = x_i \leq x_{di} \equiv x_{di+1} \leq q_{i+1} = x_{i+1}, \forall i = 1, 3, 5 \tag{3.2}$$

3.2.2.1 Inverse Kinematic Model

Regarding the parallel module of ARROW PKM, the position and orientation of the moving platform in the 3T-1R DoFs are represented by the 4-dimensional coordinate vector $\mathbf{X}_{PM} = (x\ y\ z\ \theta_z)^T$. The six actuators positions are parametrized by the 6-dimensional coordinate vector $\mathbf{q}_{PM} = (q_1\ q_2\ q_3\ q_4\ q_5\ q_6)^T$. IKM represents the transformation from the cartesian space ($\mathbf{X}_{PM}$) to the joint space ($\mathbf{q}_{PM}$) for any configuration in the workspace.

Knowing the end-effector posture $\mathbf{X}_{PM}$ in the base frame and the coordinates of D_i in the platform frame $\mathbf{PD}_i^{plat}$, we get $\mathbf{D}_i$ as follows:

$$\mathbf{D}_i = \mathbf{P} + \mathbf{R}(\mathbf{PD}_i^{plat}), \text{ for } i = 1, .., 6 \tag{3.3}$$

where $\mathbf{P} = (x\ y\ z)^T$ is the $3T$ coordinates of the end-effector, and $\mathbf{R} \in \mathbb{R}^{3\times3}$ is the basic rotation matrix of the platform frame around the z-axis. Note that $\mathbf{B}_i$, $\mathbf{C}_i$ and $\mathbf{D}_i$ are in the same horizontal plane (xy), whereas $\mathbf{A}_i$, $\mathbf{B}_i$ and $\mathbf{C}_i$ remain in the same vertical plane as illustrated in Fig. 3.9. Thus, $\mathbf{A}_i^{\dagger}$, the projection of $\mathbf{A}_i$ on the xy plane, is collinear with $\mathbf{B}_i$ and $\mathbf{C}_i$. Then, a right triangle is formed at $\mathbf{C}_i$ in the horizontal plane and we get the following equation:

$$||\mathbf{A}_i^{\dagger}\mathbf{D}_i||^2 = ||\mathbf{A}_i^{\dagger}C_i||^2 + ||\mathbf{C}_i\mathbf{D}_i||^2, \ for\ i = 1, .., 6 \tag{3.4}$$

$||\mathbf{A}_i^\dagger \mathbf{C}_i||$ can be calculated using the collinearity property of $\mathbf{A}_i^\dagger$, $\mathbf{B}_i$ and $\mathbf{C}_i$ as follows:

$$\begin{cases} ||\mathbf{A}_i^\dagger \mathbf{C}_i|| = ||\mathbf{A}_i^\dagger \mathbf{B}_i|| + ||\mathbf{B}_i \mathbf{C}_i||, \ \forall i = 1, .., 6 \\ ||\mathbf{B}_i \mathbf{C}_i|| = r_i, \ \text{constant value} \\ ||\mathbf{A}_i^\dagger \mathbf{B}_i|| = \sqrt{||\mathbf{A}_i \mathbf{B}_i||^2 - ||\mathbf{A}_i \mathbf{A}_i^\dagger||^2}, \ \text{right triangle } \overset{\triangle}{\mathbf{A}_i \mathbf{A}_i^\dagger \mathbf{B}_i} \\ \qquad\qquad\quad = \sqrt{L_i^2 - (z_i - z_i^\dagger)^2} \\ \qquad z_i^\dagger = z_{bi} = z_{di} \end{cases} \tag{3.5}$$

Knowing that $||\mathbf{C}_i \mathbf{D}_i|| = r_i'$ is a constant value, one can substitute (3.5) in (3.4) and get the length $||\mathbf{A}_i^\dagger \mathbf{D}_i||$. Then, replace the value of $||\mathbf{A}_i^\dagger \mathbf{D}_i||$ in the following equation:

$$\begin{cases} ||\mathbf{A}_i^\dagger \mathbf{D}_i|| = \sqrt{(x_i^\dagger - x_{di})^2 + (y_i^\dagger - y_{di})^2 + (z_i^\dagger - z_{di})^2}, \ \forall i = 1, .., 6 \\ \qquad\qquad\quad = \sqrt{(q_i - x_{di})^2 + (y_i - y_{di})^2} \\ z_i^\dagger = z_{bi} = z_{di}, \ x_i^\dagger = x_i = q_i \end{cases} \tag{3.6}$$

Based on the assembly mode condition mentioned in (3.2), the IKM of ARROW PKM parallel module is derived by solving (3.6) and it can be written as follows:

$$q_{PMi} = x_{di} + (-1)^i \sqrt{||\mathbf{A}_i^\dagger \mathbf{D}_i||^2 - (y_i - y_{di})^2}, \ \forall i = 1, .., 6 \tag{3.7}$$

As discussed before, the turntable adds the 5th DoF to ARROW robot allowing the rotation around x-axis. IKM of the turntable delivers the joints position, knowing the value of the rotational angle is θ_x, as follows:

$$\begin{cases} q_{T1} = \theta_x \\ q_{T2} = -\theta_x \end{cases} \tag{3.8}$$

3.2.2.2 Differential Kinematic Model

The differential Kinematic model of the parallel module can be defined as a relation between the moving platform's velocity, $\dot{\mathbf{X}}_{PM}$, and the actuated joints velocity, $\dot{\mathbf{q}}_{PM}$. The matrix links both velocities known as Jacobian is derived in this paragraph.

As $\mathbf{A}_i \mathbf{B}_i$ is supposed to be a rigid body doesn't change its length whatever is the configuration of the robot, we have:

$$\mathbf{A}_i\mathbf{B}_i^T v_{A_i} = \mathbf{A}_i\mathbf{B}_i^T \mathbf{v}_{B_i} \ \forall i = 1, .., 6 \tag{3.9}$$

with the velocities of $\mathbf{v}_{A_i}$ and $\mathbf{v}_{B_i}$ being respectively:

$$\mathbf{v}_{A_i} = \dot{q}_i \mathbf{e}_x \ \forall i = 1, .., 6 \tag{3.10}$$

$$\begin{cases} \mathbf{v}_{B_i} = \mathbf{v}_{D_i} + \mathbf{w}_{hp} \times \mathbf{D}_i\mathbf{B}_i, \ \forall i = 1, .., 6 \\ \mathbf{v}_{D_i} = \mathbf{v}_P + \mathbf{w}_{Plat} \times \mathbf{PD}_i^{plat} \end{cases} \tag{3.11}$$

where $\mathbf{w}_{hp} = \dot{\psi}_i \mathbf{e}_z \ \forall i = 1, .., 6$ and $\mathbf{w}_{Plat} = \dot{\theta}_i \mathbf{e}_z \ \forall i = 1, .., 6$ are the angular velocities of the horizontal plane $B_i C_i D_i$ and the moving platform respectively (see Fig. 3.9), knowing that both axes of rotation are parallel to the z-axis in the base frame. Substituting (3.10) and (3.11) in (3.9) yields to a relation where $\dot{\mathbf{q}}_{PM}$ and $\dot{\mathbf{X}}_{PM}$ are coupled to $\dot{\psi}$. So, there is need to find another equation that leads to uncoupled relation between $\dot{\mathbf{q}}_{PM}$ and $\dot{\mathbf{X}}_{PM}$. This equation is expressed by writing $\mathbf{v}_{A_i}$ in terms of $\mathbf{v}_{B_i}$ and $\mathbf{w}_{arm}$ knowing that each arm $\mathbf{A}_i\mathbf{B}_i$ is performing two rotations: one around an axis parallel to the z-axis in the base frame and the other is around the axis μ_i defined in Fig. 3.9. The expression will be as follows:

$$\begin{cases} \mathbf{v}_{A_i} = \mathbf{v}_{B_i} + \mathbf{w}_{arm} \times (-\mathbf{A}_i\mathbf{B}_i), \ \forall i = 1, .., 6 \\ \mathbf{w}_{arm} = \mathbf{w}_{A_i B_i} = \dot{\varphi}_i \mu_i + \dot{\psi}_i e_z \\ \mu_i = \frac{\mathbf{r}_i \times \mathbf{e}_z}{|}|\mathbf{r}_i \times \mathbf{e}_z|| \end{cases} \tag{3.12}$$

From the relations in (3.9) and (3.12), with some computations and simplifications, the Inverse Differential Kinematic Model (IDKM) for the parallel module is derived and the form of the inverse Jacobian matrix, $\mathbf{J}_m$, is obtained:

$$\begin{cases} \mathbf{J}_{qPM}\dot{q}_{PM} = \mathbf{J}_{XPM}\dot{X}_{PM} \implies \dot{\mathbf{q}}_{PM} = \mathbf{J}_m\dot{\mathbf{X}}_{PM} \text{ with } \mathbf{J}_m = \mathbf{J}_{qPM}^{-1}\mathbf{J}_{XPM} \\ \mathbf{J}_{qPM} = \begin{pmatrix} (\mathbf{m}_1^T \mathbf{e}_x)(\mathbf{e}_z^T(L_1 \times \mu_1)) & & \\ & \ddots & \\ & & (\mathbf{m}_6^T e_x)(\mathbf{e}_z^T(L_6 \times \mu_6)) \end{pmatrix} \\ \mathbf{J}_{XPM} : \begin{cases} J_X(i,1) = (\mathbf{e}_z^T(L_i \times \mu_i))(\mathbf{e}_x^T\mathbf{m}_i) \\ J_X(i,2) = (\mathbf{e}_z^T(L_i \times \mu_i))(\mathbf{e}_y^T\mathbf{m}_i) \\ J_X(i,3) = (\mathbf{e}_z^T(n_i \times \mu_i))(\mathbf{e}_z^T\mathbf{m}_i) \\ J_X(i,4) = -(\mathbf{e}_z^T(L_i \times \mu_i))(\mathbf{e}_z^T(\mathbf{m}_i \times \mathbf{PD}_i^{plat})) \\ \mathbf{n}_i = \mathbf{A}_i\mathbf{C}_i, \ \forall i = 1, .., 6 \end{cases} \\ \mathbf{m}_i = \mathbf{A}_i\mathbf{D}_i, \ \forall i = 1, .., 6 \end{cases} \tag{3.13}$$

Note that $\mathbf{J}_m$ is not invertible in the case of redundant actuated PKMs, so the pseudo-inverse of the inverse Jacobian is used as a solution widely adopted in robotics. Thus, the Forward Differential kinematic Model (FDKM) of the parallel module is obtained as follows:

$$\dot{\mathbf{X}}_{PM} = \mathbf{J}_m{}^{*}\dot{\mathbf{q}}_{PM} \tag{3.14}$$

where $\mathbf{J}_m{}^{*} \in \mathbb{R}^{4\times 6}$ is the pseudo-inverse of the inverse Jacobian matrix.

In regard to the turntable, the IDKM can be obtained by differentiating the equations of (3.8) as the following:

$$\begin{cases} \dot{q}_{T_1} = \dot{\theta}_x \\ \dot{q}_{T_2} = -\dot{\theta}_x \end{cases} \tag{3.15}$$

3.2.2.3 Dynamic Model

To establish the dynamics of the parallel module of ARROW PKM, two assumptions have been taken into consideration, similar for the hypothesis applied usually on Delta-like robots, are mentioned below:

1. The dry and viscous friction forces are neglected in all passive and active joints.
2. The rotational inertia of the arms is neglected, due to their small masses in comparison with the other components. The mass of each arm is split up into two equivalent parts, one part is added to the mass of the corresponding linear actuator while the other part is considered with the moving platform.

Applying the law of motion on the linear actuators gives the following equation:

$$\begin{cases} \mathbf{M}_a \ddot{q}_{PM} = \Gamma_{PM} - \mathbf{J}_{q_{PM}}^{T} f \\ \mathbf{M}_a = diag(m_{as}, m_{as}, m_{ap}, m_{ap}, m_{as}, m_{as}) \end{cases} \tag{3.16}$$

where $m_{as} \in \mathbb{R}$ is the mass including the actuator, moving cart and the half-mass of the simple arm, while $m_{ap} \in \mathbb{R}$ is the mass including the actuator, moving cart and the half-mass of the parallelogram arm. M_a is called the actuators' inertia matrix. $\Gamma_{PM} \in \mathbb{R}^6$ is the vector representing the six actuation forces provided by the linear motors. $\mathbf{J}_{q_{PM}} \in \mathbb{R}^{6\times 6}$ is the joint Jacobian matrix defined in (3.13). $\mathbf{f} \in \mathbb{R}^6$ is the vector of the applied forces on the arms, at $\mathbf{A}_i$, resulting from the acceleration and gravitational forces acting on the platform.

The dynamics of the moving platform is described by Newton-Euler's method as follows:

$$\begin{cases} \mathbf{M}_P \ddot{\mathbf{X}}_{PM} + \Lambda_c \dot{\mathbf{X}}_{PM} = \mathbf{J}_{X_{PM}}^T f + m_P \mathbf{g} \\ \mathbf{M}_P = \begin{pmatrix} m_P & 0 & 0 & -b\, m_P \sin\theta_z \\ 0 & m_P & 0 & b\, m_P \cos\theta_z \\ 0 & 0 & m_P & 0 \\ -b\, m_P \sin\theta_z & b\, m_P \cos\theta_z & 0 & I_{pzz} \end{pmatrix} \\ \Lambda_c = \begin{pmatrix} 0 & 0 & 0 & -b\, m_P \cos\theta_z \\ 0 & 0 & 0 & -b\, m_P \sin\theta_z \\ 0 & 0 & 0 & 0 \\ 0 & 0 & 0 & 0 \end{pmatrix} \end{cases} \tag{3.17}$$

with $\mathbf{M}_P$ and Λ_c are the mass matrix and the centrifugal and Coriolis effects of the moving platform respectively. $\mathbf{J}_{X_{PM}} \in \mathbb{R}^{6\times 4}$ is the Cartesian Jacobian matrix defined in (3.13). $m_P \in \mathbb{R}$ symbolizes the total mass of the traveling plate including the added half-masses of the arms, and $\mathbf{g} = (0\ 0\ \ -9.8\,\text{m/s}^2\ \ 0)^T$ is the gravity acceleration acting on it. The term "b" refers to the x-coordinate belonging to the center of mass of the cluster formed by the platform and the added point masses on B_i, whereas "I_{pzz}" refers to the total moment of inertia about the z-axis through TCP of the same formed cluster.

The mass m_P and the moment of inertia I_{pzz} are given as follows:

$$m_P = (mass\ of\ platform) + \frac{1}{2}\sum_{i=1}^{6}(mass\ of\ arm)_i \tag{3.18}$$

$$I_{pzz} = I_{platform} + \sum_{i=1}^{6} I_i \tag{3.19}$$

for which $I_{platform}$ is the moment of inertia about the z-axis passing through TCP only of the platform's body, and I_i is that of the point mass formed at B_i after making assumption (2) around the same axis.

Differentiating the IDKM obtained in (3.13) delivers the acceleration kinematic relation as: $\ddot{\mathbf{q}}_{PM} = \mathbf{J}_m \ddot{\mathbf{X}}_{PM} + \dot{\mathbf{J}}_m \dot{\mathbf{X}}_{PM}$. By the use of this relation and equations of (3.16) and (3.17), one can formulate the Direct Dynamic Model (DDM) of the parallel module as follows:

$$\ddot{\mathbf{X}}_{PM} = \mathbf{H}\Gamma_{PM} - \Lambda \dot{\mathbf{X}}_{PM} + \mathbf{A}_G \tag{3.20}$$

where $\mathbf{H} \in \mathbb{R}^{4\times 6}$, $\Lambda \in \mathbb{R}^{4\times 4}$ and $\mathbf{A}_G \in \mathbb{R}^4$ are expressed by:

$$\begin{cases} \mathbf{H} = \left(\mathbf{M}_P + \mathbf{J}_m^T \mathbf{M}_a \mathbf{J}_m\right)^{-1} \mathbf{J}_m^T, \ dim(\mathbf{H}) = 4 \times 6 \\ \Lambda = \mathbf{H}\mathbf{M}_a \dot{\mathbf{J}}_m + \left(\mathbf{M}_P + \mathbf{J}_m^T \mathbf{M}_a \mathbf{J}_m\right)^{-1} \Lambda_c, \ dim(\Lambda) = 4 \times 4 \\ \mathbf{A}_G = \left(\mathbf{M}_P + \mathbf{J}_m^T \mathbf{M}_a \mathbf{J}_m\right)^{-1} m_P \mathbf{g}, \ dim(\mathbf{A}_G) = 4 \times 1 \end{cases} \tag{3.21}$$

The Inverse Dynamic Model (IDM) of the parallel module can be reformulated from (3.20) by getting the actuator forces vector Γ_{PM} as function of the moving platform's position $\mathbf{X}_{PM}$ and its derivatives. In case of actuation redundancy, there is no unique Γ_{PM} vector for a specific desired acceleration. Considering the minimum norm solution of Γ_{PM} and maintaining the condition of $|\Gamma_{PM_i}| \leq \Gamma_{PM_{max}}$ $\forall i =$ 1, .., 6 (all actuators are similar) (Shayya et al. 2014), the IDM of the parallel module is then arranged in joint space as follows:

$$\Gamma_{PM} = \mathbf{H}^* \left(\ddot{\mathbf{X}}_{PM} + \Lambda \dot{\mathbf{X}}_{PM} - \mathbf{A}_G\right) \tag{3.22}$$

with $\mathbf{H}^* \in \mathbb{R}^{6\times4}$ being the pseudo-inverse of $\mathbf{H}$.

Considering the dynamics of the turntable part, the varying inertia of the object to be machined is neglected due to the low acceleration needed for the turntable rotation. The total moment of inertia of turntable with respect to the axis of rotation is given as follows:

$$I_T = 2I_{act} + I_{txx} \tag{3.23}$$

where I_{act} is the rotative motor's inertia and I_{txx} is the inertia of the table rotating around its proper axis which is x-axis. Assuming that the gravitational effect is compensated for by the springs (see Fig. 3.5), meaning that the effect of gravity and spring cancel each other, the IDM of the turntable is given by the following equations:

$$\begin{cases} \Gamma_{T_1} = (I_T \ddot{\theta}_x)/2 \\ \Gamma_{T_2} = -\Gamma_{T_1} \end{cases} \tag{3.24}$$

with Γ_{T_1} and Γ_{T_2} being the torques provided by the rotative actuators driving the turntable into its rotational operation.

3.2.3 Singularity Analysis

In this branch, a summarized singularity analysis is stated asserting that within the feasible workspace of ARROW PKM, there is no existence for all types of singularities (Shayya 2015). Two cases have been studied: constraint and classical singularity.

At first, let the desired workspace be defined as follows:

$$\begin{cases} (x) : -1.5\,\text{m} \le q_i \le +1.5\,\text{m} \\ (y) : -0.15\,\text{m} \le y - y_0 \le +0.15\,\text{m} \\ (z) : -0.15\,\text{m} \le z \le +0.15\,\text{m} \\ (\theta_z) : -45^\circ \le \theta_z \le +45^\circ \\ (\theta_x) : -45^\circ \le \theta_x \le +45^\circ \end{cases} \tag{3.25}$$

with y_0 being the y offset of the center of the desired workspace knowing that the axis of rotation of the turntable passes through $\mathbf{P}_0 = (0\ y_0\ 0)^T$ and parallel to the x direction.

3.2.3.1 Constraint Singularities

Constraint singularities occur when the pair of complex chains (III) and (IV) perform undesired rotation for the platform. Normally, the angular velocity of the platform is represented by the vector $\mathbf{w}_P = (w_x = 0, w_y = 0, w_z)^T$. Only the two complex arms are considered in this study with modifying the position of P $\equiv$ TCP to be confounded with $\mathbf{D}_3 \equiv D_4$ for more simplicity. Figure 3.10 shows the virtual equivalent chains with the modification, and defines new notations and subscripts ij with $i = 3, 4$ and $j = 1, 2$.

Investigating the constraint singularities starts from analysing the following equation:

$$\begin{cases} \mathbf{v}_{A_{ij}} = \mathbf{v}_{A_i} = \mathbf{v}_{B_{ij}} + \mathbf{w}_{a_{ij}} \times (-\mathbf{L}_{ij}),\ \forall i = 3, 4; \forall j = 1, 2 \\ \mathbf{w}_{a_{ij}} = \mathbf{w}_P + \dot{\psi}_{ij}\mathbf{e}_z + \dot{\varphi}_{ij}\mu_{ij} \end{cases} \tag{3.26}$$

Recalling that in the absence of any constraint singularity we should have $w_x = w_y = 0$ and normal generation for all necessary velocities ($v_x, v_y, w_x, w_y, \dot{\psi}_{ij}, \dot{\varphi}_{ij}, \forall i = 3, 4; \forall j = 1, 2$), (3.26) yields to the condition below:

$$h\big(\mathbf{m}_{3x}\mathbf{m}_{4y} - \mathbf{m}_{3y}\mathbf{m}_{4x}\big)\big(\mathbf{L}_{3x}\mu_{3y} - \mathbf{L}_{3y}\mu_{3x}\big)\big(\mathbf{L}_{4x}\mu_{4y} - \mathbf{L}_{4y}\mu_{4x}\big) \neq 0 \tag{3.27}$$

From (3.27), one can distinguish three cases lead to constraint singularity:

1. $h = 0$. This is not possible because h is restricted to be opposite to zero by manufacturing process.
2. $\big(\mathbf{m}_{3x}\mathbf{m}_{4y} - \mathbf{m}_{3y}\mathbf{m}_{4x}\big) = 0$. This means the xy projections of $\mathbf{m}_3$ and $\mathbf{m}_4$ are collinear. This practically is preceded by collision between platform and slider's wall, or between the actuators. So it is on the borders or outside the accessible workspace.

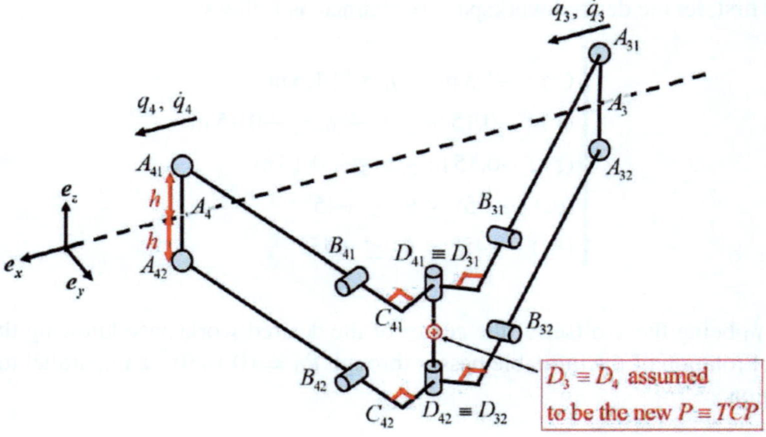

$$
\begin{cases}
\boldsymbol{L}_{ij} = \boldsymbol{A}_{ij}\boldsymbol{B}_{ij},\ \boldsymbol{r}_{ij} = \boldsymbol{B}_{ij}\boldsymbol{C}_{ij},\ \boldsymbol{r}_{p_{ij}} = \boldsymbol{C}_{ij}\boldsymbol{D}_{ij} \\
\boldsymbol{n}_{ij} = \boldsymbol{A}_{ij}\boldsymbol{C}_{ij} = \boldsymbol{L}_{ij} + \boldsymbol{r}_{ij},\ \boldsymbol{m}_{ij} = \boldsymbol{A}_{ij}\boldsymbol{D}_{ij} = \boldsymbol{L}_{ij} + \boldsymbol{r}_{ij} + \boldsymbol{r}_{p_{ij}} \\
\boldsymbol{p}_{ij} = \boldsymbol{PD}_{ij},\ \boldsymbol{\mu}_{ij} = \dfrac{\boldsymbol{r}_{ij} \times \boldsymbol{e}_z}{\| \boldsymbol{r}_{ij} \times \boldsymbol{e}_z \|},\ \varphi_{ij} = \left(\widehat{\boldsymbol{B}_{ij}\boldsymbol{C}_{ij}, \boldsymbol{B}_{ij}\boldsymbol{A}_{ij}}\right) = \left(\widehat{\boldsymbol{r}_{ij}, -\boldsymbol{L}_{ij}}\right) \\
\psi_{ij} = \left(\widehat{\boldsymbol{e}_y, \boldsymbol{D}_{ij}\boldsymbol{C}_{ij}}\right) = \left(\widehat{\boldsymbol{e}_y, -\boldsymbol{r}_{p_{ij}}}\right),\ \psi_{r_{ij}} = \left(\widehat{\boldsymbol{e}_{yp}, \boldsymbol{D}_{ij}\boldsymbol{C}_{ij}}\right) = \left(\widehat{\boldsymbol{e}_{yp}, -\boldsymbol{r}_{p_{ij}}}\right)
\end{cases}
$$

Fig. 3.10 The complex chains (III) and (IV) with notations (Shayya 2015)

3. $(\mathbf{L}_{3x}\mu_{3y} - \mathbf{L}_{3y}\mu_{3x}) = 0$ or $(\mathbf{L}_{4x}\mu_{4y} - \mathbf{L}_{4y}\mu_{4x}) = 0$. These two terms are simultaneously zero or non-zero due to the structural symmetry of the mechanism. If both are zero, this means that $\mathbf{L}_3//\mathbf{L}_4//\mathbf{e}_z$ which comes after a collision in the exterior region of the allowed workspace.

Therefore, the analysis outcomes guarantee that there are no constraint singularities within the operational workspace.

3.2.3.2 Classical Singularities

Two cases can be defined for the classical singularities to be existing: First case is described by knowing $\dot{\mathbf{q}}_{PM}$ but we cannot definitely determine $\dot{\mathbf{X}}_{PM}, \dot{\varphi} = (\dot{\varphi}_1, \ldots, \dot{\varphi}_6)^T$ or $\dot{\psi} = (\dot{\psi}_1, \ldots, \dot{\psi}_6)^T$, second case is when knowing $\dot{\mathbf{X}}_{PM}$ but we cannot definitely determine $\dot{\mathbf{q}}_{PM}, \dot{\varphi}$ or $\dot{\psi}$. Mathematically, the matrices $\mathbf{J}_{q_{PM}}, \mathbf{J}_{X_{PM}}, \mathbf{J}_{\varphi}$ and $\mathbf{J}_{\psi}$ should not be rank deficient in order to avoid the aforementioned two cases. $\mathbf{J}_{q_{PM}}, \mathbf{J}_{X_{PM}}$ were given in (3.13), and $\mathbf{J}_{\varphi}, \mathbf{J}_{\psi}$ are derived also from Eqs. (3.9) and (3.12) as follows:

$$\begin{cases} \mathbf{J}_{\varphi}\dot{\varphi} = \mathbf{J}_{X_{PM}\varphi}\dot{X}_{PM} \\ \mathbf{J}_{\varphi} = \begin{pmatrix} (\mathbf{e}_z^T(\mathbf{L}_1 \times \mu_1)) & & \\ & \ddots & \\ & & (\mathbf{e}_z^T(\mathbf{L}_6 \times \mu_6)) \end{pmatrix} \\ \mathbf{J}_{X_{PM}\varphi} = \begin{pmatrix} -\mathbf{e}_z{}^T & 0 & 0 & 0 \\ -\mathbf{e}_z{}^T & 0 & 0 & 0 \\ -\mathbf{e}_z{}^T & 0 & 0 & 0 \\ -\mathbf{e}_z{}^T & 0 & 0 & 0 \end{pmatrix} \end{cases} \tag{3.28}$$

$$\begin{cases} \mathbf{J}_{\psi}\dot{\psi} = \mathbf{J}_{X\psi}\dot{X}_{PM} \\ \mathbf{J}_{\psi} = \begin{pmatrix} (\mathbf{m}_1^T\mathbf{e}_x)(\mathbf{e}_z^T(\mathbf{L}_1 \times \mu_1)) & & \\ & \ddots & \\ & & (\mathbf{m}_6^T\mathbf{e}_x)(\mathbf{e}_z^T(\mathbf{L}_6 \times \mu_6)) \end{pmatrix} \\ \mathbf{J}_{X_{PM}\psi} = \begin{pmatrix} 0 & \mathbf{e}_z{}^T(\mathbf{L}_1 \times \mu_1) & -(\mathbf{e}_x{}^T\mu_1)(\mathbf{e}_z{}^T\mathbf{L}_1) & (\mathbf{e}_x^T\mathbf{P}_1)(\mathbf{e}_z^T(L_1 \times \mu_1)) \\ \vdots & \vdots & \vdots & \vdots \\ 0 & \mathbf{e}_z{}^T(\mathbf{L}_6 \times \mu_6) & -(\mathbf{e}_x{}^T\mu_6)(\mathbf{e}_z{}^T\mathbf{L}_6) & (\mathbf{e}_x^T P_6)(\mathbf{e}_z^T(L_6 \times \mu_6)) \end{pmatrix} \\ \mathbf{P}_i = \mathbf{P}\mathbf{D}_i^{plat}, \ \forall i = 1, .., 6 \end{cases} \tag{3.29}$$

The rank deficiencies of $\mathbf{J}_{qPM} = \mathbf{J}_{\psi}$ and $\mathbf{J}_{\varphi}$ are given by the following singularities respectively:

$$det(\mathbf{J}_{qPM}) = det(\mathbf{J}_{\psi}) = 0 \iff \exists i_0 \in \{1, .., 6\}; \quad \mathbf{e}_x^T m_{i_0} = 0 \\ \text{or } \mathbf{e}_z^T(L_{i_0} \times \mu_{i_0}) = 0 \tag{3.30}$$

$$det(\mathbf{J}_{\varphi}) = 0 \iff \exists i_0 \in \{1, .., 6\}; \ \mathbf{e}_z^T(L_{i_0} \times \mu_{i_0}) = 0 \tag{3.31}$$

Notice that (3.30) covers also the singularity of $\mathbf{J}_{\varphi}$. The case of $\mathbf{e}_x^T m_{i_0} = 0$ means that m_{i_0} is perpendicular to the x-axis of the base frame which takes place when m_{i_0} lays in the yz plane. Still this configuration is outside the regional workspace and it provokes prior collisions. For the other case, when $\mathbf{e}_z^T(L_{i_0} \times \mu_{i_0}) = 0$, then $\mathbf{L}_{i_0} // \mathbf{e}_z$ which is similar to the what have been discussed in the constraint singularities of collision between sliders and platform.

On the other hand, the rank deficiency of $\mathbf{J}_{X_{PM}}$ is studied after performing some linear operations achieving a simpler form. The changing of $\mathbf{P}$ to be confounded with $\mathbf{D}_1$ was for that purpose. Having a deficient rank of $\mathbf{J}_{X_{PM}}$ turns mathematically into the following equations:

$$\begin{cases} \mathbf{e}_z^T(L_3 \times \mu_3) = \mathbf{e}_z^T(L_4 \times \mu_4) = 0 \\ \text{or} \\ \mathbf{e}_z^T(m_3 \times r_c) = \mathbf{e}_z^T(m_4 \times r_c) = 0 \\ \text{with } \mathbf{r}_c = \mathbf{D}_1 D_3 \equiv \mathbf{D}_1\mathbf{D}_4 \end{cases} \tag{3.32}$$

Or

$$\begin{cases} (\mathbf{e}_z^T(\mathbf{L}_1 \times \mu_1))(\mathbf{m}_1^T\mathbf{e}_x) = 0 \\ \text{and} \\ (\mathbf{e}_z^T(\mathbf{L}_5 \times \mu_5))(\mathbf{m}_5^T\mathbf{e}_x) = 0 \end{cases} \tag{3.33}$$

Or

$$\begin{cases} \mathbf{e}_z^T(L_1 \times \mu_1) = 0 \text{ or } \mathbf{e}_z^T(n_5 \times \mu_5) = 0 \text{ or } \mathbf{L}_5^T e_z = 0 \\ \text{and} \\ \mathbf{e}_z^T(L_5 \times \mu_5) = 0 \text{ or } \mathbf{e}_z^T(n_1 \times \mu_1) = 0 \text{ or } \mathbf{L}_1^T e_z = 0 \end{cases} \tag{3.34}$$

Regarding the conditions in (3.32), for $\mathbf{e}_z^T(L_3 \times \mu_3) = \mathbf{e}_z^T(L_4 \times \mu_4) = 0$ it was proved in the previous paragraph that it is not possible to occur without a prior collision. However, the second case of condition (3.32) (i.e $\mathbf{e}_z^T(m_3 \times r_c) = \mathbf{e}_z^T(m_4 \times r_c) = 0$) can be interpreted by having the four vectors $\mathbf{e}_z, \mathbf{m}_3, \mathbf{m}_4$ and $\mathbf{r}_c$ in the same plane which is either xy or yz plane. This can be take into consideration after a collision between the platform and the slider's plane or between the actuators numbered 3 and 4. Thus, it is not possible to have such condition within the operating workspace. Therefore, condition (3.32) has no concern with singularities inside ARROW's workspace.

Furthermore, both cases of condition (3.33) have been interpreted in the previous parts (particularly condition (3.30)).

Moreover, the possibility of having $\mathbf{e}_z^T(L_1 \times \mu_1) = 0$ or $\mathbf{e}_z^T(n_5 \times \mu_5) = 0$ in condition (3.34) has been verified in the precedent parts that it is outside the executable range of the platform. For the case of $\mathbf{e}_z^T(L_5 \times \mu_5) = 0$ or $\mathbf{e}_z^T(n_1 \times \mu_1) = 0$, it can be analysed as having the three corresponding vectors laying in the same vertical plane. Actually, we have $\mu_i \perp \mathbf{n}_i$ and $\mu_{iz} = 0$, so the only incidents lead to the aforementioned condition is having $\mathbf{n}_1 // \mathbf{e}_z$ or $\mathbf{n}_5 // \mathbf{e}_z$ which occur after falling in collisions. Besides, the situations of $\mathbf{L}_1^T e_z = 0$ and $\mathbf{L}_5^T e_z = 0$ take place just on the boundaries of the workspace, but it can't happen simultaneously because of the assembly condition (3.1) that states $L_z \neq a$. So, as well as for conditions (3.32) and (3.33), condition (3.34) has no possibility to lead ARROW robot falling in any singularity in the feasible workspace.

Lastly, it was shown that ARROW PKM can not fall into any type of singularities inside the functional workspace providing more flexibility and less danger of structural damaging.

3.3 Motion Control

In this section, the proportional-integral-derivative with computed feedforward control solution implemented on ARROW robot is explained. An improved PID controller is used in our case to deal with the produced internal forces due to redundancy. The integral term overshoot and oscillations in case of saturated actuators have been avoided thanks to the anti-windup strategy added to the adopted PID controller.

3.3.1 PID Control in Case of Non-redundancy

The Proportional-Integral-Derivative (PID) control law proposed in Ziegler and Nichols (1942) is the simplest and easiest algorithm that can be adopted for motion control of parallel robots. Although PID is non-robust against nonlinearity effects, changing parameters and uncertainties which are abundant in parallel robots, it is still the most implemented control strategy in the industrial robotized applications.

In a typical non-redundant parallel robot, the PID control law employed in joint space can be defined in discrete form as follows:

$$\begin{cases} \Gamma_{PID}(k) = \Gamma_P(k) + \Gamma_I(k) + \Gamma_D(k) \\ \Gamma_P(k) = \mathbf{K}_P \widetilde{\mathbf{q}}(k) \\ \Gamma_I(k) = \Gamma_I(k-1) + \mathbf{K}_I T_s \widetilde{\mathbf{q}}(k) \\ \Gamma_D(k) = \mathbf{K}_D \left(\frac{\widetilde{\mathbf{q}}(k) - \widetilde{\mathbf{q}}(k-1)}{T_s} \right) \end{cases} \tag{3.35}$$

where $\Gamma_P(k)$, $\Gamma_I(k)$ and $\Gamma_D(k)$ are the corresponding output torques or forces of the proportional, the integral and the derivative controllers at time step k, while $\mathbf{K}_P$, $\mathbf{K}_I$ and $\mathbf{K}_D$ are their diagonal positive definite matrices gains respectively. $\widetilde{\mathbf{q}} = \mathbf{q}_d - \mathbf{q}$ is the joint position error vector defined by the difference between the desired position trajectory and the measured one. T_s is the sampling period. The schematic diagram of the PID controller for non-redundant parallel manipulators is shown in Fig. 3.11.

However, this control scheme doesn't work correctly for the redundant parallel manipulators since there are more control inputs than the DoFs of the moving platform, which will be discussed in the next subsection.

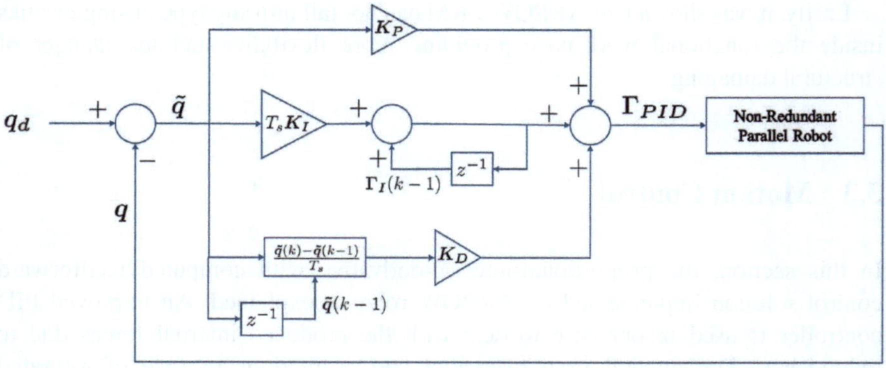

Fig. 3.11 Schematic view of the PID control for non-redundant parallel robot in joint space

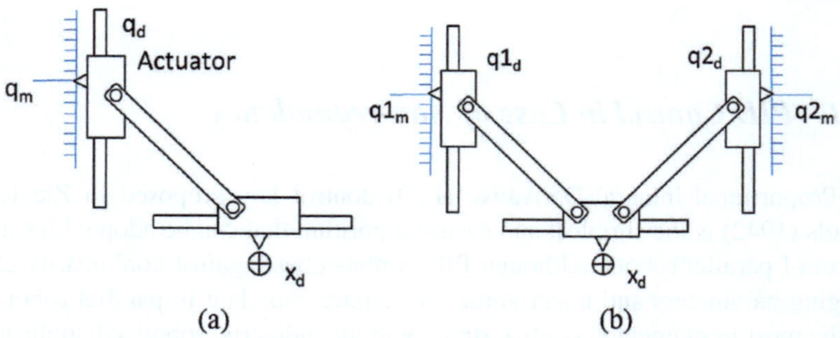

Fig. 3.12 Actuator measurements and the moving platform's position. **(a)** Ideal non-redundant case. **(b)** Ideal redundant case (Yang 2012)

3.3.2 PID Control in Case of Redundancy

Beside the various advantages of the actuation redundancy in parallel robots such as providing more accuracy, improved stiffness and more dexterity, it may declines the control performance of PKMs in the presence of undesired internal generated forces. These forces are mainly due to the "fight against" actuators efforts since there are more inputs than required to accomplish a specific end-effector motion.

3.3.2.1 Actuation Redundancy Effects

Figure 3.12 shows the one DoF of a moving platform which is actuated by one actuator for the non-redundant case and by two actuators in the redundant case. In the ideal case shown in Fig. 3.12, the actuators and moving platform can always track the desired trajectory in both cases of redundancy and non-redundancy. Indeed, geometric errors always work out from various sources such as inaccuracies in the model geometry, measurement errors, assembly errors, non-synchronized control of the actuators, backlashes, thermal expansion, etc (Muller 2009). In the shade of those errors, applying the classical scheme of PID control shown in Fig. 3.11 may lead to an error with the moving platform position tracking while still able to follow the desired actuators position signal. On the other hand, the two actuators in the redundantly actuated systems can hardly reach the desired positions at the same time conflicting to each other, which may generates antagonistic internal forces that may destroy the whole mechanism of the robot.

The position vector $\widetilde{\mathbf{q}}$ can be considered as a vector from the actuator velocity space which always be in $C(\mathbf{J}_m)$[3] in case of non-redundant actuated mechanisms even with the existence of geometric errors. On the contrary, in case of redundantly actuated mechanisms, $\widetilde{\mathbf{q}}$ may has some portions from the null space and not only formed of $C(\mathbf{J}_m)$. These errors may accumulate in the integral term of PID control generated as an input control torques or forces to the actuators. Thus the internal forces of the mechanism will be more and more significant becoming destructive to the mechanisms.

3.3.2.2 Adopted Solutions

Several approaches are proposed in Yang (2012) by following a general idea that is to get rid of the null space portions of $\widetilde{\mathbf{q}}$ in the control loop. One of the propositions is to perform regularization for the input of the PID block as well as for its output.

Regularization Before PID Block

The regularization for the input, $\widetilde{\mathbf{q}}$, consists of two steps described as follows:

1. Transforming the error vector from the joint space to the Cartesian space using the velocity mapping which guarantees the elimination of any residual parts out of $C(\mathbf{J}_m)$.

$$\widetilde{\mathbf{X}} = \mathbf{J}_m^* \widetilde{\mathbf{q}} \tag{3.36}$$

2. Calculating again the joint space error from the Cartesian one by the inverse velocity mapping relationship, which results a regularized joint space error vector.

$$\widetilde{\mathbf{q}}_{Reg} = \mathbf{J}_m \widetilde{\mathbf{X}} \tag{3.37}$$

[3] Columns of $\mathbf{J}_m$.

The transformations in (3.36) and (3.37) are combined as follows:

$$\begin{cases} \widetilde{\mathbf{q}}_{Reg} = R_V \widetilde{\mathbf{q}} \\ \mathbf{R}_V = \mathbf{J}_m \mathbf{J}_m^* \end{cases} \tag{3.38}$$

with $\mathbf{R}_V$ being the regularization matrix in the velocity space.

Regularization After PID Block

As it has been explained in Muller and Hufnagel (2011), the internal forces are caused by the control inputs which are in the null space of the inverse Jacobian matrix. The regularization matrix projects the generated control inputs into the range space of $\mathbf{J}_m$ in two steps described as follows:

1. Using the force mapping relationship, the output forces in the actuation space are transformed to the wrench in the Cartesian space.

$$\mathbf{w} = \mathbf{J}_m^T \Gamma \tag{3.39}$$

2. Calculating again the actuation space forces from the wrench by the inverse force mapping relationship, which results a regularized output forces vector.

$$\Gamma_{Reg} = (\mathbf{J}_m{}^T)^* \mathbf{w} \tag{3.40}$$

The transformations in (3.39) and (3.40) are combined as follows:

$$\begin{cases} \Gamma_{Reg} = \mathbf{R}_\Gamma \Gamma \\ \mathbf{R}_\Gamma = (\mathbf{J}_m^T)^* \mathbf{J}_m{}^T \end{cases} \tag{3.41}$$

with $\mathbf{R}_\Gamma$ being the regularization matrix in the force or torque space. It was shown in Yang (2012) that the two regularization matrices in velocity and force spaces are equal, $\mathbf{R}_\Gamma = \mathbf{R}_V \equiv \mathbf{R}_m$.

Regularization for the Integral Term

The output $\Gamma_{PID}(k)$ generated from the PID given in (3.35) can be shown as a linear function of the current input to PID block $\widetilde{\mathbf{q}}(k)$, the previous input $\widetilde{\mathbf{q}}(k-1)$, and the previous output $\Gamma_{PID}(k-1)$. This linear relation is addressed in the following equation after several manipulations as follows:

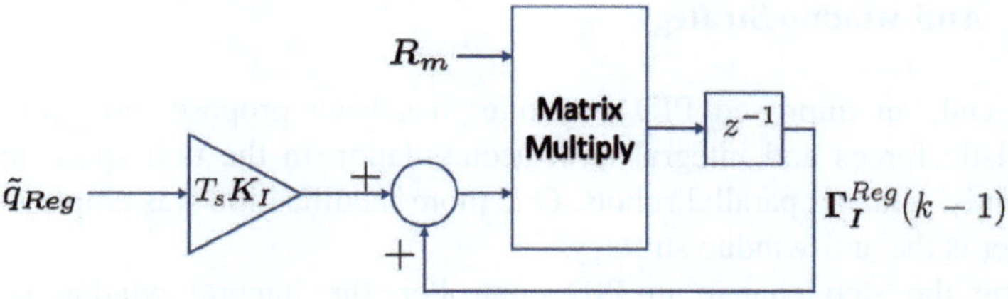

Fig. 3.13 Regularization algorithm for the integral term of PID controller in discrete form

$$\begin{cases} \Gamma_{PID}(k) = \alpha_1 \widetilde{\mathbf{q}}(k) + \alpha_2 \widetilde{\mathbf{q}}(k-1) + \alpha_3 \Gamma_{PID}(k-1) \\ \alpha_1 = \mathbf{K}_P + T_s \mathbf{K}_I + \frac{1}{T_s} \mathbf{K}_P \\ \alpha_2 = -\frac{1}{T_s} \mathbf{K}_D \\ \alpha_3 = I \end{cases} \tag{3.42}$$

To be sure that both the input and output of PID control remain in the same subspace, we shall confirm that $\widetilde{\mathbf{q}}(k-1)$ and $\Gamma_{PID}(k-1)$ have the same subspace as $\widetilde{\mathbf{q}}(k)$.

For a PID block having a regularized input position error vector, $\widetilde{\mathbf{q}}_{Reg}$, the $C(\mathbf{J}_m)$ may conserve its dimensions but may not stay in the same subspace during operation. This may influence the output of PID block, $\Gamma_{PID}(k)$ may not stay always in $R(\mathbf{J}_m)$.[4] Moreover, $\Gamma_{PID}(k-1)$ which is calculated in a recursive way may accumulate undesired quantities in the null spaces. So there is a need to an independent regularization process on the $\Gamma_{PID}(k-1)$ to ensure the space conservation of the input and output of PID block.

As well as the PID block with after regularization needs to a regularized $\Gamma_{PID}(k-1)$ since the generated regularized output Γ_{Reg} eliminates the antagonistic forces from the mechanism but doesn't solve the problem of the increasing value of the integral term with time.

A solution for the aforementioned problems is proposed in Yang (2012) which requires a regularization for the integral term at each time step using the algorithm below (see Fig. 3.13).

At each step time k:

1. Regularize the generated force from the applied integration on the regularized error joint position vector $\widetilde{\mathbf{q}}_{Reg}(k)$.

$$\Gamma_I^{Reg}(k) = \mathbf{R}_m \left(\Gamma_I^{Reg}(k-1) + \mathbf{K}_I T_s \widetilde{\mathbf{q}}_{Reg}(k) \right) \tag{3.43}$$

2. Update the regularized value for the next time step.

$$\Gamma_I^{Reg}(k-1) = \Gamma_I^{Reg}(k) \tag{3.44}$$

[4]Rows of $\mathbf{J}_m$.

3.3.2.3 Anti-windup Strategy

To this end, an improved PID controller has been proposed to deal with the antagonistic forces and integral term accumulation in the null space in case of redundantly actuated parallel robots. One more modification was employed to PID controller is the anti-windup strategy.

During the step change in PID controller, the integral windup occurs for saturation of actuator. Hence the system error decreases more slowly than ideal case causing the value of integral term to be increased, which may lead to significant large overshoot and settling time. This loss of performance provoked by the integral windup in systems of saturated actuators guides mostly to remove the integral term missing out its advantages in omission of the stable residual errors coming from friction or external loads.

The aim behind using the anti-windup strategy is to ensure conservation of the controller stability avoiding the oscillatory behaviour when the actuators saturated. The back-calculation approach of anti-windup strategy proposed in Kumar and Negi (2012) is used here for the PID controller block. It requires feeding back the difference between saturated and unsaturated signals which may reduce the integral value. The integral output of the PID block with the three regularization treatments mentioned before and anti-windup strategy is then written as follows:

$$\begin{cases} \Gamma_I^{Reg}(k) = \mathbf{R}_m\left(\Gamma_I^{Reg}(k-1) + \mathbf{K}_I T_s\left(\tilde{\mathbf{q}}_{Reg}(k) - \mathbf{K}_{AWP}\Delta\Gamma_{Reg}\right)\right) \\ \Delta\Gamma_{Reg} = \Gamma_{Reg}^{sat} - \Gamma_{Reg} \end{cases} \tag{3.45}$$

where Γ_{Reg}^{sat} and Γ_{Reg} are the generated output forces or torques from the PID block after and before saturation block respectively. $\mathbf{K}_{AWP}$ is a diagonal positive definite matrix representing the anti-windup feedback gain. The schematic diagram illustrating the regularization technique of the integral term and the anti-windup strategy for the PID block is shown in Fig. 3.14.

3.3.3 Proposed Control Solution for ARROW PKM

PKMs are well known with their highly nonlinear dynamics which increase considerably when operating at high speed leading to instability. These nonlinearities shall be treated carefully in control algorithms in order to compensate their effects, knowing that linear controllers fail to attain this mission such as PID controller.

In fact, a closed-loop control algorithm full enough with knowledge about the dynamics of the PKM brings out good robustness against nonlinearities and disturbances. A PID with FeedForward control law (PIDFF) is one of the most efficient non-adaptive model-based controllers that exist for parallel manipulators. PDFF control law was firstly suggested in (Liégeois et al. 1980) as an alternative to the

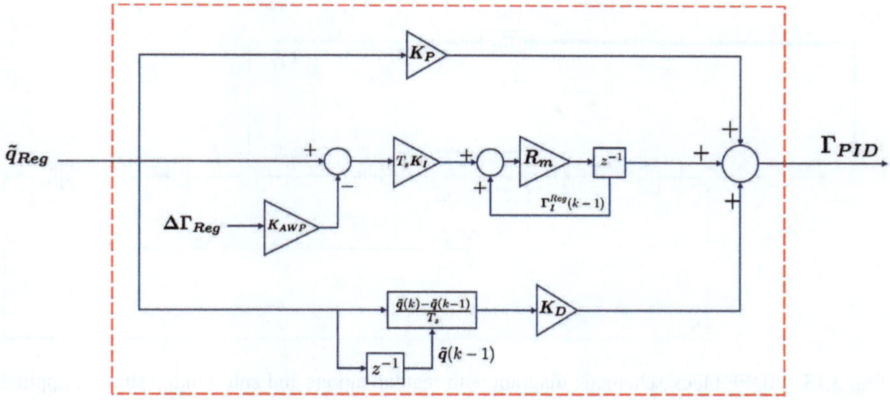

Fig. 3.14 PID block schematic diagram with regularized integral term and anti-windup strategy in the discrete form

on-line computation requirements of others model-based motion control schemes, particularly the computed torque control method. PIDFF control law consists of a linear PID feedback plus a feedforward part results from the dynamics of the robot evaluated using the desired motion trajectory. Consequently, the feedforward dynamics can be computed offline reducing the control computation efforts and making this type of model-based controllers easy and simple for implementation as well as PID controller.

For ARROW PKM, the implemented control law is the PIDFF controller using the PID block recently obtained in Fig. 3.14, individually for each part of ARROW: the parallel module and the turntable. The PIDFF control implemented for ARROW PKM can be written in joint space as follows:

$$\begin{cases} \mathbf{U}_{PM;T} = \mathbf{R}_m\left(\mathbf{K}_P\tilde{q} + \Gamma_I^{Reg} + K_D\dot{\tilde{q}}\right) + \Gamma_{PM;Tff} \\ \Gamma_{PMff} = \mathbf{H}_d^*\left(\ddot{X}_d^{PM} + \Lambda_d\dot{X}_d^{PM} - A_{Gd}\right) \\ \Gamma_{Tff} = \begin{pmatrix} I_T/2 \\ -I_T/2 \end{pmatrix}\theta_{xd} \end{cases} \tag{3.46}$$

where Γ_I^{Reg} is the modified integral term including the regularization and anti-windup strategy obtained in (3.45). $\mathbf{X}_d^{PM}$, $\dot{\mathbf{X}}_d^{PM}$ and $\ddot{\mathbf{X}}_d^{PM}$ are the desired position, velocity and acceleration of the moving platform respectively. $\mathbf{H}_d$, Λ_d and $\mathbf{A}_{Gd}$ are the matrices used in the dynamic model of the parallel module in (3.22) calculated based on the desired generated trajectory. $\mathbf{U}_{PM;T}$ is the control input forces vector applied to the linear actuators or torques vector applied to the rotative motors. θ_{xd} is the desired generated rotation in the Cartesian space around x-axis. The proposed control solution for ARROW PKM is described as a schematic view in Fig. 3.15.

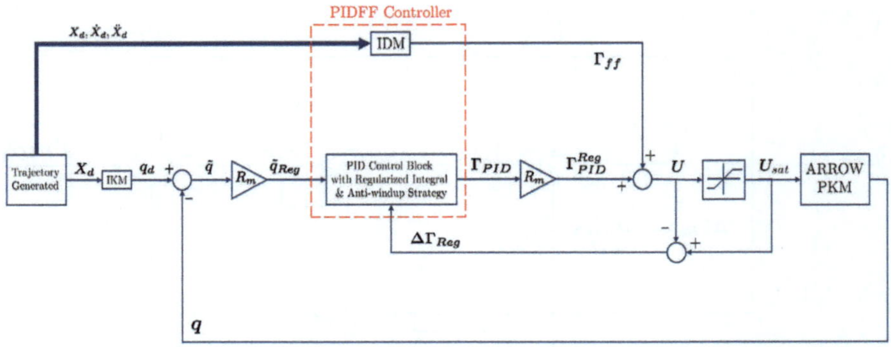

Fig. 3.15 PIDFF block schematic diagram with regularizations and anti-windup strategy applied on ARROW PKM

It is visible in (3.46) that feedforward dynamics are composed from the desired trajectories making this controller preferable for real-time implementation. In other meaning, PIDFF controller enhances the general performance and increases the accuracy with the same processing time cost of a simple PID. As well as it avoids the noise estimations of velocity and acceleration for the actuators knowing that most of the real manipulators are equipped only with position sensors (i.e encoders). Nevertheless, such kind of model-based controllers needs a precise knowledge about the dynamics of the parallel robot, which is a difficult development task, in order to achieve high level of performances.

The stability analysis of this model-based controller has been studied in the literature. In Reyes and Kelly (2001), it has been reported that the position error of a PD control with computed feedforward will vanish asymptotically in a local sense after selecting properly the feedback matrices. Furthermore, it has been proved in Santibanez and Kelly (2000) that this controller is able to yield a globally asymptotically stable closed-loop system with an experimental validation.

3.4 Motion Generation

A strong requirement in the motion generation for a machining tool is the constant feed rate. The tool have to maintain a constant velocity along the path. To ensure this, the Gcode position data series coming from a CAD/CAM software are used as input for a Spline based definition of path.

3.4.1 Spline Description

We can extract, from the Gcode instructions, the tool positions along the machining path. The first idea in creating a trajectory through points is to generate straight

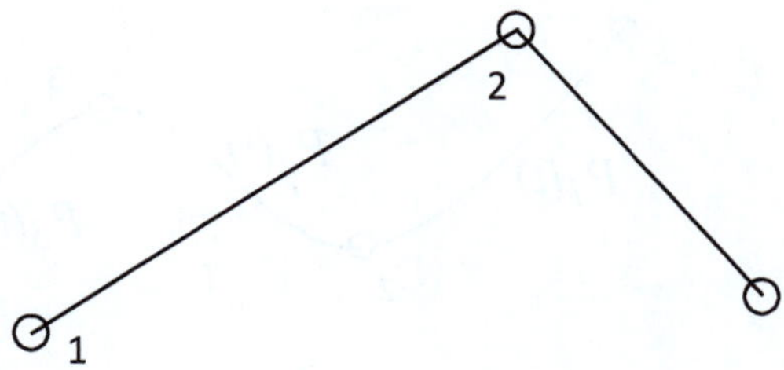

Fig. 3.16 Straight lines path with C_0 position continuity

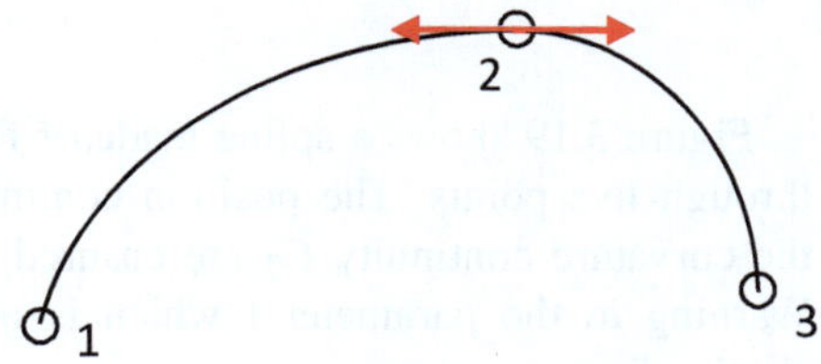

Fig. 3.17 Path with a tangency continuity C_1

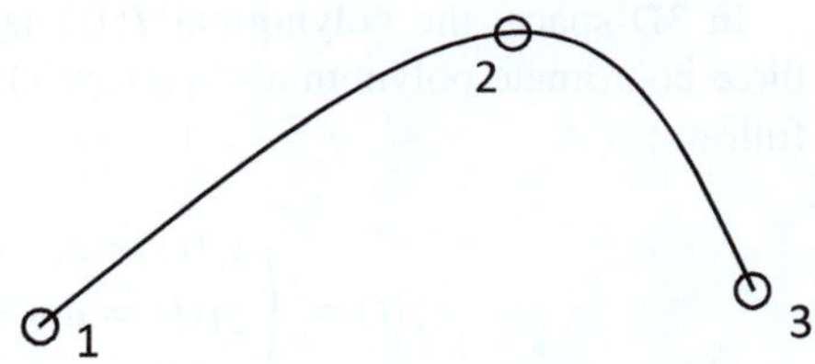

Fig. 3.18 Path with a curvature continuity C_2

lines between points such as in Fig. 3.16. This kind of trajectory ensures a position continuity at each junction point called C_0 continuity.

Off course, this kind of trajectories will generate velocity discontinuities along the path which is not suitable to generate a proper machining motion. A second manner to generate a trajectory is to create polynomials between points and ensure that the successive polynomials are tangent at junction points (see Fig. 3.17). This kind of continuity is called C_1 and is characterized by an equality of the first derivatives of successive polynomials at a junction point.

In the context of high speed machining, the continuity C_1 is not sufficient. Indeed, the path curvature from one side to the other of a point isn't continuous (ex: point 2 in Fig. 3.17). To ensure the continuity of curvature, called C_2 continuity, the second derivative of successive polynomials have to be equal at the junction points (ex: point 2 in Fig. 3.18).

This natural cubic spline technique is used to create a path which is optimal for the machining process. A spline is a function defined piecewise by polynomials between each successive pair of points describing the path. In the case of natural cubic spline, those polynomials are of degree three with the following form:

$$P(t) = a + bt + ct^2 + dt^3 \tag{3.47}$$

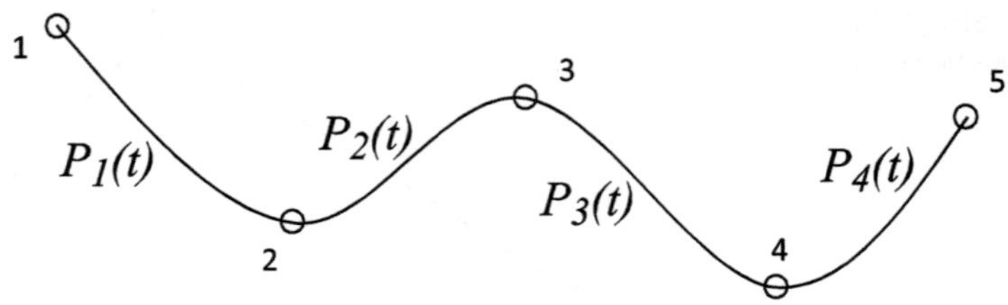

Fig. 3.19 Spline passing through four points (defined by four polynomials)

Figure 3.19 shows a spline made of four polynomials ($P_1(t)$, .., $P_4(t)$) passing through five points. The position continuity C_0, the tangency continuity C_1 and the curvature continuity C_2 are ensured by the use of natural cubic spline curve. Warning to the parameter t which is not a time variable but a sort of "spline abscissa".

In 3D space, the polynomial $P_i(t)$ is represented by a concatenated vector of three coordinate polynomials $x_i(t)$, $y_i(t)$ and $z_i(t)$ with a similar form to (3.47) as follows:

$$P_i(t) = \begin{pmatrix} x_i(t) = a_{ix} + b_{ix}t + c_{ix}t^2 + d_{ix}t^3 \\ y_i(t) = a_{iy} + b_{iy}t + c_{iy}t^2 + d_{iy}t^3 \\ z_i(t) = a_{iz} + b_{iz}t + c_{iz}t^2 + d_{iz}t^3 \end{pmatrix} \tag{3.48}$$

The Matlab curve fitting toolbox provides the "cscvn" function allowing to get the natural cubic spline definition. This function takes as input the n Gcode path points series and gives as output the spline structure. It is specified by a "break sequence" along the "spline abscissa" for each Gcode path point and the "coefficient array" of polynomials composing the spline shown respectively as follows:

$$breaks = [0\ break_1\ break_2\ \ldots\ break_n] \tag{3.49}$$

$$coefs = \begin{pmatrix} coef\ for\ P_1 = \begin{cases} d_{1x}\ c_{1x}\ b_{1x}\ a_{1x} \\ d_{1y}\ c_{1y}\ b_{1y}\ a_{1y} \\ d_{1z}\ c_{1z}\ b_{1z}\ a_{1z} \end{cases} \\ coef\ for\ P_2 = \begin{cases} d_{2x}\ c_{2x}\ b_{2x}\ a_{2x} \\ d_{2y}\ c_{2y}\ b_{2y}\ a_{2y} \\ d_{2z}\ c_{2z}\ b_{2z}\ a_{2z} \end{cases} \\ \vdots \qquad\qquad \vdots \\ coef\ for\ P_n = \begin{cases} d_{nx}\ c_{nx}\ b_{nx}\ a_{nx} \\ d_{ny}\ c_{ny}\ b_{ny}\ a_{ny} \\ d_{nz}\ c_{nz}\ b_{nz}\ a_{nz} \end{cases} \end{pmatrix} \tag{3.50}$$

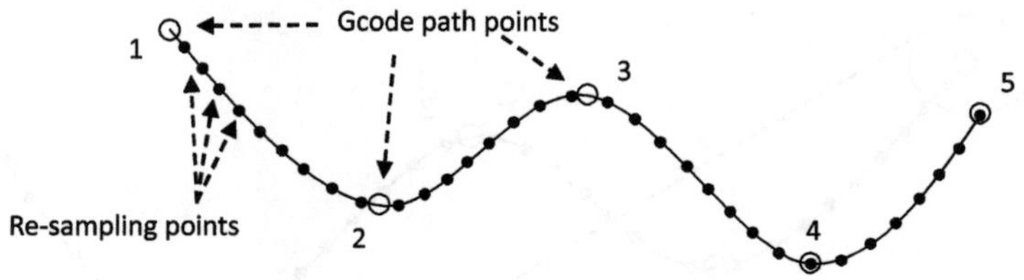

Fig. 3.20 Trajectory re-sampling at each sampling period

Thus, the polynomial $P_1(t)$ gives the position of Gcode point 1 when parameter $t = 0$ and gives the position of point 2 when parameter $t = break_1$. The spline $P_2(t)$ gives the position of Gcode point 2 when parameter $t = 0$ and gives the position of point 3 when parameter $t = break_2$.

3.4.2 Spline Re-sampling During Constant Velocity Phase

During the real-time execution of the machining motion, the sampling period is fixed at 0.2 ms (5 KHz). The Gcode path points are generated from CAD/CAM geometrical specifications and are not distributed in terms of sampling period. The spline re-sampling will consist in generating the path at each sampling period as illustrated in Fig. 3.20.

As the velocity has to be constant along the path during a machining process, we have made the assumption that the distance between two successive re-sampled points is constant. Indeed, we consider that distance between two points along the path is very close to the cord distance between those two points. The determination of re-sampling points position is done by the following iterative process:

1. From Gcode point 1, small iterations δt are done on the parameter t and the corresponding position on the curve is computed from spline polynomial $P_1(t)$ as shown in Fig. 3.21.
2. This is repeated till the distance between the Gcode point 1 and the actual position on the curve exceeds the desired distance between two successive re-sampling points.
3. The last position reached via this iterative process is considered as the first re-sampling point.

The process is re-iterated for the following successive re-sampling points (black points in Fig. 3.21) till the reaching of Gcode points 2, 3 and so on. The more tiny δt is considered, the more precise process is obtained. During this process, the increasing of parameter t along the break sequence allows to know which polynomial has to be used to compute the positions on the curve (ex: (t) is used between Gcode points 1 and 2, $P_2(t)$ between Gcode points 2 and 3).

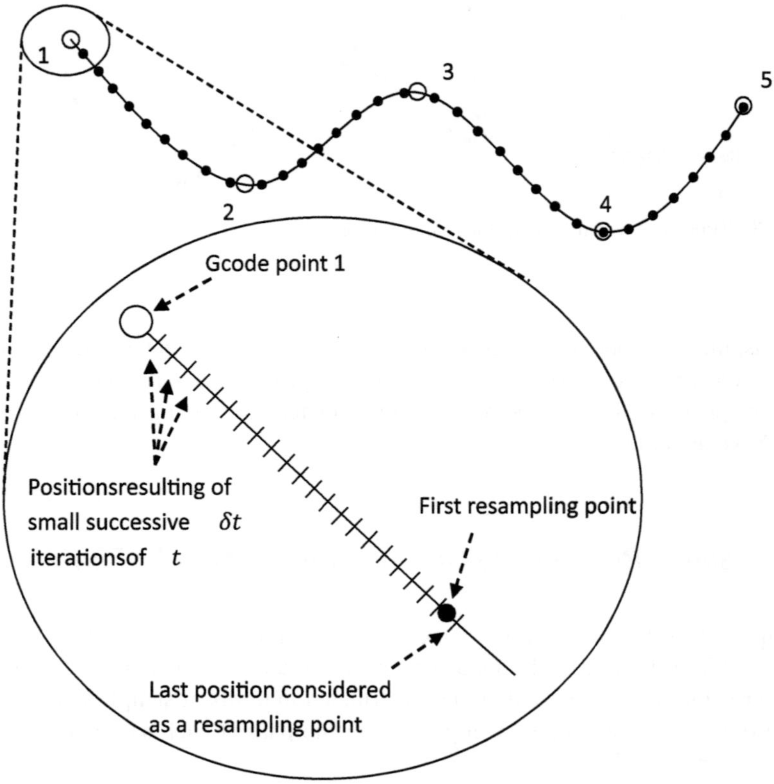

Fig. 3.21 Iterative process for re-sampling point determination

During the constant velocity phase, the use of spline curve guarantees an optimal trajectory in terms of velocity and acceleration discontinuities for machining motion.

3.4.3 Acceleration and Deceleration Phases

The previous section deals with the motion generation along the spline curve when the velocity is constant. In a fast machining process, the reaching of the constant velocity phase has to be done in a smooth way aiming at increasing durability of mechanical parts. The motion has to start with an acceleration phase and to finish with a deceleration phase. A suitable acceleration procedure is used in integrating a constant jerk profile to get a proper acceleration and velocity profile. The Jerk profile (Fig. 3.22a) is introduced as follows:

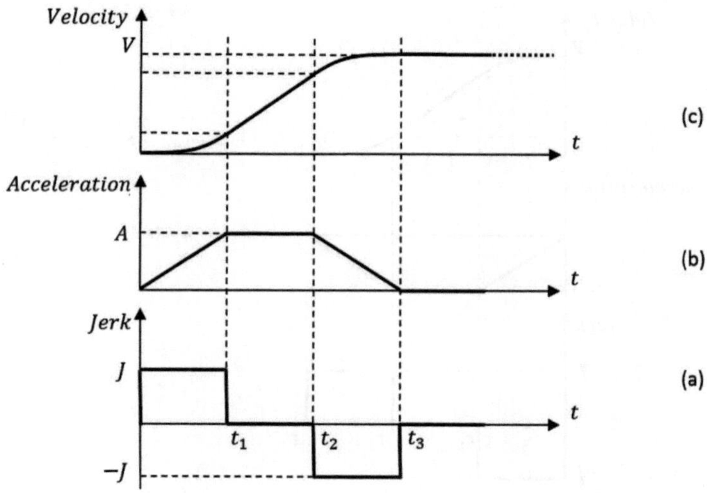

Fig. 3.22 Acceleration phase

$$Jerk = \begin{cases} J, & 0 < t < t_1 \\ 0, & t_1 < t < t_2 \\ -J, & t_2 < t < t_3 \end{cases} \tag{3.51}$$

The integration of Jerk versus time t gives the acceleration profile shown in Fig. 3.22b such as:

$$Acceleration = \begin{cases} Jt, & 0 < t < t_1 \\ A, & t_1 < t < t_2 \\ A - J(t - t_2), & t_2 < t < t_3 \end{cases} \tag{3.52}$$

A second integration of acceleration versus time t gives the Velocity profile shown in Fig. 3.22c such as:

$$Velocity = \begin{cases} \frac{Jt^2}{2}, & 0 < t < t_1 \\ \frac{Jt_1^2}{2} + A(t - t_1), & t_1 < t < t_2 \\ \frac{Jt_1^2}{2} + A(t_2 - t_1) + A(t - t_2) - \frac{J(t-t_2)^2}{2}, & t_2 < t < t_3 \end{cases} \tag{3.53}$$

The deceleration phase is only an inversion of acceleration phase as shown in Fig. 3.23. Thus, the Velocity profile can be given as follows:

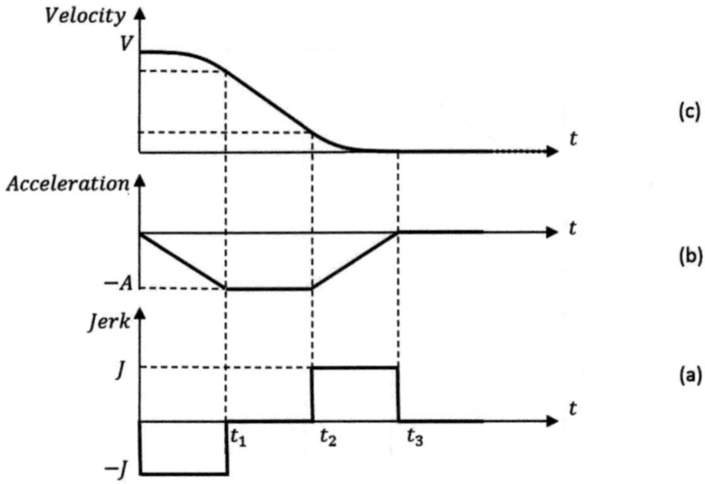

Fig. 3.23 Deceleration phase

$$Velocity = \begin{cases} V - \left(\frac{Jt^2}{2}\right), & 0 < t < t_1 \\ V - \left(\frac{Jt_1^2}{2} + A(t - t_1)\right), & t_1 < t < t_2 \\ V - \left(\frac{Jt_1^2}{2} + A(t_2 - t_1) + A(t - t_2) - \frac{J(t-t_2)^2}{2}\right), & t_2 < t < t_3 \end{cases} \tag{3.54}$$

Those velocity profiles are used in determination of re-sampling points spacing along the spline curve during acceleration and deceleration phases. Now, we have gotten a complete motion description at a certain sampling period.

3.5 Real-Time Experiments

This section exposes the real-time evaluation for the machining performance of the introduced ARROW robot. It covers the description of the experimental testbed consists of ARROW, the generated machining trajectory and the results concerning precision and performance.

3.5.1 Experimental Testbed of the ARROW Robot

The mechanical structure and general mechanism of ARROW PKM has been discussed previously a lot in this chapter. The manufactured ARROW robot is shown in Fig. 3.24, where you can see the turntable with the object to be machined, the arms and the actuators.

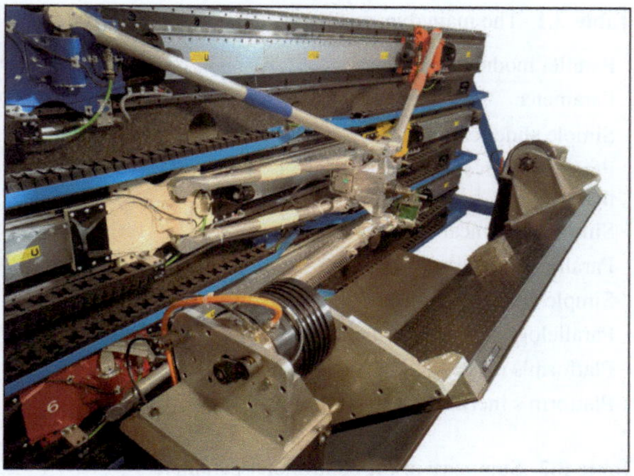

Fig. 3.24 The manufactured ARROW robot

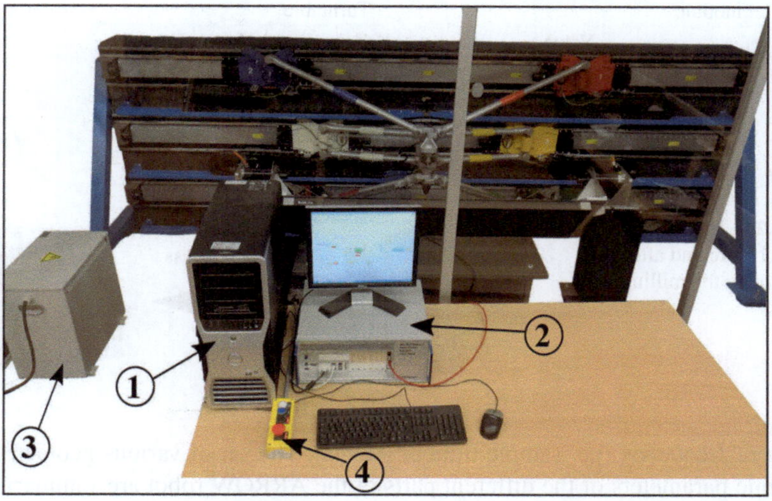

Fig. 3.25 The experimental setup of ARROW robot. 1: Master computer, 2: xPC Target, 3: Power supply, 4: Emergency button

The parallel module actuators are all identical of type Ironless ETEL ILM12-060, providing each one a maximum force of 2500 N and reaching speed up to 15 m/s.

The actuators of the turntable are of type TMB0140-100-3RBS ETEL direct-drive motors. They can provide a maximum peak torque of 127 Nm and they are able to reach 550 rpm of maximum speed. Each actuator is equipped with a non-contact incremental optical encoder providing a total number of 5000 pulses per revolution.

The experimental setup of the ARROW robot is displayed in Fig. 3.25. Simulink and Real-Time Workshop from Mathworks Inc. are used to implement the control scheme and to generate the real-time execution code. The target PC is running under

Table 3.1 The main dynamic parameters of ARROW robot

Parallel module	
Parameter	Value
Simple slider cart's mass	11.1 kg
Parallelogram slider cart's mass	11.34 kg
Simple arm linear mass	1.744 kg/m
Parallelogram arm linear mass	3.488 kg/m
Simple arms' length	0.96 m
Parallelograms arms' length	0.61 m
Platform's mass	10.2 kg
Platform's inertia	0.414 kg.m^2

Turntable	
Parameter	Value
Turntable actuator's inertia	0.004 kg.m^2
Total inertia of the turntable	1.204 kg.m^2

Table 3.2 Summary of the controllers' parameters

Parallel module	
Gain	Value
k_P	5860796.8
k_D	8070.24
k_I	20425584.8
k_{AWP}	166×10^{-5}

Turntable	
Gain	Value
k_P	5550
k_D	25.9
k_I	7400
k_{AWP}	37×10^{-6}

Fig. 3.26 The machining object before and after performing the milling process

5 kHz of frequency (i.e. sample time of 0.2 ms). The set of various geometric and dynamic parameters of the different parts of the ARROW robot are summarized in Table 3.1.

The control tuning gains that are specified for both control architectures of parallel module and turntable are shown in Table 3.2.

The chosen machining trajectory leads to obtain the pyramid squared shape seen in Fig. 3.26 after the milling operation is performed on a cubed shape. A 3D plot of the adopted trajectory is illustrated in Fig. 3.27 from the side view of the parallel module. The chosen machined shape estimates the five DoFs of ARROW robot and allows us to evaluate the whole performance in the accessible workspace.

3.5.2 Experimental Results

In order to evaluate the performance of the proposed control solution, PIDFF control law, on the whole trajectory in both Joint and Cartesian spaces, the Root Mean

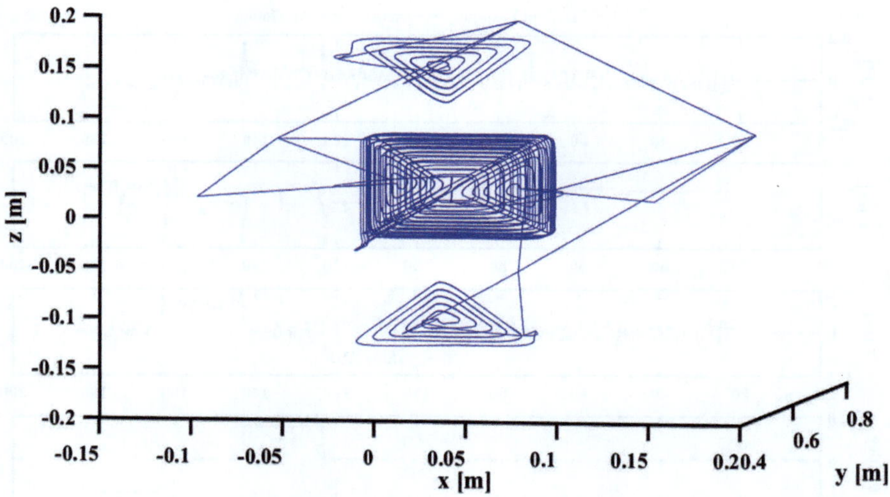

Fig. 3.27 3D plot for the machining trajectory as if seen from the parallel module side view

Square of the tracking Error in Translational (RMSET) and Rotational (RMSER) motions are calculated as follows:

$$RMSET_C = \Big(\frac{1}{N}\sum_{i=1}^{N}\big[\widetilde{x}^2(i) + \widetilde{y}^2(i) + \widetilde{z}^2(i)\big]\Big)^{1/2} \tag{3.55}$$

$$RMSER_C = \Big(\frac{1}{N}\sum_{i=1}^{N}\big[\widetilde{\theta}_z^2(i) + \widetilde{\theta}_x^2(i)\big]\Big)^{1/2} \tag{3.56}$$

$$RMSET_J = \Big(\frac{1}{N}\sum_{i=1}^{N}\sum_{j=1}^{6}\widetilde{q}_j^2(i)\Big)^{1/2} \tag{3.57}$$

$$RMSER_J = \Big(\frac{1}{N}\sum_{i=1}^{N}\sum_{j=1}^{2}\widetilde{q}_j^2(i)\Big)^{1/2} \tag{3.58}$$

where N is the number of the time-samples.

The actual operating motion of the moving platform compared to the desired trajectory shows a high performance in terms of accuracy. Both trajectories are plotted in Fig. 3.28. The evolution of the tracking Cartesian error is shown in Fig. 3.29 zoomed to the interval [90,95] seconds for clarification purposes. The evaluation of the Cartesian tracking error is done by calculating the root mean square error over the whole trajectory and is shown in Table 3.3. The root mean square error values show a very small error over the whole trajectory which validates the high expected precision of the designed ARROW robot.

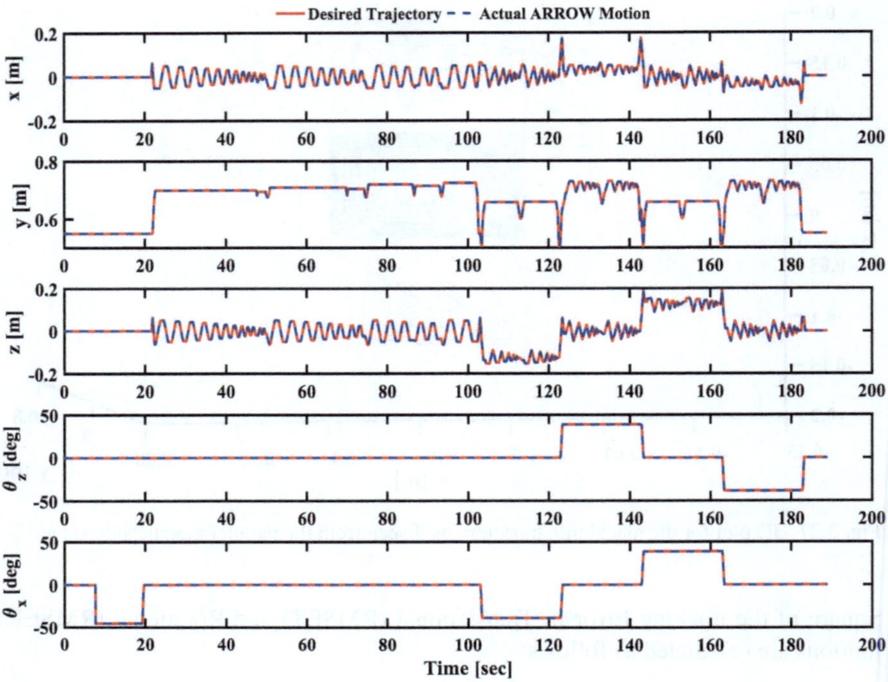

Fig. 3.28 Evolution of the tracking Cartesian coordinates while following the machining trajectory

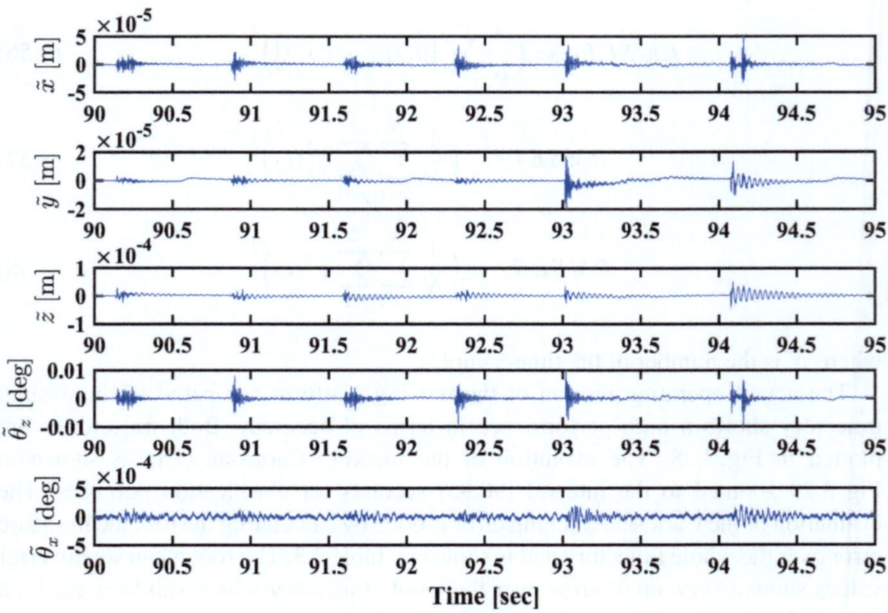

Fig. 3.29 Evolution of the Cartesian tracking error while following the machining trajectory

Table 3.3 Root mean square of Cartesian and joint tracking errors

Cartesian space		Joint space	
RMSET [m]	RMSER [deg]	RMSET [m]	RMSER [deg]
7.0574×10^{-6}	0.0170	6.8796×10^{-6}	0.0240

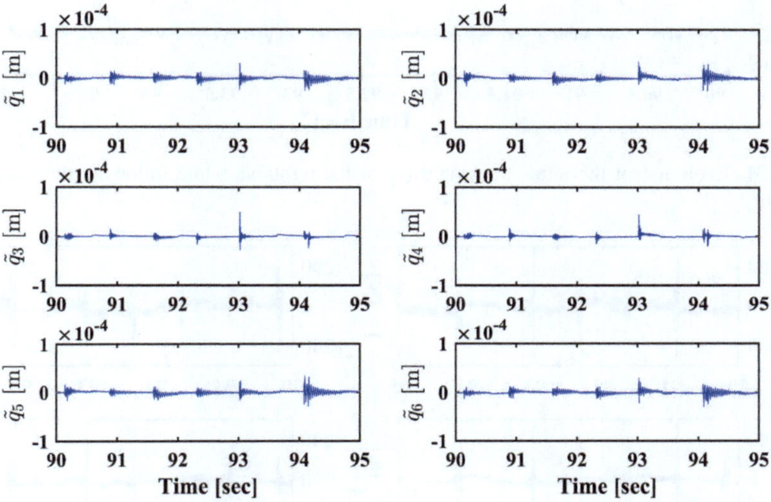

Fig. 3.30 Evolution of the joints tracking error of the parallel module while following the machining trajectory

The tracking joint errors of the parallel module and the turntable are shown in Figs. 3.30 and 3.31 respectively. Also it verifies a very high performance of trajectory tracking machining task, and the evaluations of the root mean square error are mentioned in Table 3.3 approving the good precision. Indeed, the monitoring of Cartesian coordinates is available through measuring the joints positions using encoders, and then transforming these measurements into Cartesian space by kinematic mapping (Jaccobian). This is used in the most of the parallel manipulators. As shown, both joint and Cartesian tracking trajectories achieve very high accuracy which is an extra evidence to the accurate kinematic model designed for ARROW robot.

The evolution of the control input forces and torques applied to their corresponding linear and rotative motors are shown in Figs. 3.32 and 3.33 respectively. It is clear that all the control inputs are in the safe range of the allowed capability for the actuators. Moreover, it seems that at some instants, the actuators reached the saturation zone, but the control maintained the stability and performance thanks to the anti-windup strategy that eliminates overshoots and prevent undesired oscillatory behaviour which might be dangerous for the structure.

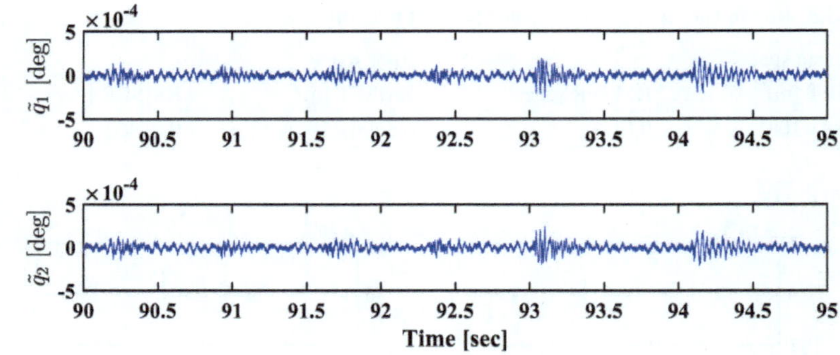

Fig. 3.31 Evolution of the joints tracking error of the turntable while following the machining trajectory

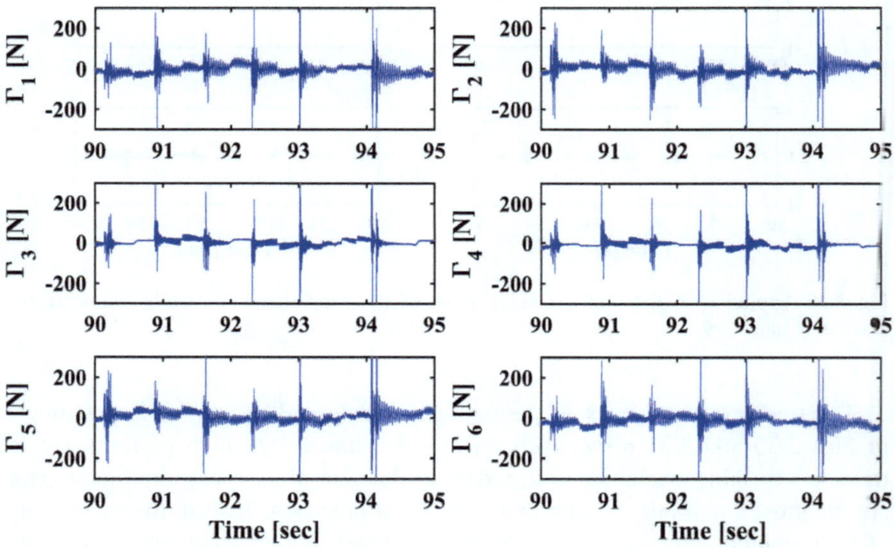

Fig. 3.32 Evolution of the control input forces of the parallel module linear motors

3.6 Conclusions and Future Work

In this work, a 5-DoFs (3T-2R) machining tool named ARROW robot based on redundantly actuated parallel structure was presented. ARROW robot is featured with a special mechanism providing a combination between rapidity and precision with large operational workspace without any type of interior singularities. A detailed explanation for the kinematics and dynamics associated to ARROW PKM were introduced in this chapter, in addition to the singularity analysis. In order to follow up accurately a machining trajectory, an improved PID with computed feedforward control law has been proposed. Regularization techniques were used to

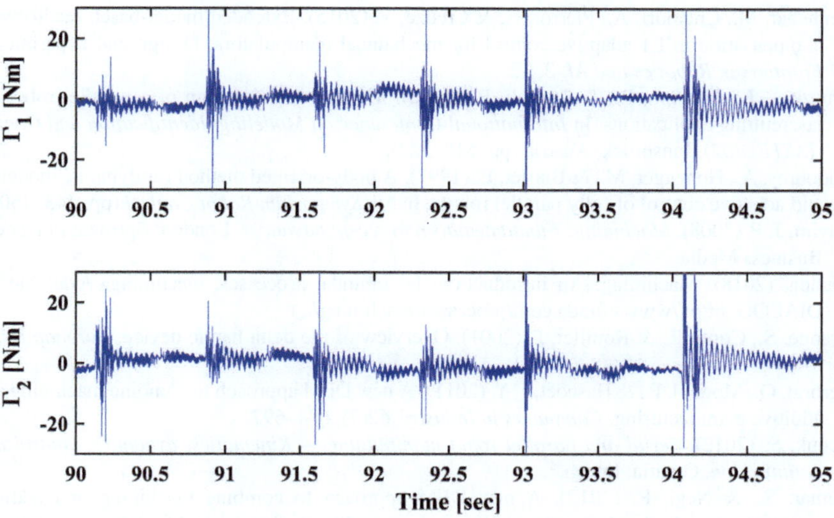

Fig. 3.33 Evolution of the control input torques of the turntable rotative motors

eliminate the undesired antagonistic forces resulting from redundancy and to deal with the increasing value of the integral term in the PID control block. Thanks to the adopted anti-windup strategy which preserves the stability of the controller in case of saturation. Real-time experiments were conducted on ARROW PKM making a certain machining task. The general performance was evaluated using the root mean square error criteria justifying a good sufficient accuracy for a machining tasks.

For the future work, one can enhance more the control performance in terms of precision, motion speed and robustness. Implementing some adaptive model-based controllers based on the online dynamic calibration technique provides more robustness against parameters variation and disturbances. From a control point of view, considering more aspects in a PKM to be enclosed in the control-loop system can improve the general performance, such as motor drives, actuator dynamics, transmission system and friction in the articulations.

Acknowledgements This work has been supported by the ARPE ARROW project.

References

Bennehar, M., Chemori, A., & Pierrot, F. (2014a). A new extension of desired compensation adaptive control and its real-time application to redundantly actuated PKMs. In *Intelligent Robots and Systems (IROS)*, Chicago, IL, USA.

Bennehar, M., Chemori, A., & Pierrot, F. (2014b). A novel RISE-based adaptive feedforward controller for redundantly actuated parallel manipulators. In *Intelligent Robots and Systems (IROS)*, Chicago, IL, USA.

Bennehar, M., Chemori, A., Pierrot, F., & Creuze, V. (2015). Extended model-based feedforward compensation in L1 adaptive control for mechanical manipulators: Design and experiments. *Frontiers in Robotics and AI, 2*, 32.

Bruzzone, L., Molfino, R., & Razzoli, R. (2002). Modelling and design of a parallel robot for lasercutting applications. In *International Conference on Modeling, Identification and Control (IASTED'02)*, Innsbruck, Austria (pp. 518–522).

Codourey, A., Honegger, M., & Burdet, E. (1997). A body-oriented method for dynamic modeling and adaptive control of fully parallel robots. In *5th Symposium Robot Control* (pp. 443–450).

Davim, J. P. (2008). *Machining: Fundamentals and recent advances*. London: Springer Science & Business Media.

eFunda. (2018). Machining: An introduction. In: eFunda, processes, machining. Available via DIALOG. http://www.efunda.com/processes/machining/

Grange, S., Conti, F., & Rouiller, P. (2001). Overview of the delta haptic device. *Eurohaptics, 1*, 5–7.

Kerbrat, O., Mognol, P., & Hascoët, J. Y. (2011). A new DFM approach to combine machining and additive manufacturing. *Computers in Industry, 62*(7), 684–692.

Kucuk, S. (2012). *Serial and parallel robot manipulators – Kinematics, dynamics, control and optimization*. Croatia: Intech.

Kumar, S., & Negi, R. (2012). A new DFM approach to combine machining and additive manufacturing. In *2nd International Conference on Power, Control and Embedded Systems*.

Li, Y., & Xu, Q. (2007). Design and development of a medical parallel robot for cardiopulmonary resuscitation. *IEEE/ASME Transaction on Mechatronics, 12*(3), 265–273.

Liégeois, A., Fournier, A., & Aldon, M. J. (1980). Model reference control of high-velocity industrial robots. In *Joint Automatic Control Conference*, San Francisco.

Merlet, J. P. (2006). Introduction (chapter 1). Structural synthesis and architectures (chapter 2). In Gladwell, G. (Ed.), Parallel robots (Solid mechanics and its applications, 2nd ed.). The Netherlands: Springer.

Muller, A. (2009). Effects of geometric imperfections to the control of redundantly actuated parallel manipulators. In *IEEE International Conference on Robotics and Automation (ICRA'09)* (pp. 1782–1787).

Muller, A., & Hufnagel, T. (2011) A projection method for the elimination of contradicting control forces in redundantly actuated PKM. In *IEEE International Conference on Robotics and Automation (ICRA'11)* (pp. 3218–3223).

Natal, G. S., Chemori, A., & Pierrot, F. (2012). Dual-space adaptive control of redundantly actuated parallel manipulators for extremely fast operations with load changes. In *IEEE International Conference on Robotics and Automation (ICRA)*, Saint Paul, MN, USA.

Natal, G. S., Chemori, A., & Pierrot, F. (2015). Dual-space control of extremely fast parallel manipulators: Payload changes and the 100G experiment. *IEEE Transaction on Control Systems Technology, 23*(4), 1520–1535.

Reyes, F., & Kelly, R. (2001). Experimental evaluation of model-based controllers on a direct-drive robot arm. *Mechatronics, 11*, 267–282.

Santibanez, V., & Kelly, R. (2000). PD control with feedforward compensation for robot manipulators: Analysis and experimentation. *Robotica, 19*, 11–19. Cambridge University Press.

Shayya, S. (2015). *Towards rapid and precise parallel kinematic machines*. Ph.D. thesis, Université Montpellier (Ex UM2).

Shayya, S., Krut, S., & Company, O. (2014). Dimensional synthesis of 4 dofs (3t-1r) actuatedly redundant parallel manipulator based on dual criteria: Dynamics and precision. In *IEEE/RSJ International Conference on Intelligent Robots and Systems (IROS'14)* (pp. 1716–1723).

Shoham, M., Burman, M., & Zehavi, E. (2003). Bone-mounted miniature robot for surgical procedures: Concept and clinical applications. *IEEE Transaction on Robotics and Automation, 19*(5), 893–901.

Stewart, D. (1965). A platform with six degrees of freedom. *Proceedings of the Institution of Mechanical Engineers, 180*, 371–386. ARCHIVE.

Su, Y., Duan, B., & Zheng, C. (2004). Nonlinear pid control of a six-dof parallel manipulator. *IEEE Proceedings-Control Theory and Applications, 151*, 95–102.

Yang, H. (2012). *Agile mobile manufacturing for large workpieces*. Ph.D. thesis, Université Montpellier.

Youssef, H. A., & El-Hofy, H. (2008). *Machining technology: Machine tools and operations*. Boca Raton: Taylor & Francis.

Zhang, Y. X., Cong, S., & Shang, W. W. (2007). Modeling, identification and control of a redundant planar 2-dof parallel manipulator. *International Journal of Control, Automation and Systems, 5*, 559–569.

Ziegler, J., & Nichols, N. (1942). Optimum settings for automatic controllers. *Transaction of the American Society of Mechanical Engineer, 64*, 759–768.

Chapter 4
Cable-Driven Parallel Robot Modelling for Rehabilitation Use

Hachmia Faqihi, Maarouf Saad, Khalid Benjelloun, Mohammed Benbrahim, and M. Nabil Kabbaj

Abstract The aim of this chapter is to presents a Kinematic analysis of a Cable-Driven Robot for rehabilitation use of human lower limb, by taking into account the constraints required by the entrainment system and the mobile platform (human leg). The proposed approach is focused on optimizing the manipulability and the human performance of the human leg, as being a physiologically constrained three-link arm. The obtained forward kinematic model leads to define the feasible workspace of the human leg in the considered configuration. Using an effective optimization-based human performance measure that incorporates a new objective function of musculoskeletal discomfort, and the mapping relation between articular joints actuator, length cables and articular joint mobile platform, the optimal inverse kinematic (IK) model is obtained.

Keywords Cable-Driven Robot · Rehabilitation · Lower limb · Optimization · Inverse Kinematics · Trajectory generation · Minimum Jerk

---

H. Faqihi (✉) · K. Benjelloun
Automatic and Industrial Informatics Laboratory (LAII), Ecole Mohammadia d'Ingenieurs, Mohammed V University, Rabat, Morocco
e-mail: hachmiafaqihi@research.emi.ac.ma; bkhalid@emi.ac.ma

M. Saad
Electrical Engineering Department, Ecole de Technologie Superieure, Montreal, QC, Canada
e-mail: maarouf.saad@etsmtl.ca

M. Benbrahim · M. N. Kabbaj
Faculty of Sciences, Integration of Systems and Advanced Technologies Laboratory (LISTA), University of Fez, Fez, Morocco
e-mail: mohammed.benbrahim@usmba.ac.ma; n.kabbaj@usmba.ac.ma

N. Derbel et al. (eds.), *New Developments and Advances in Robot Control*, Studies in Systems, Decision and Control 175,
https://doi.org/10.1007/978-981-13-2212-9_4

4.1 Introduction

Recently, the number of patients with neurological disorders is in perpetual growth. These disorders are carried out from the neuromuscular diseases, stroke, and spinal cord injury cerebellar disorders, or the impaired functions of the member musculature after a serious injury, illness or surgery, which lead among others to the problems in the lower limb joint and generate atypical gait movement, thereby reducing patient's quality of life.

Rehabilitation process based on physical therapy are considered as the best method to restore patient's strength, and the mobility fitness. Traditionally physical therapy sessions were carried out manually, and require a highly qualified staff. However, due to the increase in the number of patients who need such care, and the poor performances reached in traditionally rehabilitation process in term of quickiness, and efficience many researchers have been encouraged to require the robotic, where a good repeatability, and a precisely controllable assistance, providing quantitative measures of the subject's performance and reducing the required labor of physical therapists, and the financial burden on patients (Pennycott et al. 2012).

Several robotic solutions have been proposed for rehabilitation use. Their designs change depending on the desired level of sophistication. However, the major robotic solutions developed for neurologically impaired patients, use rigid links and mechanical joints attached to human limbs and joints, which adds extra weight and inertia to the human limbs, and thereby changes the natural motion dynamics of the patient. Furthermore, these devices require accurate alignment between its mechanical joints and the patient limbs, which is difficult or impossible to accomplish due to the complex geometry of the human body.

To overcome that, a special kind of parallel robots defined as Cable-Driven Parallel Robots (CDPRs) are increasingly used, especially in physical rehabilitation.

Generally, their entrainment system is the flexible cables which are attached to fixed base, to move the mobile platform. By changing the lengths of the cables actuated by the fixed motors and winches, the displacement of the mobile platform is reached.

In review of the literature, some researchers are been motivated to develop and design CDPRs to physical rehabilitation use, especially for lower human limb. Each of them adopts some specific configuration and geometry according to the type of defficience to restore. These types of devices can be classified as end-effector-based devices. In this case, one talk about multi-body cable-driven mechanisms, which is an extension of the basic cable robots where the moving platform is replaced by the human limbs, as being multi-body system (MBS) (Bryson et al. 2013). Thereby, the use of cables ensure to this type of robots a low inertia, ability to reach high speeds, relatively a large workspace, low fabrication costs, a possibility to reconfiguration, tolerance to anatomical differences among patients, tolerance to misalignment of apparatus with patient bodies, etc, which are widely recommended in rehabilitation exercises.

For all of the proposed rehabilitation device in this order, the use of CDPRs generally, still have some challenges to be taken in consideration. Due to the compliance of cables, the stiffness analysis of CDPRs becomes a vital concern (Gouttefarde and Gosselin 2012). Indeed, defficient static stiffness can decrease the positioning accuracy of CDPRs, and bad dynamic stiffness characteristics can lead to vibration and long settling time.

Among challenges also for such mechanisms are the risk of interference between the cables and the links, or between the cables themselves (Marler et al. 2009). Though, the major challenge is ensuring their ability to have the best force distribution in all cables (positive tension), known as tensionability condition (wrench closure or force closure) (Rezazadeh and Behzadipour 2011).

The proposed CDPR for rehabilitation of lower limb in this study, has a specific design, compared to those existed. It allows to applying rehabilitative motion for lower patient limbs, following two possible embodiments. The first embodiment, with standing position of the patient, requires the control of three degrees of freedom in the sagittal plane for its right and left lower limbs. However, the second embodiment, where the patient is in an elongated position, allows the movement of one leg at a time in the seven degrees of freedom.

The study of this kind of mechanism require not only to take into account the challenges faced by the usual CDPRs, but also the constraints required in medical applications. Generally, the study of robotic real application require to resorting to the Inverse Kinematic (IK) task, to find a configuration at which the end-effector of the robot reaches given point in the task space. Specially for human body application, the IK resolution complexity is enhanced with the increased Degrees Of Freedom (DOF). The application of the classical IK methods, remains just viable mathematically, do not take into account the physiological feasibility and biofidelity of human posture, and suffer from numerical problems (Abdel-Malek 2004).

Optimization based approaches can be suitable ways to overcome the above mentioned problems. It refers to predict the realistic posture of human limb in its feasible workspace. As any optimization problem, for the posture prediction problem, the joint angles of the human leg are considered as the design variables, the constraints are considered according to physiological feasibility and motion precision, and for the objective function, the human performance measures are used.

There are many forms used in the literature to define the human performance measure, such as physical fatigue defined as reduction of physical capacity. It is mainly the result of three reasons: magnitude of the external load, duration and frequency of the external load, and vibration (Chen 2000). However, for the movements required low speed such as rehabilitation exercises, the physical fatigue is not so significant. Indeed, the required movements can lead to some human discomfort (Abdel-Malek 2004), where its evaluation may vary from person to person, such as potential energy (Mi et al. 2009), torque joints, muscle fatigue, or perturbation from a neutral position (Spong et al. 2006).

In the current work, the passive modeling approach is adopted, with the thorough description with a customization of anthropometric parameters, and taking into account, the different constraints which can be appears in this system. Arising, a unique and sophisticated biomechanical model, compared to those found in the literature. The obtained biomechanical model characterize a class of nonlinear systems with unknown dynamics.

To better illustrate these aspects, the remainder of the paper is organized as follows: In the first the robot description will be presented according to the associated configurations. In the following the human lower limb consideration will be discussed. Therefore the forward kinematics has been developed for the tow configurations where motions of the human lower limb occur with three and seven degrees of freedom. According to that, in following section, the feasible workspace have been established. Thereby, in the followed section, the new optimal posture prediction has been described and thereby applied on the human lower limb for the provided motion configuration. To check the effectiveness of the proposed approach, a simulation model has been developed using Matlab package. To sum up, the results of the study are outlined in finally section.

4.2 Modeling Robot

The use of Cable Driven Robotic Rehabilitation (CDRR) of lower limb, is based on eight cables, attached to the fixed winches, then, to the lower limb (Faqihi et al. 2016). The length of the cables is adjusted by motors to reproduce the desired rehabilitation movement. For a given position, the robot must calculate and adjust the length of the necessary cables so that the leg reaches the desired place in space. These calculations must be carried out dynamically and continuously during a movement.

The robot is designed to apply rehabilitation exercises according to two possible configurations:

4.2.1 First Configuration

In the first configuration, the patient is in a standing position. This configuration is adopted to imitate the walking movement, in each legs can move according to three degrees of freedom, in the sagittal plane, as illustrated in Fig. 4.1.

With the eight used cables, the robot in this configuration has one Degree Of Redundancy (DOR) for each legs at a time, which will thereafter serve as a necessary condition to maintain the tensionability of the mechanism.

The provided motion of the human leg can be modeled by three revolute degrees of-freedom: A hip extension-flexion degree-of-freedom, a knee extension-flexion degree-of-freedom, and an ankle dorsiflexion-plantar degree-of-freedom.

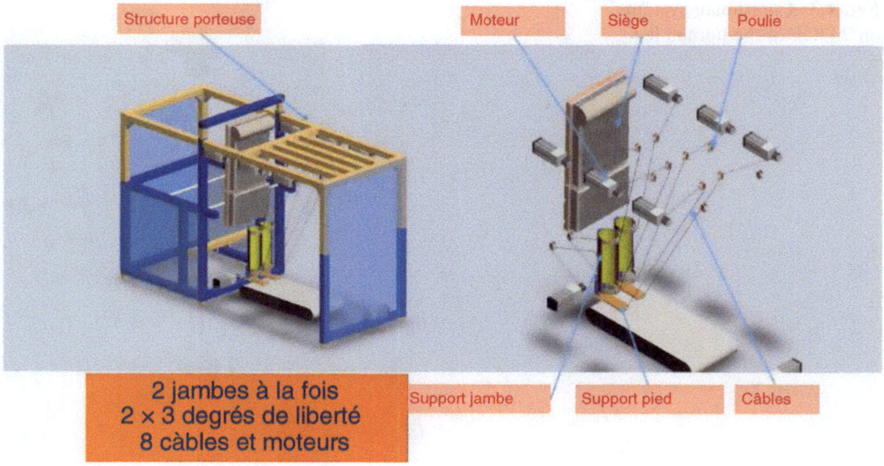

Fig. 4.1 First configuration of CDRR Legend in Frensh !!!

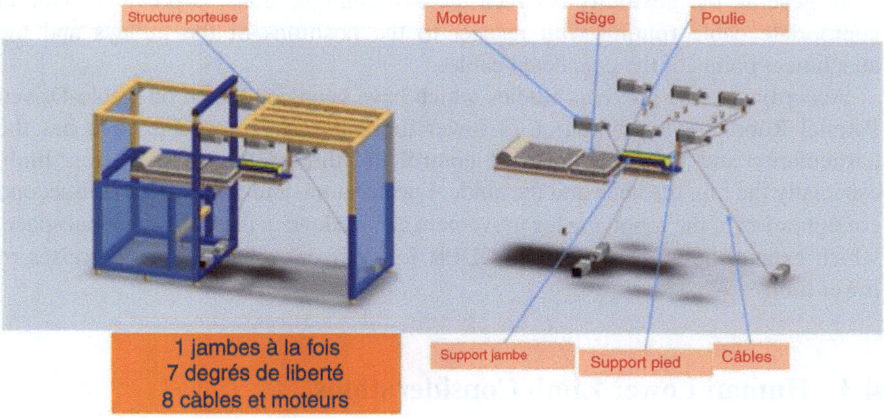

Fig. 4.2 Second configuration of CDRR Legend in Frensh !!!

4.2.2 Second Configuration

For this configuration, the patient is in an elongated position, which ensure a single leg movement at the time in the three dimensional space, as illustrated in Fig. 4.2.

With this configuration also, the robot has one DOR.

The motion of the human leg is provided for sevens degree-of-freedom, defined as: 3 DOF hip (extension-flexion degree-of-freedom, abduction-adduction degree-of-freedom, and inversion-eversion degree-of-freedom), 1 DOF knee (extension-flexion degree-of-freedom), 3 DOF ankle (extension-flexion

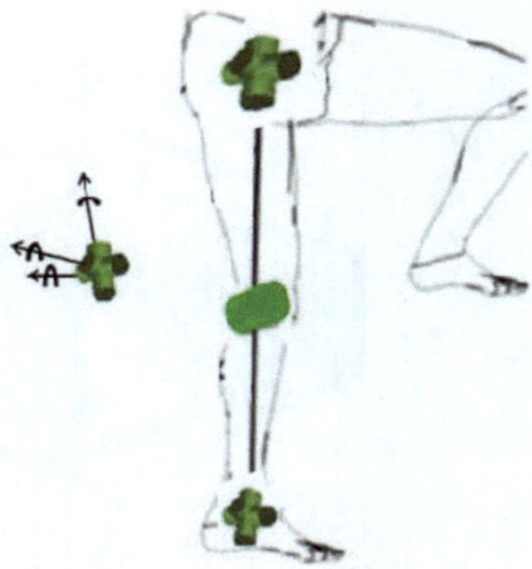

Fig. 4.3 Coordinate systems on 7 DOF of the human leg

degree-of-freedom, abduction-adduction degree-of-freedom, and inversion-eversion degree-of-freedom), as depicted in Fig. 4.3.

In general, the flexibility between the two configurations, is provided with an appropriate cable routing with respect to the positions of the pulleys and the attachment points of the connected cables.

According to the previous studies which have been conducted on Cable-Driven Parallel Robot for rehabilitation of lower limbs use, the proposed robot has the particularity and the uniqueness of control the three joints of the lower limb, especially the hip, the knee and the ankle. Furthermore, and according to the second configuration of the robot, the leg movement can be done in three dimensional space, which extend the spectrum of use CDPR for different rehabilitation exercises of lower limbs.

4.3 Human Lower Limb Considerations

As described above, the studied CDRR lead to move human lower limb by entrainment system of cables according to tow possible configurations. In order to give a credible model of this robot, it is necessary to give a precision of its mobile platform.

Generally, the human body is a complex system, its biomechanical modeling represents a simplification of its real operating. The introduction of assumptions is necessary in this order, which are selected according to the desired performances.

4.3.1 Modeling Hypothesis

The model adopted for the lower limb based on three segments to model its anatomical structure (bone structure): thigh, shank and foot considered as the length between ankle and metatarsal.

Since, exercised movement on the lower limb, during the rehabilitation sessions, is of small variation of acceleration. In this order, some hypothesis were fixed, and which will be valid for all structures:

H1: The body segments are considered to be non-deformable rigid bodies. This assumption enables the application of the mechanical rigid body approach. As the body segments are rigid, their center of mass and their moments of inertia does not change during movement.
H2: The joints assumed to be perfect mechanical connections.
H3: The actions that generate movement are localized in the joint centers.
H4: The body mass is assumed concentrated on the center of mass.
H5: The length of each segment, remains constant during the period of executing movement.
H6: To ignore artifacts soft tissue, the movement marrow masses (skin and fat) is negligible and does not affect the inertial properties during the execution of the movement.
H7: The joints are defined with viscoelastic behavior in anatomical axes of the joint. In addition, these axes are fixed, thus the translations of a bone on the other (knee) are also negligible.
H8: The muscular action is due to a single muscle group. Therefore, there are no antagonist muscles that oppose the movement created by the agonists.

The connection of all three segments is ensured naturally by ligaments and muscles, and should be kinematically redundant to ensure biofidelity of the human leg motion (Winter 2009).

For a Kinematic analysis, the human leg is modeled by a kinematic chain of rigid bodies, interconnected by kinematic joints, which can be either simple or complex according to required physiological behavior, and thus the degree of freedom associated with the possible joints (9).

According to the second configuration, the leg motion is conduct with seven degree of freedom, distributed as depicted in Fig. 4.3.

4.3.2 Anthropometric Parameters

The modeling of body segments must take into account some anthropometric parameters. In order to customize the model, accurate measurements of the anthropometric parameters are required and can be obtained from statistical tables proposed in Winter (2009). It refers to adopt a proportional anthropometric model, based on statistical regression equations to estimate these segmental inertial parameters (PIS), from a comprehensive knowledge of the body such as: size and body weight. The Table 4.1, provides the necessary information to calculating inertial parameters. For this study, physical length segments will be used. They can be computed using total body height (H), where W is the total body weight, and L is the total body length;

The segment moment of inertia around its center of mass I_{cm}, can be expressed by:

$$I_{cm} = W_s \times (L_s \times R)^2 \tag{4.1}$$

4.3.3 Range of Motion

To ensure users security during rehabilitation session, physiological safety of humans lower limb must be considered in the design of robot. Indeed, the suspension of the human leg according to used cable, allows to maintain the patients balance during training, when he is unable to do it. In the other hand, defining ROM of the human lower limb model, is not limited to the designed mechanical structure, but also to the human physiological factors, such as the age, body build, gender, health condition (Chaffin et al. 1992), so that the Maximum Range Of Motion (MROM) and Maximum Speed (MS) must be taked into account, to increase the physiological safety.

These specifications are used as constraints of the desired trajectory, ensuring that MROM and MS will not be exceeded during motion exercise. The designed robot, can assist patients to perform joint training, gait training, or other desired trajectory tracking in hip, knee and ankle joints, according to the controller's performance.

Due to the biomechanical constraints of human body motion, the bounds of the joint variables are fixed, which define the Range of Motion (ROM).

Generally, the ROM of human legs, based on a previous study by Hernandez et al. (2011) is given in Chaffin et al. (1992).

4.3.4 Comfort Zone

To ensure the comfort motion of the human body, each joint variable can be defined by its comfort zone, which must belong to the range of motion (ROM) of the associated joint variable.Referring to the literature, the comfort zone represents 35% of the range of motion (ROM). The center of the comfort zone, q_{ic}, is calculated by the following expression (Glowinski and Krzyzynski 2016):

$$q_i^C = 0.5(q_{icz}^u + q_{icz}^l) + q_i^h \tag{4.2}$$

where q_{icz}^u, and q_{icz}^l are respectively, the upper and lower angles of the comfort zone, for the ith joint variable associated. q_i^h is the home position angle of the ith joint variable. Generally, the home position angle can differ from tested tasks (standing, recumbent, seating,...).

4.4 Kinematic Analysis of CDRR

The studied robot is a mechanical system that transforms the movements of its actuators on movement of the platform (lower limb). In order to describe the kinematics of the cable mechanism, the following notation is adopted:

- The joint coordinates of the actuators (angles of rotation of the motors) are denoted by $q = [q_1, \ldots q_m]$, where m is the number of actuators.
- The articulare coordinates of the platform are noted $\theta = [\theta_1, \ldots \theta_d]$, where d is the Degree Of Freedom of the platform.
- The operational coordinates describing the platform position, are defined in R^n and denoted by $x = [x_1 \ldots x_n]$, where n the number of DDLs of the platform.

In other words, the robotic system transforms joint forces: the cable tensions, noted $t = [t_1, \ldots t_m] \in R^m$, in operational forces $f = [f_1, \ldots f_n] \in R^n$. These efforts represent the cable effects on the platform.

Since the platform represents a multi-body, we will therefore have $j = 1 \ldots m$ referencial marks R_{Pj} associated to each segment of the platform.

In general, CDPR with m cables, whose geometry is illustrated in Fig. 4.4, has m output points of the cables, denoted A_i on the base and m points of attachment of the cables on the platform, B_i.

The coordinates of these points are grouped in the vectors a_i and b_i, respectively. Where a_i is expressed in the fixed frame of the robot, while b_i is expressed in the mobile frame attached to the platform. The vector p gives the Cartesian position of the platform of geometric center P. The orientation of the fixed reference with respect to the fixed coordinate system is given by the orthogonal matrix of rotation Q. The length of Cable i is denoted by l_i. By writing e_i a unit vector, we can write:

Fig. 4.4 General Architecture of CDPR

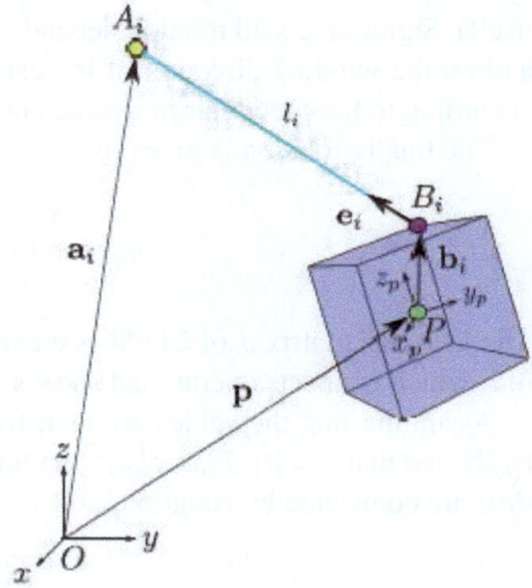

$$A_i B_i = l_i e_i \tag{4.3}$$

The problem of inverse geometry model ifor CDPRs consists to determine the lengths of the cables l, required for a precise displacement of the actuators described by their articular coordinates q, from the vector describing the placement (position and orientation) x of the mobile platform such as:

$$q = f(x) \tag{4.4}$$

In the kinematic chain that connects the base to the platform, only the m cables connect the platform to the fixed base of the robot. From Eq. 4.3, we can write directly:

$$l_i = \|A_i B_i\| = \|x + Qb_i - a_i\| \tag{4.5}$$

It is assumed that there exists a linear relationship between the cable length l_i, $i = 1 \ldots m$ and the rotation angle of the motor q_i expressed in meters and radians, respectively (El-Ghazaly et al. 2014) :

$$l_i = \pm r_i q_i \tag{4.6}$$

where r_i is the ratio expressed by:

$$r_i = \kappa_i \sqrt{e_i^2 + \frac{\rho_i^2}{2\pi}} \tag{4.7}$$

with κ_i is the transmission ratio of the ith winder with the ith motor, e_i and ρ_i the radius and the pitch of the ith winder respectively. Note that κ_i and ρ_i are constants.

The relation (4.7) is true in the case of an inextensible cable whose winding system guarantees that e_i remains constant over time (the cable does not curl up on itself). Sign of the said relation depends on the direction of rotation of the ith motor and on the winding direction of the cable. The sign must be therefore established according to the particular mechanics of the robot.

The finally relation is given by:

$$q_i = \frac{1}{r_i}\|x + Qb_i - a_i\| \tag{4.8}$$

The Jacobian matrix J of CDPR is the matrix formed by the gradient composants l_i of l, which connects operational speeds $dotx$ of the platform to joint speeds $\dot{q}$.

Assuming that the cables are taut, for a robot with m DDL, the J is expressed by: Noted that $\dot{l} = [\dot{l}_1, \dot{l}_2, \ldots \dot{l}_m]^T$ the linear velocity vector $dotq$ and cable lengths $dotl$ are connected by equation:

$$\dot{l} = R\dot{q} \tag{4.9}$$

where R is diagonal matrix defined by $diag\{r_i\} = \{r_1, \dots r_m\}$. According to the operational platform speeds:

$$\dot{l} = J(x)\dot{x} \tag{4.10}$$

Hence the relationship between joint speeds (speeds of rotation of the motors) $dotq$, and the operational speeds $dotx$ of the platform as follows:

$$\dot{q} = R^{-1}J(x)\dot{x} \tag{4.11}$$

In the studied CDPR the mobile plate-form is MBS, characterized by their articulation coordinates θ_i. For safety and comfort purpose of patient it would be necessary to take into account the constraints of θ_i to determine $f(x)$ defined in Eq. 4.4.

Therefore, the main goal is to find the displacement of the actuators q_i for a given position x of the end-effector, by defining the associated articular joints of the plateforme θ_i according to the discussed constraints, and thereby associated cable lengths l_i .

In this order, the remain of this section focuses on introducing a general optimization-based formulation to predict the θ_i from given end-effector position x, and deduce therefrom the articulation coordinates θ_i. The new proposed idea, is to use an objective function incorporating three factors that contribute to musculoskeletal discomfort as human performance measure.

4.4.1 Forward Kinematic Analysis

The forward kinematic (FK) model in n plane, defined by g function, can determine the pose of the end-effector (x_j, $j = 1 \dots n$), from given articular joints variables of the paltforme (θ_i, $i = 1 \dots DOF$). It is a necessary step in the kinematic analysis process.

$$x_j = g(\theta_i) \tag{4.12}$$

For the rigid bodies robotic systems, several methods can be used to resolve this problem.

Despite of the human body complexity, for all practical purposes, it has been shown that approximated modeling of gross human motion, in order to ensure human motion simulation, ergonomic analysis, or rehabilitation process, can be achieved using homogenous transformation matrices method, and the Denavit Hartenberg (DH) representation, based on appropriate kinematic coordinates. Indeed, the DH method provides an adequate, and systematic method for embedding the local coordinate systems for each link.

Table 4.1 DH human leg parameters for 3 DOF

Link(i)	α_i	a_i	d_i	q_i
1	0	a_1	0	q_1
2	0	a_2	0	q_2
3	0	a_3	0	q_3

Table 4.2 DH human leg parameters for 7 DOF

Joint(i)	α_i	a_i	d_i	q_i
1	$-\frac{\pi}{2}$	0	0	q_1
2	$\frac{\pi}{2}$	0	0	q_2
3	0	a_1	0	q_3
4	0	a_2	0	q_4
5	0	a_3	0	q_5
6	$-\frac{\pi}{2}$	0	0	q_6
7	$\frac{\pi}{2}$	0	0	q_7

The forward kinematic human leg model is developed from the DH parameters (Dombre and Khalil 2010) where each degree-of-freedom can be modeled as a revolute joints. The DH parameters are depicted in Table 4.1 for the first configuration and Table 4.2 for the second configuration, from the defined kinematic coordinates.

From the provided D-H parameters, the forward kinematic model can be computed using the transfomation matrix, given in Eq. 4.13. Generally, the transformation matrix is the relationship expression between two consecutive frames $i-1$ and i, which depends on the described parameters (θ_i,α_i,a_i,d_i) given in the above tables.

$$^{i-1}T_i = \begin{pmatrix} c\theta_i & -s\theta_i c\alpha_i & s\theta_i c\alpha_i & \alpha_i c\theta_i \\ s\theta_i & c\theta_i s\alpha_i & -c\theta_i s\alpha_i & \alpha_i s\theta_i \\ 0 & s\alpha_i & c\alpha_i & d_i \\ 0 & 0 & 0 & 1 \end{pmatrix} \tag{4.13}$$

The global transformation matrix for the first configuration can be expressed by:

$$^{0}T_3 =^0 T_1.^1T_2.^2T_3 \tag{4.14}$$

and for the second one:

$$^{0}T_7 =^0 T_1.^1T_2.^2T_3.^3T_4.^4T_5.^5T_6.^6T_7 \tag{4.15}$$

From Eqs. 4.14 and 4.15, the forward kinematic model is given by the following equations respectively for the first and second configurations:

$$\begin{pmatrix} x \\ y \end{pmatrix} = \begin{pmatrix} c_{123}a_3 + a_2c_{12} + a_1c_1 \\ s_{123}a_3 + a_2s_{123} + a_1s_1 \end{pmatrix} \tag{4.16}$$

where: $c_{123} = cos(q_1 + q_2 + q_3)$, $s_{123} = sin(q_1 + q_2 + q_3)$, $c_{12} = cos(q_1 + q_2)$, $s_{12} = sin(q_1 + q_2)$, $c_1 = cos(q_1)$, and $s_1 = sin(q_1)$.

$$\begin{pmatrix} x \\ y \\ z \end{pmatrix} = \begin{pmatrix} a_2c_{1234} + a_1c_{123} + a_3c_{1234}c_4 - a_3s_{1234}s_5 \\ a_2s_{1234} + a_1s_{123} + a_3s_{1234}c_4 + a_3c_{1234}s_5 \\ -a_1s_{23} - a_2s_{234} - a_3s_{234}c_4 - a_3c_{234}s_5 \end{pmatrix} \tag{4.17}$$

where: $c_{1234} = \cos(\theta_1 + \theta_2 + \theta_3 + \theta_4)$, $s_{1234} = \sin(\theta_1 + \theta_2 + \theta_3 + \theta_4)$, $c_{123} = \cos(\theta_1 + \theta_2 + \theta_3)$, $s_{123} = \sin(\theta_1 + \theta_2 + \theta_3)$, $c_{12} = \cos(\theta_1 + \theta_2)$, $s_{12} = \sin(\theta_1 + \theta_2)$, $c_4 = \cos(\theta_4)$, $s_4 = \sin(\theta_4)$, and $s_5 = \sin(\theta_5)$.

In terms of velocity and acceleration, the forward kinematic is given by:

$$\dot{x}_j = J\dot{\theta}_i, \quad \ddot{x}_j = J\ddot{\theta}_i + \dot{J}\dot{\theta}_i \tag{4.18}$$

where $\dot{x}_j$, $\ddot{x}_j$, represent respectively the end-effector velocity, the end-effector acceleration in task space, J and $\dot{J}$, represent the Jaccobian matrix of the system and its derivative. Finally, and $\dot{\theta}_i$ and $\ddot{\theta}_i$ represent respectively the end-effector velocity and acceleration in joint space.

4.4.2 *Feasible Workspace*

In order to analyze the feasible workspace associated to the human lower limb, where we plot the different possible positions of foot which can be achieved, the direct kinematic model $f(\theta_i)$ is used. Thus, the workspace can be defined as the set of all the possible positions in the task space according to the ROM, as following:

$$E_p = \left\{\theta_i \in ROM / \sum_j x_j = f(\theta_i)\right\} \tag{4.19}$$

Using the appropriate forward kinematic model given in Eq. 4.17, and the ROM described, the feasible workspace can be plotted for the motion provided according to the used configuration.

4.4.3 *Proposed Optimization Approach*

4.4.3.1 Problem Formulation

The optimal posture prediction is considered to be a constrained optimization problem (CO), using a constraint to find a realistic configuration.

Generally, the CO problem can have equality and/or inequality constraints according to the described problem, and the objective function, which requires some assumptions according to the continuity and differentiability. In that fact, in the following the optimization problem model are described.

4.4.3.2 Design Variables

The design variables represent in this case, the joint variables θ_i, $i = 1 \ldots DOF$, following the used configuration. However, the initial guess consists of determining initial values, for the design variables. Indeed it should be suitable to use a feasible point. In this study, an initial guess is used to satisfy the ROM.

4.4.3.3 Constraints

The first constraints consider the difference between the current end-effector position, velocity, and acceleration, and the given target position, velocity and acceleration respectively in cartesian space, as following:

$$||x_j^{computed}(\theta_i) - x_j^{desired}(\theta_i)|| \leq \varepsilon_1$$

$$||\dot{x}_j^{computed}(\theta_i) - \dot{x}_j^{desired}(\theta_i)|| \leq \varepsilon_2 \tag{4.20}$$

$$||\ddot{x}_j^{computed}(\theta_i) - \ddot{x}_j^{desired}(\theta_i)|| \leq \varepsilon_3 \tag{4.21}$$

where $||.||$ define the Euclidean norm. The end-effector position $x_j^{computed}$ hits a predetermined target point $x_j^{desired}$ in cartesian space, within a specified tolerance ε_1 a small positive number that approximates zero, similarly to end-effector velocity and acceleration.

It should be noted that, determining the end-effector position, velocity and acceleration are ensured using the forward kinematic model.

On the other hand, each joint variable is constrained to lower and upper limits, represented by θ_i^l and θ_i^u, respectively. These limits ensure that the human posture does not assume an unrealistic position to achieve the target point.

To more rigorous biofidelity end, we can choose that each joint variable is constrained to lie between upper and lower angles of the comfort zone, designed by θ_{icz}^u and θ_{icz}^l respectively.

$$\theta_{icz}^l \leq \theta_i \leq \theta_{icz}^u \tag{4.22}$$

Finally, the constraints in term of velocity and acceleration limits are used, as following.

$$\dot{\theta}_i^l \leq \dot{\theta}_i \leq \dot{\theta}_i^u \tag{4.23}$$

$$\ddot{\theta}_i^l \leq \ddot{\theta}_i \leq \ddot{\theta}_i^u \tag{4.24}$$

where $\dot{\theta}_i^l$ and $\ddot{\theta}_i^l$ are the lower limit of velocity and acceleration, $\dot{\theta}_i^u$, and $\ddot{\theta}_i^u$ are the upper limits of velocity and acceleration respectively. These limits are fixed according to biomedical studies, where the physiological and health state of patient are taken into account.

4.4.3.4 Cost Function

As described previously, the posture prediction requires a human performance measure, where its rigorous choice ensures the optimal realistic posture.

To this end, the modeling musculoskeletal discomfort is used as human performance measure, which can be somewhat ambiguous, as it is a subjective quantity, thus its evaluation may vary from one person to another (Abdel-Malek 2004).

According to the last described researches (Abdel-Malek 2004; Mi et al. 2009), in this order, different forms of human performance measures have been adopted, but it often results in postures with joints extended to their limits, and thus to some uncomfortable positions.

As remedy, we can add factors associated with moving while joint variables are near their respective limits in terms of position, velocity and acceleration. In this respect, it is possible to incorporate different factors that contribute to discomfort. The first factor, is referred to the tendency to move different segments of the body sequentially. The second factor, is referred to the tendency to gravitate to a reasonably comfortable position. Finally, the discomfort associated to the motion while joints are near their ROM in term of position, velocity and acceleration, expresses the third factor.

According to the previous studies, the proposed objective function is similar to that adopted by Yang et al. (2004) applied for the upper limb. However, there is currently no research focused on prediction of human leg posture, by applying this form of discomfort, with the restriction particularity of 15%, taking into account the end-effector position, velocity and acceleration.

In order to incorporate the first factor, we can find several strategies which induce motion in a certain order, or with higher weighted joints than others.

Consider q_{ic} the comfortable position of ith joint variable, measured from the home configuration defined by $q_i^h = 0$. Then, conceptually, the displacement from the comfortable position for a particular joint position is given by: $|q_i - q_{ic}|$. However, to avoid numerical difficulties and non-differentiability, we can use: $(q_i - q_{ic})^2$.

Generally, the terms should be combined using a weighted sum w_i, to emphasize the importance of particular joints depending on the characteristics of each patient. Thereby, the joint displacement function is given as follows:

$$f_{displacment} = \sum_i w_i (q_i - q_{ic})^2 \tag{4.25}$$

The weights are used to approximate the lexicographic approach (Marler et al. 2009).

In order to incorporate the tendency to gravitate to a reasonably comfortable neutral position, each term in Eq. (4.25) is normalized, as described in the following:

$$\Delta q_i = \frac{q_i - q_{ic}}{q_i^u - q_i^l} \tag{4.26}$$

Each term of $(\Delta q_i)^2$ of this normalization scheme is considered as a fitness function with each individual joint and has normalized values, which lie between zero and one.

The principal limitation of this approach often results in postures with joints limits extended, thereby an uncomfortable joint. In this order, the third factor is introduced, which defines the discomfort of moving while joints are near their respective limits. This factor requires to add some designed penalty terms to increase significantly the discomfort where joint values are close to their limits.

– Barrier Penalty Function

Generally, the new designed penalty term $P(d)$ is a barrier penalty function (Eschenauer and Thierauf 1989), of d argument, expressed by:

$$P(d) = [\sin(ad + b)]^p \tag{4.27}$$

The $P(d)$ function is adapted to penalize any number d, considered as normalized parameters, which is approaching zero at some number value.

The proposed idea is that the penalty term remains zero until the d value reaches $d \leq 0.15$, which defines the desired curve data.

Thereby, the parameters a, b, and p of the basic structure of barrier penalty function are fitted to reach the desired curve data.

The penality function $P(d)$ is computed, according to the desired curve data as explained previously, by using Curve Fitting Package in Matlab, as depicted in Fig. 4.5.

The obtained parameters of the penalty function are given by : $a = 2.5$, $b = 7.855$, $p = 100$. Thereby the final expression of the penality term is given by:

$$P(d) = [\sin(2.5d + 7.855)]^{100} \tag{4.28}$$

According to the three described factors, the consequent discomfort function is obtained as follows:

$$\begin{aligned} f_{dicomfort} = \sum_{i=1}^{DOF} [w_i(\Delta q_i)^2 + P(R_{pui}) + P(R_{pli}) \\ + P(R_{vui}) + P(R_{cli}) + P(R_{aui}) + P(R_{ali})] \end{aligned} \tag{4.29}$$

where $P(R_{pli})$ and $P(R_{pui})$, $P(R_{vli})$ and $P(R_{vui})$, $P(R_{ali})$ and $P(R_{aui})$ are the penality terms with joint values that approach their lower limits, and their upper limits, respectively for end-effector position, velocity and acceleration.

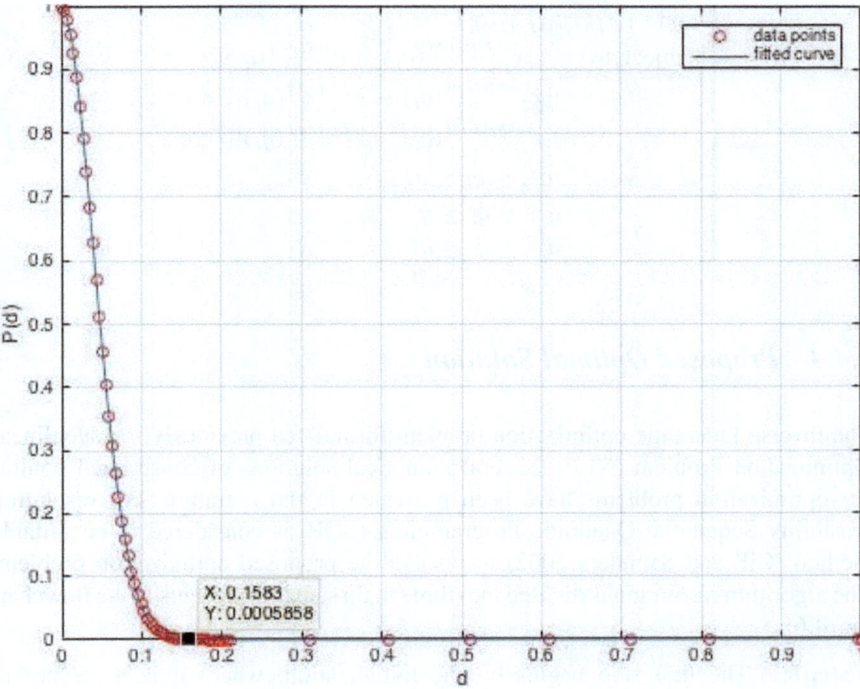

Fig. 4.5 Penality term P(d)

$$R_{pui} = \frac{q_i^u - q_i}{q_i^u - q_i^l}; \quad R_{pli} = \frac{q_i - q_i^l}{q_i^u - q_i^l} \tag{4.30}$$

$$R_{vui} = \frac{\dot{q}_i^u - \dot{q}_i}{\dot{q}_i^u - \dot{q}_i^l}; \quad R_{vli} = \frac{\dot{q}_i - \dot{q}_i^l}{\dot{q}_i^u - \dot{q}_i^l}$$

$$R_{aui} = \frac{\ddot{q}_i^u - \ddot{q}_i}{\ddot{q}_i^u - \ddot{q}_i^l}; \quad R_{ali} = \frac{\ddot{q}_i - \ddot{q}_i^l}{\ddot{q}_i^u - \ddot{q}_i^l}$$

The penalty term that depends on parameter d, remains zero as long as the upper or lower joint value does not reach 15% of its range, as depicted in Fig. 4.5.

4.4.3.5 Constrained Optimization Model

From the described design variables, constraints and cost function, the final optimization problem can be formulated, as the following:

$$\begin{cases} \min \ : f_{Discomfort}(q_i) \\ \text{subject to}: \quad ||x_j^{computed}(q_i) - x_j^{desired}(q_i)|| < \varepsilon \\ \qquad ||\dot{x}_j^{computed}(q_i) - \dot{x}_j^{desired}(q_i)|| < \varepsilon \\ \qquad ||\ddot{x}_j^{computed}(q_i) - \ddot{x}_j^{desired}(q_i)|| < \varepsilon \\ \qquad q_{icz}^l \le q_i \le q_{icz}^u \\ \qquad \dot{q}_i^l \le \dot{q}_i \le \dot{q}_i^u \\ \qquad \ddot{q}_i^l \le \ddot{q}_i \le \ddot{q}_i^u \end{cases}$$

4.4.4 Proposed Optimal Solution

The inverse kinematic optimization problem formulated previously is a Nonlinear Optimization Problem (NLP). Several numerical solutions of constrained nonlinear optimization problems have been presented in the literature. For resolution feasibility Sequential Quadratic Programming (SQP) is considered to be suitable method (Gill and Saunders 2002) to resolve the proposed optimization problem. The algorithm resolution is divided into three main steps, as presents in the flowchart Fig. 4.6:

- **Step1:** The first step begins by the initialization, where it is necessary to determine the total body height of a subject and to calculate the length of thigh, shank, and foot segments. Then, we fix the initial joint position, velocity and acceleration variables, and calculate the initial guess as being initial effector position, velocity and acceleration by using forward kinematics, according to the used configuration. Next, taking into account the joints constraints, the workspace and the comfort zone of each joint are computed.
- **Step2:** Giving the end position, velocity and acceleration coordinates, in the second step, we check if the position is in the workspace, else it is necessary to find new coordinates.
- **Step3:** SQP optimization technique is applied in this step to find the optimal postures. This is performed using Matlab fmincon constrained function. The obtained solution is a matrix with position, velocity and acceleration angles. Taking into account the comfort zone of each joint, the obtained angles are the most comfortable.

4.4.5 Trajectory Generation

In rehabilitation robots, the reference trajectory must be predefined as a human limb motion practiced during activities of daily live. Indeed, motion therapy can be carried out in different modes including passive, active, active-resistive,

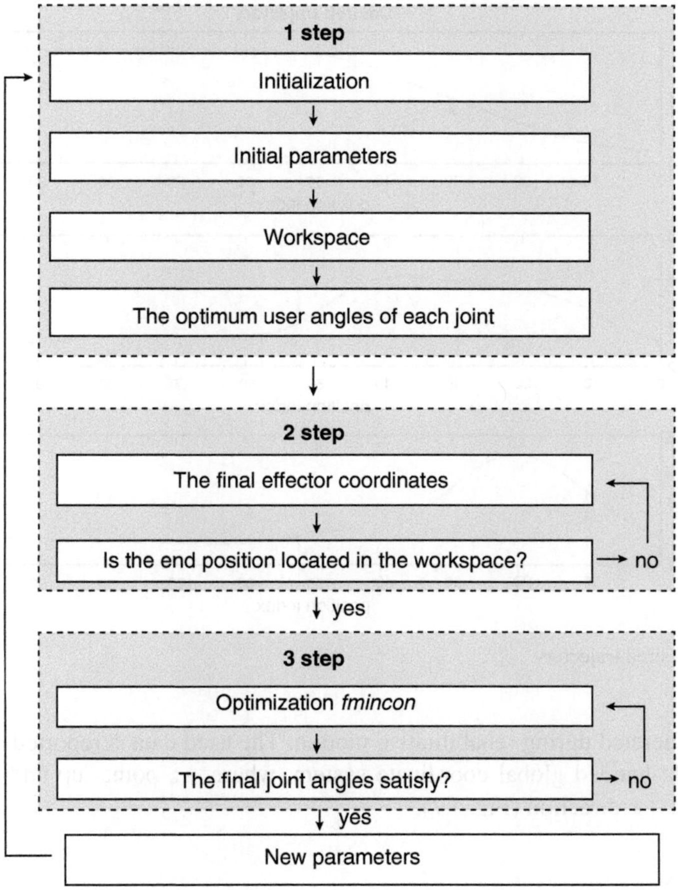

Fig. 4.6 Proposed flowchart solution

active-assistive, and bilateral exercises, which differ depending on the degree of patient involvements. Selecting the proper mode strategy requires an appropriate rehabilitation robot choice, with concerned patients.

To determine the appropriate trajectory for the movement of the rehabilitation robot, there are several methods such as a prerecorded trajectory obtained by gait analysis, and a prerecorded trajectory during therapist assistance, which require data use, and modelling the trajectory based on normative movements which can be based on kinematics and/or dynamics constraints during the path motion in terms of fitting more realistic motion (Rastegarpanah and Mozafar 2016). Generally, it is desirable to use reference trajectories ensuring the feasibility and biofidelity of rehabilitation session.

According to passive rehabilitation exercises, the desired trajectory used to check the effectiveness of the proposed approach is defined from Clinical Gait Database

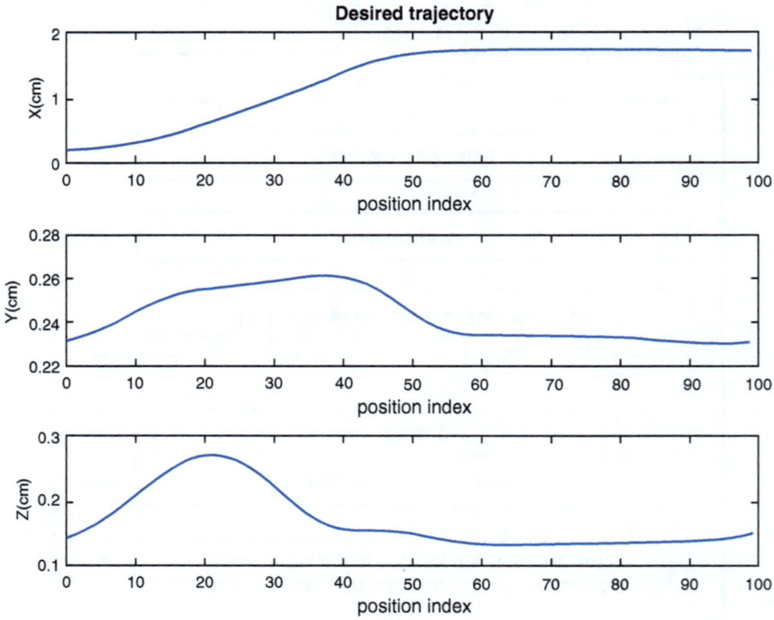

Fig. 4.7 Desired trajectory

(CGD) generated during rehabilitative motion. The used data is reported in relation to the right-handed global coordinate system, where $+z$ points up and subject is walking in $+x$ direction (Fig. 4.7).

4.5 Simulation Results and Discussion

In order to show the effectiveness of the proposed approach, this section presents a simulation results of the developed Kinematic analysis of the Cables-Driven Robot for Rehabilitation use, for the tow discussed configurations.

The validation test is held on a patient of 1.80 m height. Using the anthropometric associated data, the developed forward kinematic model and the described ROM, the associated workspaces are plotted, as shown in Figs. 4.8 and 4.9.

As described above, the main goal is to find the associated joint coordinates of the actuators q_i defining the length cables l_i, which lead to move the end-effector to the given position x_i.

To define the mapping between the end-effector position x_i and the associated articular joints θ_i the developed optimal posture algorithm is used. Indeed, desired end-effector position for the two configurations are fixed to lie their associated workspace.

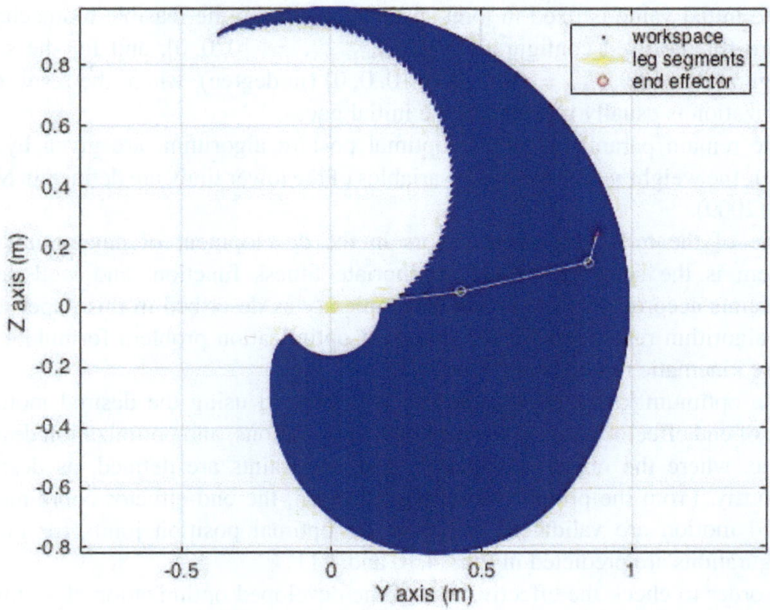

Fig. 4.8 First configuration workspace

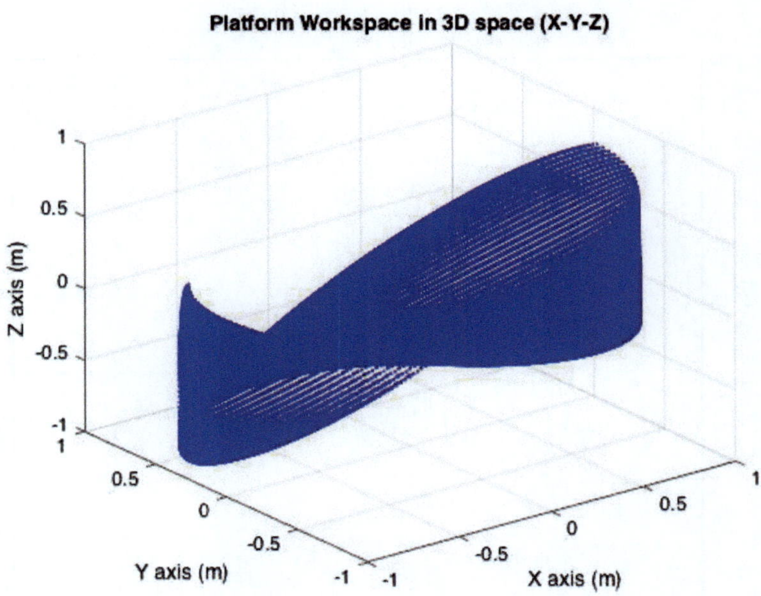

Fig. 4.9 Second configuration workspace

The initial value is fixed in joint space according to the feasible biomechanical posture for the used configuration as $q_{1initguess} = [0, 0, 0]$, and for the second configuration $q_{2initguess} = [0, 0, 0, 0, 10, 0, 0]$ (in degree), where the result of the optimization is usually sensitive to the initial guess.

The remain parameters of the optimal posture algorithm, are given by $\varepsilon = 0.0001$, the weight w_i for the joints variables of the lower limb, are defined in Marler et al. (2009).

One of the most important factors in the development of any optimization problem is the selection of the appropriate fitness function, and well defined constraints according to the problem complexity as described in this paper. Using SQP algorithm resolution for the proposed optimization problem formulation the inverse kinematic model is obtained.

The optimum kinematic parameters are obtained using the desired motion in terms of end effector position for the two configurations, and optimization designed routine, where the objective function, and constraints are defined, as described previously. From the provided checking function, the end-effector coordinates of desired motion are validated, and then the optimal position joints for the tow configurations are predicted in Figs. 4.10 and 4.11.

In order to check the effectiveness of the developed optimization algorithm for the first configuration in sagittal plane, another objective function, already used in the same context (Glowinski and Krzyzynski 2016), is considered. The results of running optimization routine using tow objective functions are shown in Fig. 4.12.

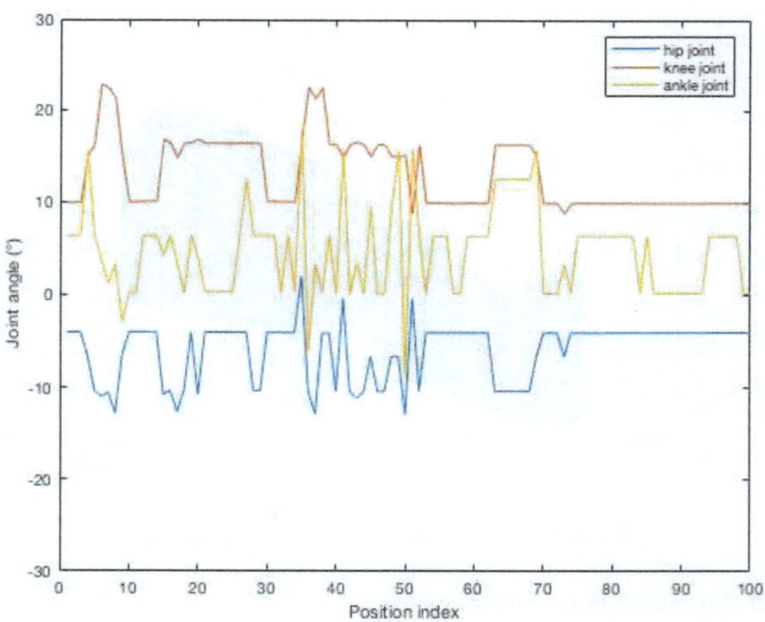

Fig. 4.10 Obtained joint-space position (First configuration)

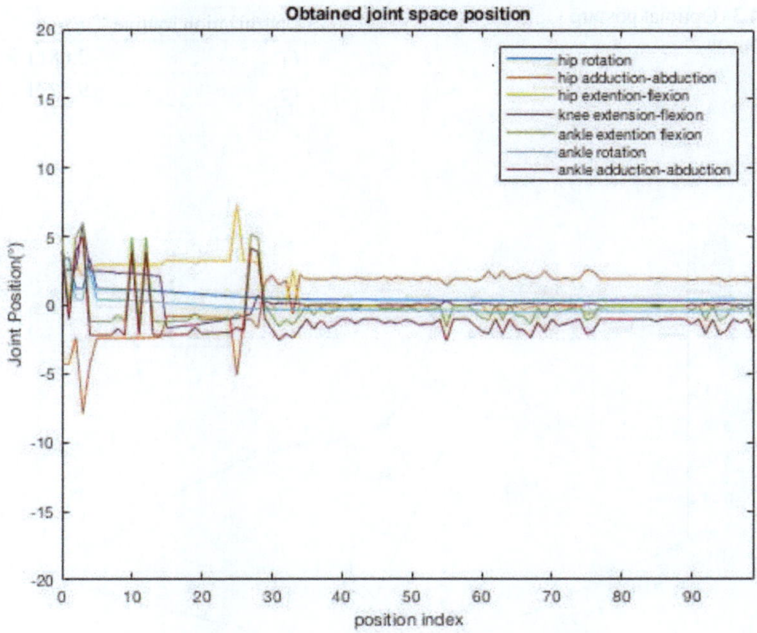

Fig. 4.11 Obtained joint-space position (Second configuration)

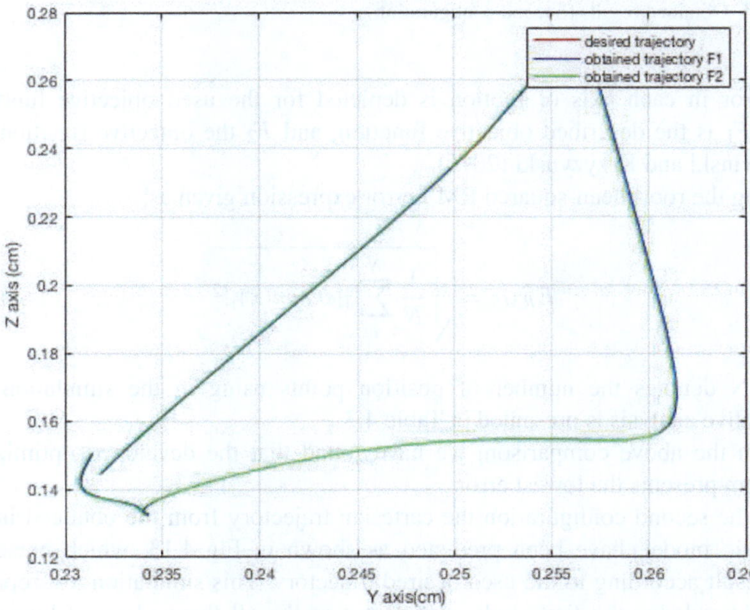

Fig. 4.12 Obtained results (First configuration)

Table 4.3 Optimal posture comparaison

Optimization routine	E_{RMS}
F_1	2.0811×10^{-4}
F_2	9.5271×10^{-4}

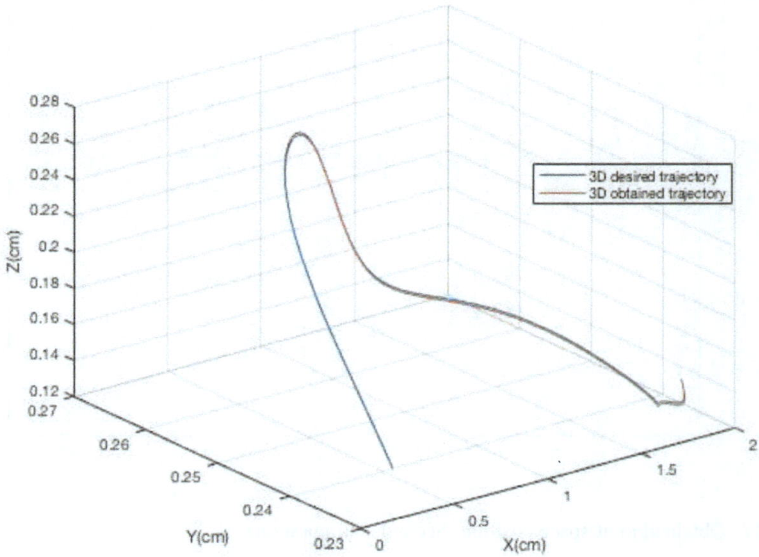

Fig. 4.13 Obtained results (Second configuration)

The error in each axis of motion is depicted for the used objective functions, where F_1 is the described objective function, and F_2 the objective function used in Glowinski and Krzyzynski (2016).

Using the root-mean-squared RMS error expression given as:

$$E_{RMS} = \sqrt{\frac{1}{N}\sum_{i}^{N} ||error(x)||^2} \tag{4.31}$$

where N denotes the number of position points using in the simulation. The quantitative analysis is presented in Table 4.3.

From the above comparison, we have found that the developed optimization algorithm presents the lowest error.

For the second configuration the cartesian trajectory from the obtained inverse kinematic model, have been predicted as shown in Fig. 4.13, which presents a good result according to the used desired trajectory. This simulation test represent to our knowledge the first study carried out in the all three plane of human leg configuration,

4.6 Conclusion

A new general optimization-based formulation for optimal kinematic analysis for the human lower limb has been developed. The proposed method is developed specially in three dimensional space, in terms of the kinematic parameters, using an objective function incorporating three factors that contribute to musculoskeletal discomfort as human performance measure.

The performance measure is referred to the tendency to move sequentially all segments of the human leg, the tendency to move leg while joints are near their ROM, and the discomfort associated with gravitating around a reasonably comfortable position.

This new form of objective function is developed principally for rehabilitation use, where physiological patient constraints need to be taken into consideration. In this order, according to the physiological constraints and the developed forward kinematic model, the feasible workspace is presented. To validate the feasibility and the effectiveness of the proposed kinematic method to predict the inverse kinematic model of 3D space human leg configuration, a reference trajectory is generated to be suitable in rehabilitation case, and thereby, applied in the proposed algorithm solution.

The simulation results, were present an optimal joint space parameters defining the inverse kinematics.

However, this study still valid for the static purpose, and can be improved according to a large definition of discomfort according to dynamical parameters in term of muscle, fatigue... which can be developed in the future research.

References

Abdel-Malek, K., Yang, J., Yu, W., & Duncan, J. (2004). *Human performance measures: Mathematics. Iowa*: University of Iowa.

Bryson, J. T., & Agrawal, S. K. (2013). Methodology to identify and analyze optimal cable configurations in the design of cable-driven serial manipulators. In *International Design Engineering Technical Conference and Computers and Information in Engineering Conference, 37th ASME Conference Mechanisms and Robotics IDETC/CIE.*

Chaffin, D. B., Andersson, G. B. J., & Martin, B. J. (1992). *Occupational biomechanics*. New York: Wiley.

Chen, Y. (2000). Changes in lifting dynamics after localized arm fatigue. *International Journal of Industrial Ergonomics, 25*(6), 611–619.

Dombre, E., & Khalil, W. (2010). *Modeling, performance analysis and control of robot manipulators*. London: Wiley.

El-Ghazaly, G., Gouttefarde, M., & Creuze, V. (2014). Adaptive terminal sliding mode control of a redundantly-actuated cable-driven parallel manipulator: Cogiro. In *Proceedings of the Second International Conference on Cable-Driven Parallel Robots*. Springer.

Eschenauer, H. A., & Thierauf, G. (1989). Discretization methods and structural optimization, procedures and applications. In *Proceedings of a GAMM-Seminar October 5–7, 1988, Siegen, FRG* (Lecture notes in engineering). Springer.

Faqihi, H., Saad, M., Ben Jelloun, K., Benbrahim, M., & Kabbaj, M. N. (2016). Tracking trajectory of a cable-driven robot for lower limb rehabilitation. *International Journal of Electrical, Computer, Energetic, Electronic and Communication Engineering, 10*(8), 1036–1041.

Gill, P. E., Murray, W., & Saunders, M. A. (2002). Snopt: An SQP algorithm for large-scale constrained optimization. *SIAM Review, 47*(1), 99–131.

Glowinski, S., & Krzyzynski, T. (2016). An inverse kinematic algorithm for human leg. *Journal of Theoretical and Applied Mechanics, 54*(1), 53–61.

Gouttefarde, M., & Gosselin, C. M. (2012). Analysis of the wrench-closureworkspace of planar parallel cable-drivenmechanisms. *IEEE Transaction on Robotics, 22*(3), 434–445.

Hernandez-Santos, C., Soto, R., & Rodriguez, E. (2011). Design and dynamic modeling of humanoid biped robot e-robot. *IEEE Conference on Electronics, Robotics and Automotive Mechanics CERMA*, Cuernavaca, Mexico.

Marler, R. T., Arora, J. S., Yang, J., Kim, H. J., & Abdel-Malek, K. (2009). Use of multi-objective optimization for digital human posture prediction. *Engineering Optimization, 41*(10), 925–943.

Mi, Z., Yang, J., & Abdel-Malek, K. (2009). Optimization-based posture prediction for human upper body. *Robotica, 27*(4), 607–622.

Pennycott, A., Wyss, D., Vallery, H., Klamroth-Marganska, V., & Riener, R. (2012). Towards more effective robotic gait training for stroke rehabilitation: A review. *Journal of NeuroEngineering and Rehabilitation, 9*, 65.

Rastegarpanah, A., & Mozafar, S. (2016). Lower limb rehabilitation using patient data. *Applied Bionics and Biomechanics, 2016*, 1–10.

Rezazadeh, S., & Behzadipour, S. (2011). Workspace analysis of multibody cable-driven mechanisms. *Journal of Mechanisms and Robotics. ASME, 3*, 021005.

Spong, W. M., Hutchinson, S., & Vidyasagar, M. (2006). *Robot modeling and control*. New Delhi: Wiley.

Winter, D. A. (2009). *Biomechanics and motor control of human movement*. New York: Wiley.

Yang, J., Marler, R. T., Kim, H. J., Arora, J. S., & Abdel-Malek, K. (2004). Multi-objective optimization for upper body posture prediction. In *Proceedings of the 10th AIAA/ISSMO Multidisciplinary Analysis and Optimization Conference*, Albany, NY (pp. 1–18). American Institute of Aeronautics and Astronautics.

Chapter 5
Control of Robot Manipulators Using Modified Backstepping Sliding Mode

Yassine Kali, Maarouf Saad, and Khalid Benjelloun

Abstract In this chapter, a robust backstepping sliding mode controller with time delay estimation method is presented for high accuracy joint space tracking trajectory of robot manipulators with unknown dynamics and external disturbances. The proposed method is based on backstepping scheme, on a switching function and on a time delay estimation in the final step. This structure is used to estimate unknown dynamics and disturbances, to adapt the switching gains in term of the error and its first time derivative and to reduce the control effort and the chattering phenomenon. Lyapunov theory is used to establish sufficient condition that ensures the stability of the closed-loop system. The proposed controller is simulated on a two-link robot and implemented in real time on the 7-DOF ANAT robot to show the effectiveness regarding unknown dynamics and external disturbances and chattering reduction.

Keywords Backstepping · Sliding mode · Time delay estimation · Uncertainties · Robot manipulators · Tracking position

5.1 Introduction

During the last two decades, nonlinear control of uncertain robotic manipulators has been the topic of many researches. In general, all developed controllers are based on the mathematical model which is always an approximation of reality due to computational complexity and external disturbances which make the implementation

---

Y. Kali (✉) · K. Benjelloun
Department of Electrical Engineering, Ecole Mohammadia d'Ingénieurs, University of Mohammed V, Rabat, Morocco
e-mail: yassinekali@research.emi.ac.ma; bkhalid@emi.ac.ma

M. Saad
Department of Electrical Engineering, École de Technologie Supérieure, Montreal, QC, Canada
e-mail: maarouf.Saad@etsmtl.ca

N. Derbel et al. (eds.), *New Developments and Advances in Robot Control*, Studies in Systems, Decision and Control 175,
https://doi.org/10.1007/978-981-13-2212-9_5

difficult (Lewis et al. 1993). In literature, many control algorithms have been developed for uncertain robotic manipulators, including adaptive control (Cheah et al. 2006; Seraji 1987), intelligent controllers such as fuzzy logic control (Guo and Chung 2003; Yi and Woo 1997) and neural network (Hsia and Jung 1995), Backstepping (Slotine and Li 1991), Sliding Mode Control (SMC) (Utkin 1992; Utkin et al. 1999).

Sliding mode control where a switching control is designed to drive the systems trajectories onto a specific hyperplane in the state space, also called sliding surface, has been proposed for uncertain systems (Slotine and Sastry 1983; Utkin et al. 1999). Nonetheless, this control has many drawbacks, the major one is the well-known chattering phenomenon which is due to the choice of switching gain larger than the over-estimated bound of unknown dynamics and disturbances (Boiko and Fridman 2005; Fridman 2001).

As a solution to this aforementioned phenomenon, some works have suggested the use of continuous functions as saturation or hyperbolic tangent instead of the sign one, but the steady state error does not converge in a finite-time (Boiko 2013; Utkin et al. 1999). Another approach is to use the observer-based sliding mode (de Wit and Slotine 1991; Liu and Wang 2012; Slotine et al. 1986) where the goal is to provide exact and robust estimation in order to allow chattering reduction by a small choice of the switching gain matrix. However, the control performance can be reduced if the estimation is not accurate. Moreover, sliding mode using adaptive switching gains (Jamoussi et al. 2013) has been proposed, which provide an adaptation of the switching gain to be as small as possible and sufficient to eliminate the effect of unknown dynamics and external disturbances. However, the bounds of the uncertainties must be known.

Otherwise, a Higher Order Sliding Mode (HOSM) controller (Fridman and Levant 2002; Geng et al. 2013; Levant 1993; Ling et al. 2012; Utkin 2016) has been proposed for chattering reduction, but the increasing of required informations and the knowledge of the superior bound of the unknown dynamics and disturbances is always required. This in turn makes the implementation difficult. Otherwise, a combination of sliding mode with intelligent controller as neural-network and fuzzy logic has been considered (Dotoli 2003; Jezernik et al. 1997; Lin 2003). Even if these controllers provide a good approximation of unknown dynamics and disturbances; however, their implementation is still difficult due to the large number of parameters or fuzzy rules.

In addition, one can notice that Time Delay Estimation method (TDE) (Toumi and Ito 1990; Toumi and Shortlidge 1991), can estimate nonlinear unknown dynamics and unexpected disturbances effectively and simply. In this method, the unknown parts of dynamics are assumed continuously differentiable with respect to the time variable, and are estimated by using time delayed signals of the system state derivatives and control. The TDE is proven to be efficient because the estimation does not need an exact knowledge of the robot model. In Kali et al. (2015), a combination of sliding mode control using saturation function instead the signum one and time delay estimation method has been proposed. This controller is chattering free but leads to a large steady state error.

In this chapter, an accurate trajectory tracking control of robot manipulators with unknown dynamics and unexpected disturbances is proposed for joint space trajectory tracking by using a new method based on a combination of Backstepping Sliding Mode and TDE. The choice of using backstepping sliding mode based on exponential reaching law instead of conventional SMC is to deal with the chattering problem since the exponential function allows adaptation to the variations of the sliding surface (Fallaha et al. 2011; Rahman et al. 2013) and to eliminate the effect of TDE error.

In the design procedure, the controller will be designed recursively by designing intermediate control laws for some of the state variables considered as "virtual controls" (Chang 2011; Slotine and Li 1991), and at the end of the procedure, the switching manifold will be introduced. Finally, the term consisting on unknown dynamics and external disturbances is approximated using time delay estimation method.

The chapter is organized as follows. Section 5.1 introduces the chapter. In Sect. 5.2, the dynamics of a general n-DOF robot manipulator and problem formulation are presented. The new method is designed for high accuracy joint space tracking and a Lyapunov based stability analysis for the closed loop dynamics with sufficient condition are provided in Sect. 5.3. Simulation results are provided in Sect. 5.4, to illustrate the advantages of the proposed nonlinear control technique. In Sect. 5.5, experimental results of the proposed method applied on 7-DOF ANAT robot arm installed at GREPCI Laboratory in real-time are given to prove the effectiveness of the proposed controller. Finally, Sect. 5.6 concludes the chapter.

5.2 Mathematical Model of Robot Manipulators and Preliminaries

5.2.1 Robot Dynamics

Consider the dynamics of n-DOF robot manipulator in the following matrix equation (Spong et al. 2005):

$$M(q(t))\ddot{q}(t) + C(q(t), \dot{q}(t))\dot{q}(t) + G(q(t)) = \tau(t) + \tau_d(t) \tag{5.1}$$

where:

- $q(t), \dot{q}(t), \ddot{q}(t) \in R^n$ represent the joint space position, velocity and acceleration vectors, respectively.
- $M(q(t)) \in R^{n\times n}$ denotes the inertia matrix which is symmetric definite positive and its inverse always exists.
- $C(q(t), \dot{q}(t)) \in R^{n\times n}$ is the centrifugal and Coriolis matrix.
- $G(q(t)) \in R^n$ represents the gravitational vector.
- $\tau(t) \in R^n$ is the torque input vector.
- $\tau_d(t) \in R^n$ denotes the disturbance vector.

Introducing a diagonal matrix where the elements are strictly positive constants:

$$\overline{M} = \begin{bmatrix} \overline{m}_{11} & 0 & \cdots & 0 \\ 0 & \overline{m}_{22} & \ddots & \vdots \\ \vdots & \ddots & \ddots & 0 \\ 0 & \cdots & 0 & \overline{m}_{nn} \end{bmatrix} \tag{5.2}$$

The dynamic equation of the robot manipulator given in partitioned form in Eq. (5.1) can be rewritten as follows:

$$\overline{M}\ddot{q}(t) + N(q(t), \dot{q}(t)) + H(q(t), \dot{q}(t), \ddot{q}(t)) = \tau(t) \tag{5.3}$$

where:

$$N(q(t), \dot{q}(t)) = C_0(q(t), \dot{q}(t))\dot{q}(t) + G_0(q(t))$$

$$H(q(t), \dot{q}(t), \ddot{q}(t)) = (M(q(t)) - \overline{M})\ddot{q}(t) + C_u(q(t), \dot{q}(t))\dot{q}(t) + G_u(q(t)) - \tau_d(t)$$

with $C_0(q(t), \dot{q}(t))$ and $G_0(q(t))$ are the nominal parts while $C_u(q(t), \dot{q}(t))$ and $G_u(q(t))$ denote the uncertain parts caused by load variations, unknown parameters, unmodel dynamics and external disturbances of $C(q(t), \dot{q}(t))$ and $G(q(t))$, respectively. For simplicity, let us denote $M(t) = M(q(t))$, $N(t) = N(q(t), \dot{q}(t))$ and $H(t) = H(q(t), \dot{q}(t), \ddot{q}(t))$.

5.2.2 *Problem Formulation*

The objective is to design a robust joint space controller that ensures high accuracy tracking trajectory even in presence of unknown dynamics and unexpected disturbances. To that end, the controller will be designed and its stability analysis will be carried out based on the following properties and assumptions:

Property 1 The inertia matrix $M(q(t))$ is symmetric definite positive and bounded such as:

$$0 < m_1 \leq \|M(q(t))\| \leq m_2$$

with m_1 and m_2 are two known positive constants (Spong et al. 2005).

Property 2 The diagonal positive constant matrix $\overline{M}$ is chosen such as the following condition in Spong et al. (2005) is verified:

$$\|I_{n\times n} - M^{-1}(q(t))\overline{M}\| < 1$$

Assumption 1 The joint position and velocity states are measurable.

Assumption 2 The functions $H_i(t)$ for $i = 1, \cdots, n$ of the uncertain vector $H(t) = [H_1(t), \cdots, H_n(t)]^T$ are continuously differentiable with respect to the time variable and don't vary largely during a small L period of time (Toumi and Ito 1990).

5.3 Controller Design

In this section, a combination of backstepping sliding mode with exponential reaching law and time delay estimation method will be designed to ensure the convergence of the joint space trajectories to the known desired ones with high accuracy even in presence of uncertainties. The architecture of the closed-loop system is represented in Fig. 5.1.

The design of the proposed controller is described step-by-step as follows:
Step 1 This first step consists on defining the regulated variable to be the joint space tracking error as follows:

$$e(t) = q(t) - q_d(t) \tag{5.4}$$

where $q_d(t) \in R^n$ is the desired joint space trajectory vector. Therefore, its first time derivative is computed as:

$$\dot{e}(t) = \dot{q}(t) - \dot{q}_d(t) \tag{5.5}$$

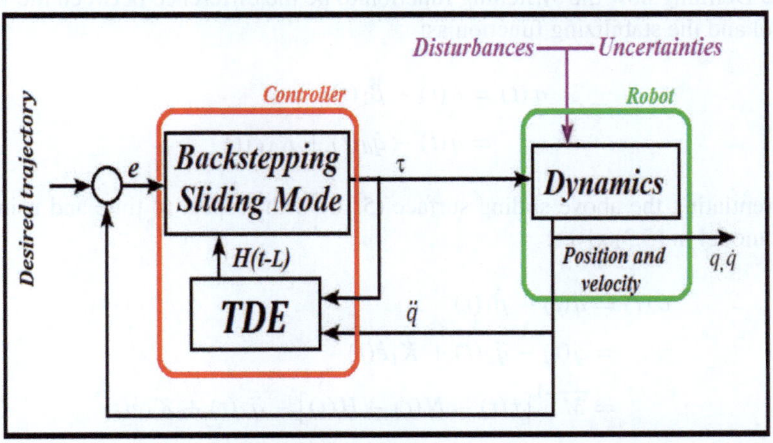

Fig. 5.1 Block diagram of the proposed controller

Our objective in this step is to force the convergence of the regulated variable to zero $e(t) \to 0$ by designing a virtual control. To reach this goal, the following Lyapunov function is selected:

$$V_1(t) = \frac{1}{2} e^T(t) e(t) \tag{5.6}$$

Its first time derivative is obtained as follows:

$$\begin{aligned} \dot{V}_1(t) &= e^T(t)\dot{e}(t) \\ &= e^T(t)(\dot{q}(t) - \dot{q}_d(t)) \end{aligned} \tag{5.7}$$

Choosing $\dot{q}(t)$ as virtual control variable. Then, an appropriate stabilizing function is selected to ensure stability as follows:

$$\begin{aligned} \dot{q}(t) &= \beta_1(t) \\ &= \dot{q}_d(t) - K_1 e(t) \end{aligned} \tag{5.8}$$

where $K_1 =$ diag $(k_{11}, k_{12}, \cdots, k_{1n})$ is a diagonal positive matrix.

Substituting the stabilizing function $\beta_1(t)$ in the first time derivative of the Lyapunov function (5.7) leads to:

$$\dot{V}_1(t) = -e^T(t) K_1 e(t) \tag{5.9}$$

From the above equation, it is clear that $\dot{V}_1(t)$ is negative definite, which proves that the convergence of the tracking error $e(t)$ to zero is ensured.

Step 2 Defining now the switching function to be the difference between the virtual control and the stabilizing function as:

$$\begin{aligned} \sigma(t) &= \dot{q}(t) - \beta_1(t) \\ &= \dot{q}(t) - \dot{q}_d(t) + K_1 e(t) \end{aligned} \tag{5.10}$$

Differentiating the above sliding surface (5.10) with respect to time and using the robot model in (5.3) gives:

$$\begin{aligned} \dot{\sigma}(t) &= \ddot{q}(t) - \dot{\beta}_1(t) \\ &= \ddot{q}(t) - \ddot{q}_d(t) + K_1 \dot{e}(t) \\ &= \overline{M}^{-1}\big[\tau(t) - N(t) - H(t)\big] - \ddot{q}_d(t) + K_1 \dot{e}(t) \end{aligned} \tag{5.11}$$

Therefore, based on Eq. (5.10), the first time derivative of the error $\dot{e}(t)$ in (5.5) can be rewritten as follows:

$$\dot{e}(t) = \sigma(t) - K_1 e(t) \tag{5.12}$$

Thus, the second Lyapunov function is selected as:

$$V_2(t) = V_1(t) + \frac{1}{2}\sigma^T(t)\sigma(t) \tag{5.13}$$

Its first time derivative is calculated as:

$$\begin{aligned}
\dot{V}_2(t) &= \dot{V}_1(t) + \sigma^T(t)\dot{\sigma}(t) \\
&= e^T(t)\dot{e}(t) + \sigma^T(t)\dot{\sigma}(t) \\
&= e^T(t)\big(\sigma(t) - K_1 e(t)\big) \\
&\quad + \sigma^T(t)\left(\overline{M}^{-1}[\tau(t) - N(t) - H(t)] - \ddot{q}_d(t) + K_1\dot{e}(t)\right) \\
&= \sigma^T(t)\left(\overline{M}^{-1}[\tau(t) - N(t) - H(t)] - \ddot{q}_d(t) + K_1\dot{e}(t) + e(t)\right) \\
&\quad - e^T(t)K_1 e(t)
\end{aligned} \tag{5.14}$$

By choosing:

$$\begin{aligned}
&\overline{M}^{-1}[\tau(t) - N(t) - H(t)] - \ddot{q}_d(t) + K_1\dot{e}(t) + e(t) = \\
&\qquad -K_2\sigma(t) - K(\sigma(t))\,\text{sign}\,(\sigma(t))
\end{aligned} \tag{5.15}$$

where $K_2 = \text{diag}\,(k_{21}, k_{22}, \cdots, k_{2n})$ is a diagonal positive matrix and $K(\sigma(t)) = \text{diag}\left[\frac{k_1}{D_1(\sigma_1(t))}, \frac{k_2}{D_2(\sigma_2(t))}, \cdots, \frac{k_n}{D_n(\sigma_n(t))}\right]$ with $k_i > 0$ and the term $D_i(\sigma_i(t))$ for $i = 1, \cdots, n$ is defined in Fallaha et al. (2011) as follows:

$$D_i(\sigma_i) = \delta_i + (1 - \delta_i)e^{-\gamma_i|\sigma_i(t)|^{p_i}} \tag{5.16}$$

where $0 < \delta_i < 1$, $\gamma_i > 0$, $p_i > 0$. Finally, the signum function is defined by $\text{sign}\,(\sigma(t)) = \left[\,\text{sign}\,(\sigma_1(t)), \cdots, \text{sign}\,(\sigma_n(t))\right]^T$ with:

$$\text{sign}\,(\sigma_i(t)) = \begin{cases} 1, & if\ \sigma_i(t) > 0 \\ 0, & if\ \sigma_i(t) = 0 \\ -1, & if\ \sigma_i(t) < 0 \end{cases} \tag{5.17}$$

Remark 1 Notice that the exponential reaching law allows chattering reduction due to the gradually variations of the term $\frac{k_i}{D_i(\sigma_i(t))}$ in Eq. (5.16) that increases and decreases between k_i and $\frac{k_i}{\delta_i}$ depending on the sliding surface variations. In other words, $\frac{k_i}{D_i(\sigma_i(t))}$ increases during the reaching phase which make the attraction of

$\sigma_i(t)$ faster (i.e, when the value of $|\sigma_i(t)|$ becomes high, then, $D_i(\sigma_i)$ converges to δ_i and $\frac{k_i}{D_i(\sigma_i)}$ to $\frac{k_i}{\delta_i}$) while decreases during the sliding phase when the error and its first time derivative converge to zero (i.e, when $\sigma_i(t)$ converges to 0, then, the term $D_i(\sigma_i)$ converges to 1 and $\frac{k_i}{D_i(\sigma_i)}$ to k_i).

Then, the Backstepping Sliding Mode Control (BSMC) is obtained by resolving Eq. (5.15) as:

$$\tau(t) = \overline{M}u(t) + N(t) + H(t) \tag{5.18}$$

where:

$$u(t) = \ddot{q}_d(t) - K_1\dot{e}(t) - e(t) - K_2\sigma(t) - K(\sigma(t)) \text{ sign } (\sigma(t)) \tag{5.19}$$

Since $H(t)$ is uncertain, the control performance will be affected. Then, based on Assumption 2 given in Sect. 5.2, $H(t)$ can be estimated using time delay estimation method (Toumi and Ito 1990) as:

$$\begin{aligned} \hat{H}(t) &\cong H(t-L) \\ &= \tau(t-L) - N(t-L) - \overline{M}\ddot{q}(t-L) \end{aligned} \tag{5.20}$$

where L is the estimation time delay. Clearly the accuracy of $\hat{H}(t)$ improves as L decreases. In practice, L can't be chosen smaller than the sampling period.

Finally, the proposed robust backstepping sliding mode with time delay estimation is obtained as:

$$\begin{aligned} \tau(t) &= \overline{M}u(t) + N(t) + \hat{H}(t) \\ &= \tau(t-L) + \overline{M}\big[u(t) - \ddot{q}(t-L)\big] + N(t) - N(t-L) \end{aligned} \tag{5.21}$$

where $u(t)$ is given in Eq. (5.19) and the time delayed acceleration $\ddot{q}(t-L)$ can be obtained by one of the following approximations:

$$\ddot{q}(t-L) = \frac{1}{L}\big[\dot{q}(t-L) - \dot{q}(t-2L)\big] \tag{5.22}$$

$$\ddot{q}(t-L) = \frac{1}{L^2}\big[q(t-L) - 2q(t-2L) + q(t-3L)\big] \tag{5.23}$$

Step 3 For the stability analysis, the proposed controller in Eq. (5.21) is replaced in the first time derivative of the selected second Lyapunov function in Eq. (5.14) as:

$$\begin{aligned} \dot{V}_2(t) = {} & \sigma^T(t)\Big(\overline{M}^{-1}[\hat{H}(t) - H(t)] - K_2\sigma(t) - K(\sigma) \text{ sign } (\sigma(t))\Big) \\ & - e^T(t)K_1e(t) \end{aligned}$$

$$
\begin{aligned}
&= -e^T(t)K_1e(t) - \sigma^T(t)K_2\sigma(t) + \sigma^T\big(\varepsilon(t) - K(\sigma)\,\mathrm{sign}\,(\sigma(t))\big) \\
&= \sum_{i=1}^{n} -\big(k_{1i}e_i^2(t) + k_{2i}\sigma_i^2(t)\big) + \sigma_i(t)\left(\varepsilon_i(t) - \frac{k_i}{D_i(\sigma_i)}\,\mathrm{sign}\,(\sigma_i)\right) \\
&= \sum_{i=1}^{n} -\big(k_{1i}e_i^2(t) + k_{2i}\sigma_i^2(t)\big) + \left(\sigma_i(t)\varepsilon_i(t) - \frac{k_i}{D_i(\sigma_i)}|\sigma_i(t)|\right) \\
&\le \sum_{i=1}^{n} -\big(k_{1i}e_i^2(t) + k_{2i}\sigma_i^2(t)\big) - |\sigma_i(t)|\left(\frac{k_i}{D_i(\sigma_i)} - |\varepsilon_i(t)|\right)
\end{aligned}
\tag{5.24}
$$

where $\varepsilon(t) = \overline{M}^{-1}[\hat{H}(t) - H(t)]$ denotes the time delay estimation error. To ensure that $\dot{V}_2(t)$ is a negative definite function, the following condition must be verified:

$$
k_i > |\varepsilon_i(t)|,\ for\ i = 1, \cdots, n. \tag{5.25}
$$

Therefore, the error $e(t)$ and its first time derivative $\dot{e}(t)$ are asymptotically bounded (i.e. the closed loop system is stable) if the TDE error $\varepsilon(t)$ is asymptotically bounded. Therefore, the asymptotic boundedness of ε will be proved as in Jin et al. (2008).

Substituting the controller given in Eq. (5.21) in the mathematical model of the robot (5.3) leads to:

$$
\varepsilon(t) = \ddot{q}(t) - u(t) \tag{5.26}
$$

Multipliying both sides of the above equation by $M(t)$ and using Eq. (5.1) gives:

$$
\begin{aligned}
M(t)\varepsilon(t) &= M(t)(\ddot{q}(t) - u(t)) \\
&= \tau(t) + \tau_d(t) - C(q(t), \dot{q}(t))\dot{q}(t) - G(q(t)) - M(t)u(t)
\end{aligned}
\tag{5.27}
$$

Then, substituting (5.21) in the above equation leads to:

$$
\begin{aligned}
M(t)\varepsilon(t) &= \overline{M}u(t) + N(t) + \hat{H}(t) + \tau_d(t) - C(q(t), \dot{q}(t))\dot{q}(t) \\
&\quad - G(q(t)) - M(t)u(t) \\
&= [\overline{M} - M(t)]u(t) + \hat{H}(t) + \tau_d(t) - C_u(q(t), \dot{q}(t))\dot{q}(t) \\
&\quad - G_u(q(t))
\end{aligned}
\tag{5.28}
$$

As $H(t) = \big[M(t) - \overline{M}\big]\ddot{q}(t) + C_u(q(t), \dot{q}(t))\dot{q}(t) + G_u(q(t)) - \tau_d(t)$. Then, $\hat{H}(t)$ can be expressed as:

$$
\begin{aligned}
\hat{H}(t) &= \big[M(t-L) - \overline{M}\big]\ddot{q}(t-L) + G_u(q(t-L)) - \tau_d(t-L) \\
&\quad + C_u(q(t-L), \dot{q}(t-L))\dot{q}(t-L)
\end{aligned}
\tag{5.29}
$$

Therefore, Eq. (5.28) becomes:

$$M(t)\varepsilon(t) = [\overline{M} - M(t)]u(t) + \left[M(t-L) - \overline{M}\right]\ddot{q}(t-L) + \Delta \tag{5.30}$$

where:

$$\begin{aligned}\Delta &= C_u(q(t-L),\dot{q}(t-L))\dot{q}(t-L) - C_u(q(t),\dot{q}(t))\dot{q}(t)\\ &\quad + G_u(q(t-L)) - G_u(q(t)) + \tau_d(t) - \tau_d(t-L)\end{aligned} \tag{5.31}$$

Obviously, for a sufficiently small L, Δ is bounded. Then, using Eq. (5.26), the following expression can be stated:

$$\ddot{q}(t-L) = \varepsilon(t-L) + u(t-L) \tag{5.32}$$

Then:

$$\begin{aligned}M(t)\varepsilon(t) &= \left[\overline{M} - M(t)\right]u(t) + \left[M(t-L) - M(t)\right]\ddot{q}(t-L) + \Delta\\ &\quad + \left[M(t) - \overline{M}\right]\ddot{q}(t-L)\\ &= \left[\overline{M} - M(t)\right]u(t) + \left[M(t-L) - M(t)\right]\ddot{q}(t-L) + \Delta\\ &\quad + \left[M(t) - \overline{M}\right]\left[\varepsilon(t-L) + u(t-L)\right]\\ &= \left[M(t) - \overline{M}\right]\varepsilon(t-L) + \left[M(t) - \overline{M}\right]\left[u(t-L) - u(t)\right]\\ &\quad + \left[M(t-L) - M(t)\right]\ddot{q}(t-L) + \Delta\end{aligned} \tag{5.33}$$

Therefore, the TDE error is expressed as:

$$\varepsilon(t) = D\varepsilon(t-L) + D\zeta_1 + \zeta_2 \tag{5.34}$$

where

$$D = I - M^{-1}(t)\overline{M}$$

$$\zeta_1 = u(t-L) - u(t)$$

$$\zeta_2 = M^{-1}(t)[M(t-L) - M(t)]\ddot{q}(t-L) + M^{-1}(t)\Delta$$

Using Assumptions 1 and 2, it is clear that ζ_1 and ζ_2 are bounded for a small L. In addition, by choosing an appropriate $\overline{M}$, the following inequation is verified (Spong et al. 2005):

$$\|D\| = \|I - M^{-1}\overline{M}\| < 1 \tag{5.35}$$

In the discrete-time domain, Eq. (5.34) is represented as:

$$\varepsilon(k) = D(k)\varepsilon(k-1) + D\zeta_1(k) + \zeta_2(k) \tag{5.36}$$

Since $\|D\| < 1$, the roots of $D(k)$ reside inside a unit circle, thus, the TDE error in Eq. (5.36) is asymptotically bounded with bounded functions ζ_1 and ζ_2.

Hence, the stability of the closed-loop system is proved.

Remark 2 In real-time, the measured signals are contaminated by noise. The noise effect might be amplified when $\ddot{q}(t-L)$ is obtained using one of the approximations (5.22) and (5.23). As a solution to solve this problem, a Low Pass Filter (LPF) may be introduced before calculating the time delayed acceleration. However, the attenuation of noise without using a LPF is possible by choosing small values for $\overline{M}$.

If a digital LPF with the cutoff frequency β is adopted, the control law can be modified as follows:

$$\tau^f(t) = \beta L(1+\beta L)^{-1}\tau(t) + (1+\beta L)^{-1}\tau^f(t-L) \tag{5.37}$$

where $\tau(t)$ denotes the calculated input before the filter while $\tau^f(t)$ denotes the filtered control input. Substituting $\tau(t)$ by its expression in (5.21), gives:

$$\begin{aligned}\tau^f(t) &= \tau^f(t-L) + \beta L(1+\beta L)^{-1}\overline{M}\big(u(t) - \ddot{q}(t-L)\big)\\ &\quad + \beta L(1+\beta L)^{-1}\big(N(t) - N(t-L)\big)\end{aligned} \tag{5.38}$$

Comparing (5.38) with the controller in (5.21), then:

$$\overline{M}' = \beta L(1+\beta L)^{-1}\overline{M} \tag{5.39}$$

Since $\beta L(1+\beta L)^{-1} < 1$, then, for very small value of $\overline{M}$, a similar result as using a digital LPF will be obtained.

5.4 Simulation Studies

In this section, three numerical simulations are presented for trajectory tracking of a two-link rigid robotic manipulator shown in Fig. 5.2 using Matlab/Simulink software with ODE 4 Solver and a sampled time equal to 0.001 s.

For this two-link manipulator, the dynamic equation (5.1) has the following matrices,

$$M(q(t)) = \begin{bmatrix} l_2^2 m_2 + 2l_1 l_2 m_2 c_2 + l_1^2(m_1+m_2) + J_1 & l_2 m_2(l_1+l_2) \\ l_2 m_2(l_1+l_2) & l_2^2 m_2 + J_2 \end{bmatrix},$$

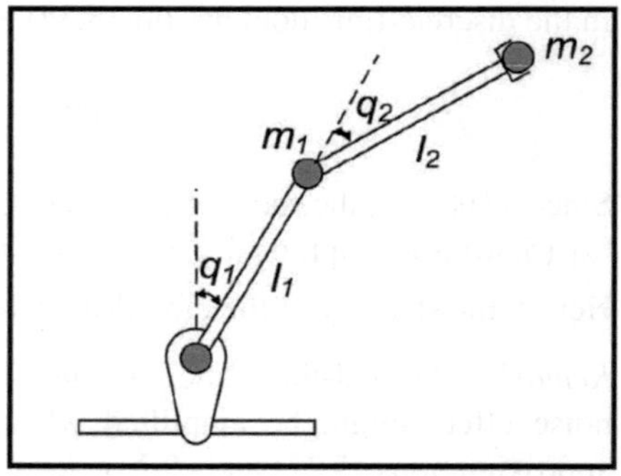

Fig. 5.2 Considered two-link robot manipulator

$$C(q(t), \dot{q}(t))\dot{q}(t) = \begin{bmatrix} -l_1 l_2 m_2 s_2 \dot{q}_2^2(t) - 2 l_1 l_2 m_2 s_2 \dot{q}_1(t)\dot{q}_2(t) \\ l_1 l_2 m_2 s_2 \dot{q}_2^2(t) \end{bmatrix},$$

$$G(q(t)) = \begin{bmatrix} l_2 m_2 g c_{12} + (m_1 + m_2) l_1 g c_1 \\ l_2 m_2 g c_{12} \end{bmatrix},$$

$$\tau_d(t) = \begin{cases} \begin{bmatrix} 2\sin t + 0.5\sin(200\pi t) \\ \cos 2t + 0.5\sin(200\pi t) \end{bmatrix}, & if\ t < 6\,\text{s} \\ \begin{bmatrix} 20\sin t + 5\sin(200\pi t) \\ 10\cos 2t + 5\sin(200\pi t) \end{bmatrix}, & else \end{cases}$$

where s_i, c_i and c_{ij} are defined by $s_i = \sin(q_i(t))$, $c_i = \cos(q_i(t))$ and $c_{ij} = \cos(q_i(t) + q_j(t))$, respectively. The parameters of the above elements are reported in Table 5.1.

Table 5.1 Parameters of the two-link robot manipulator

Symbol	Definition	Value	S.I. Unit
l_1	Length of the first link	1	(m)
l_2	Length of the second link	0.85	(m)
J_1	Moment of inertia of the first motor	5	(kg m^2)
J_2	Moment of inertia of the second motor	5	(kg m^2)
m_1	Mass of link 1	0.5	(kg)
m_2	Mass of link 2	1.5	(kg)
$\hat{m}_1$	Nominal mass of link 1	0.4	(kg)
$\hat{m}_2$	Nominal mass of link 2	1.2	(kg)
g	Gravitational constant	9.81	(m/s^2)

The control objective here is to follow the reference signals defined by:

$$q_{1d}(t) = 1.25 - (7/5)e^{-t} + (7/20)e^{-4t} \tag{5.40}$$

$$q_{2d}(t) = 1.25 + e^{-t} - (1/4)e^{-4t} \tag{5.41}$$

The initial states (joint positions velocities) are selected to be $q_1(0)$ =0.4 rad, $q_2(0) = 1.8$ rad and $\dot{q}_1(0) = \dot{q}_2(0) = 0$ rad/s.

5.4.1 Classical SMC Simulated on Robotic Arm

A complete study of classical sliding mode theory can be found in Utkin (1992). The control law for the considered robot is given as:

$$\tau(t) = M_0(q(t))\big[\ddot{q}_d(t) - \lambda\dot{e}(t) - \ \text{sign}\ (\sigma(t))\big] + C_0(q(t), \dot{q}(t))\dot{q}(t) + G_0(q(t))$$

where $M_0(q(t))$, $C_0(q(t), \dot{q}(t))$ and $G_0(q(t))$ are the nominal matrices calculated using the nominal masses $\hat{m}_1$ and $\hat{m}_2$ given in Table 5.1. The controller gains used in the simulation are:

$$\lambda = \ \text{diag}\ (2, 2), \quad K = \ \text{diag}\ (5, 5)$$

The simulation results are given in Fig. 5.3.

5.4.2 SMC with TDE Simulated on Robotic Arm

The theoretical development of the sliding mode control with time delay estimation can be found in Kali et al. (2015, 2017a). The control law for robot is given as:

$$\begin{aligned}\tau(t) = \tau(t-L) - M_0(q(t-L))\ddot{q}(t-L) + M_0(q(t))\big[\ddot{q}_d(t) - \lambda\dot{e}(t) - Ksat(\sigma(t))\big]\\ + C_0(q(t), \dot{q}(t))\dot{q}(t) + G_0(q(t)) - C_0(q(t-L), \dot{q}(t-L))\dot{q}(t-L) - G_0(q(t-L))\end{aligned}$$

In this simulation, the used parameters for sliding mode control using saturation function with time delay estimation are:

$$\lambda = \ \text{diag}\ (2, 2), \quad K = \ \text{diag}\ (5, 5), \quad L = T_s = 0.001\ \text{s}$$

Moreover, the time delayed acceleration $\ddot{q}(t - L)$ is obtained using the second approximation (5.23). The obtained simulation results are shown in Fig. 5.4.

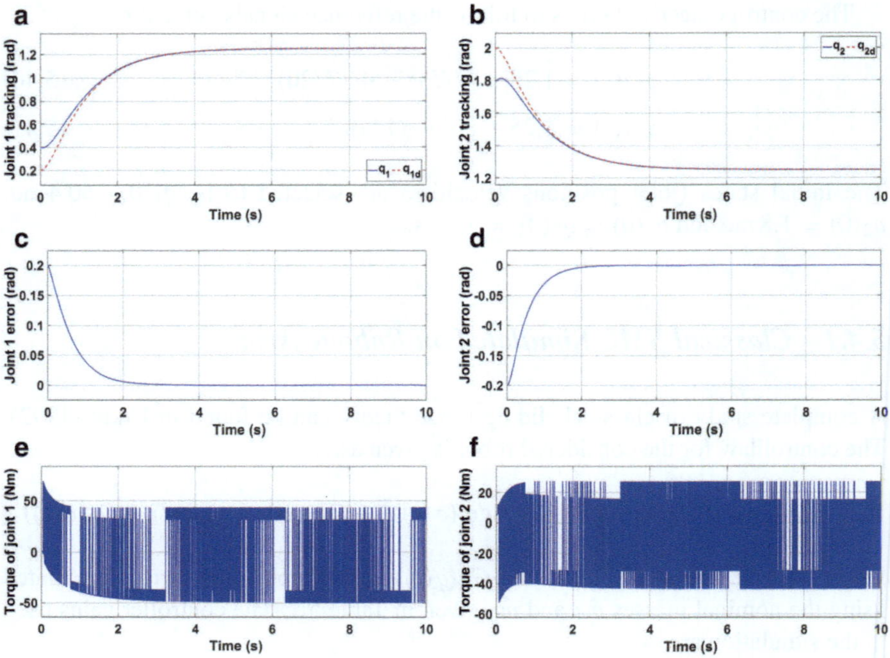

Fig. 5.3 Obtained simulation results via classical sliding mode control. (**a**) Joint 1 tracking. (**b**) Joint 2 tracking. (**c**) Joint 1 error. (**d**) Joint 2 error. (**e**) Torque of joint 1. (**f**) Torque of joint 2

5.4.3 *Proposed Controller Simulated on Robotic Arm*

The proposed modified backstepping sliding mode simulated on the two-link robot is given in Eq. (5.21) with:

$$K_1 = \text{diag}\,(2, 2),\;\; K_2 = \text{diag}\,(5, 5),\;\; k_1 = k_2 = 3$$
$$\delta_1 = \delta_2 = 0.6,\;\; \gamma_1 = \gamma_2 = 20,\;\; p_1 = p_2 = 1$$

and $L = T_s = 0{,}001$ s while the time delayed acceleration $\ddot{q}(t - L)$ is obtained using the second approximation (5.23). The obtained simulation results are depicted in Fig. 5.5.

5.4.4 *Discussion*

5.4.4.1 Classical Sliding Mode Control

The joint positions $q_1(t)$ and $q_2(t)$ shown respectively in Fig. 5.3a, b did not follow their desired references properly (the tracking error doesn't converge totally to zero).

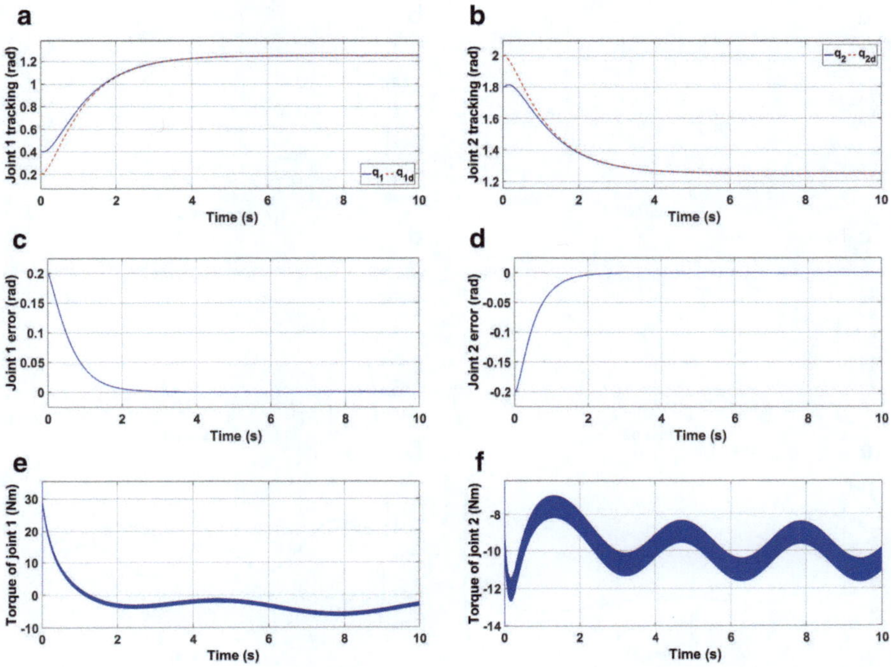

Fig. 5.4 Obtained simulation results via sliding mode control with time delay estimation. (**a**) Joint 1 tracking. (**b**) Joint 2 tracking. (**c**) Joint 1 error. (**d**) Joint 2 error. (**e**) Torque of joint 1. (**f**) Torque of joint 2

In addition, Fig. 5.3e, f show that the chattering phenomenon and the control effort are very important due to the large choice of the switching gains.

5.4.4.2 Sliding Mode Controller with Time Delay Estimation Method

Figure 5.4e, f show that the control effort and the chattering phenomenon in control torque inputs is reduced except at the beginning. However, the robot system deals with a large steady state error as depicted in Fig. 5.4c, d.

5.4.4.3 Proposed Backstepping Sliding Mode Controller with Time Delay Estimation Method

The obtained simulation results show the effectiveness of the modified backstepping sliding mode method, since the joints q_1 and q_2 shown respectively in Fig. 5.5a, b convergence to their desired trajectories in finite time. This is confirmed in the tracking error figures (Fig. 5.5c, d). In addition, the proposed controller allows chattering reduction as shown in Fig. 5.5e, f.

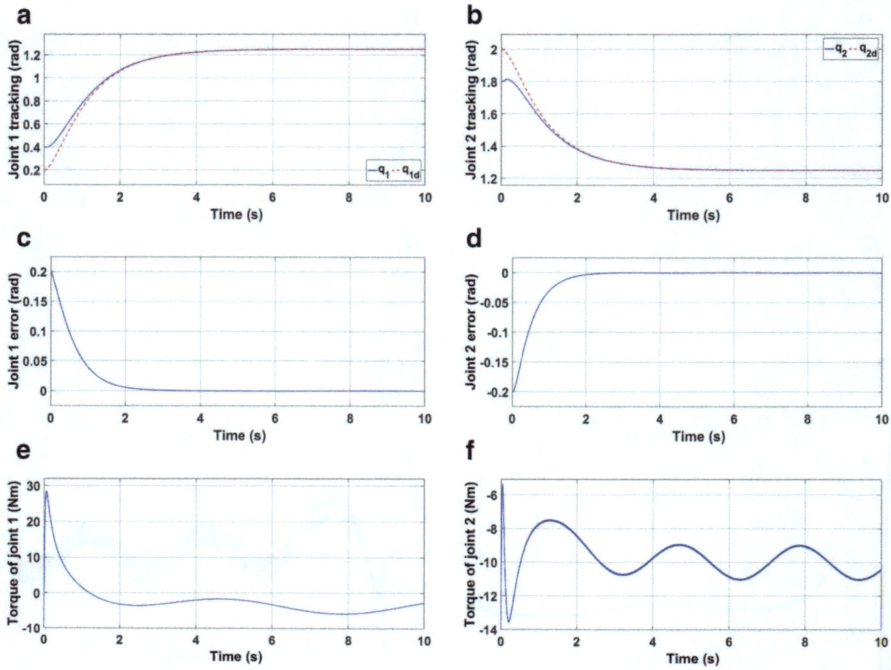

Fig. 5.5 Obtained simulation results via the proposed controller. (**a**) Joint 1 tracking. (**b**) Joint 2 tracking. (**c**) Joint 1 error. (**d**) Joint 2 error. (**e**) Torque of joint 1. (**f**) Torque of joint 2

From the above obtained results, the proposed controller shows the superiority such that the steady state error converges to zero faster due to the exponential reaching law. The chattering is still present SMC with TDE method, which can damage the controlled robot, while is reduced significantly in the proposed method.

5.5 Case Study

5.5.1 System Description

In this section, the real time implementation of the proposed optimal super-twisting with time delay estimation on the 7-DOF ANAT robot shown in Fig. 5.6 using Simulink with Real-Time Workshop is presented. ANAT (Articulated Nimble Adaptable Trunk) Robot is built by Robotics Design inc. (Canada). The first joint of the used ANAT is prismatic, followed by six rotary joints (Kali et al. 2017b).

As shown in Fig. 5.7, the frames of references are selected using the modified Denavit-Hartenberg (Craig 2005) (D-H) convention. The D-H parameters that determine the homogeneous transformation matrices are given in Table 5.2.

Fig. 5.6 7-DOF ANAT robot arm

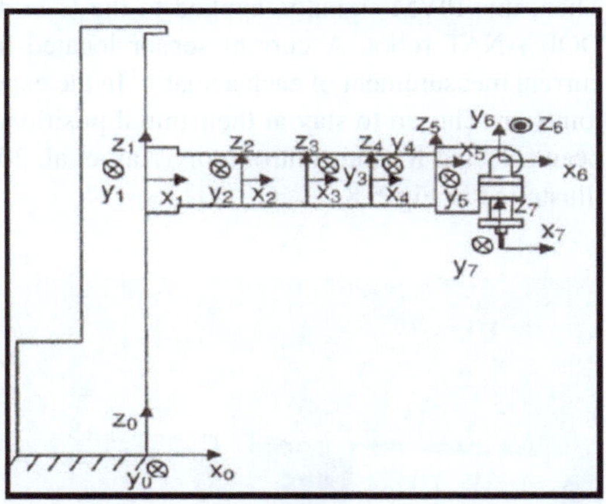

Fig. 5.7 D-H frames of the ANAT robot arm

Table 5.2 D-H parameters of the 7-DOF ANAT robot

Joints	α_{i-1} (rad)	a_{i-1} (m)	d_i (m)	q_i (rad)
1	0	0	q_1	0
2	0	L_1	0	q_2
3	0	L_0	0	q_3
4	0	L_0	0	q_4
5	0	L_0	L_2	q_5
6	$\pi/2$	L_3	0	q_6
7	$-\pi/2$	0	$-L_4$	q_7

The initial task space position of the end-effector that is placed on the last joint (7th joint) is: $x(0) = [0.6764\ 0\ \ -0.19]^T$ while the initial joint positions are 0 rad and joint velocities are 0 rad/s. The equation of motion are obtained such in Eq. (5.3) with:

$$\overline{M} = \text{diag}\,(0.23, 0.11, 0.14, 0.12, 0.12, 0.12, 0.12).$$

In addition, the ANAT model verifies the properties and the assumptions given in Sect. 5.2.

5.5.2 Real-Time Setup

All robot joints are actuated by DC motors. The actuators encoders provide position measurements. The ATMEGA 16 microcontrollers are used to send the angles positions to simulink, and using the forward kinematics, the joint space trajectory is transformed to the workspace trajectory. The microcontrollers are also used to transform the control input signals into a Pulse Width Modulation (PWM) signal. Then, this PWM signal is applied to the H-bridge drive of the actuators of the 7-DOF ANAT robot. A current sensor located in the H-bridge drive provides the current measurement of each actuator. In the experimentation, the sixth, and seventh joints are chosen to stay at their initial position. A sampling time $T_s = 0.03$ s has been selected for the control loop (Kali et al. 2017b). The above real-time setup is illustrated in Fig. 5.8.

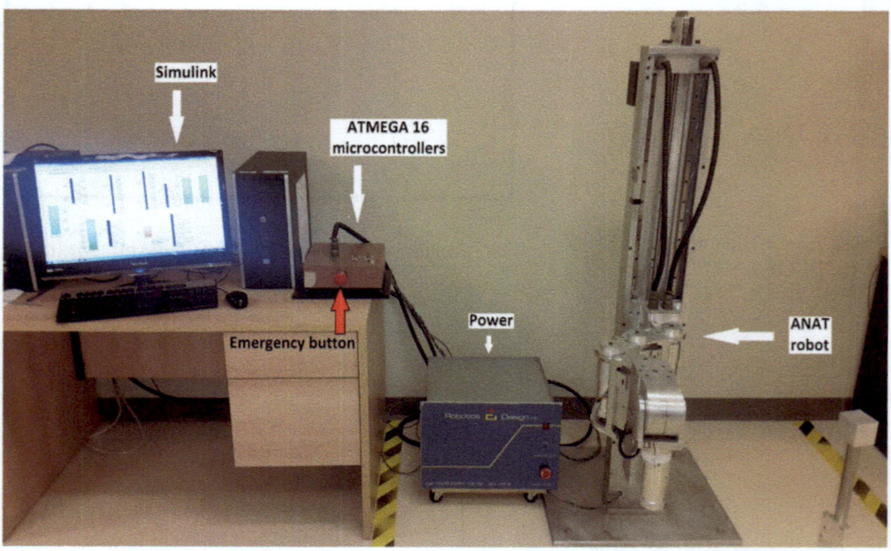

Fig. 5.8 Experimental setup

5.5.3 Controller Setting

The proposed robust backstepping sliding mode control with time delay estimation has four diagonal matrices with constant elements. This makes the controller setting simplify, which is briefly described as follows:

- Determine the matrix K_1 of the sliding surface in (5.11).
- Determine the matrix K_2 such as a fast convergence will be obtained.
- Select L as small as possible (equal to the sampling time interval T_s).
- Tuning the matrix $\overline{M}$ based on the Property 2 given in Sect. 5.2. This step is very important since $\overline{M}$ plays a key role in the stability and the noise attenuation. To that end, the diagonal elements of $\overline{M}$ are increased gradually from small positive values, while checking the control performance by trial and error.
- Finally the constants k_i for $i = 1, \cdots, n$ are chosen such as the condition of stability in (5.25) is verified, they should be further tuned to achieve the optimal performance.

5.5.4 Experimental Results

The control objective here is to follow a known spiral desired trajectory defined in the task space. For this application the desired spiral is chosen such as all kinematic singularities are avoided, the relation between the desired accelerations in the task space and the joint space is given by:

$$\ddot{q}_d(t) = J^{+}\ddot{x}_d(t) - J^{+}\dot{J}J^{+}\dot{x}_d(t) \tag{5.42}$$

where $J^{+} = J^T(JJ^T)^{-1}$ denotes the generalized inverse of the Jacobian matrix while $\dot{J}$ is its derivative, $\ddot{q}_d(t)$ is the desired joint acceleration vector, $\ddot{x}_d(t)$ and $\dot{x}_d(t)$ are the task space desired acceleration and velocity vectors, respectively. The desired velocity in the joint space $\dot{q}_d(t)$ is obtained from $\ddot{q}_d(t)$ using an integrator while the desired position in the joint space $q_d(t)$ is obtained from $\dot{q}_d(t)$ using another integrator.

In Sect. 5.3, the proposed controller that ensures the stability of the closed loop system is given in (5.21). The time delay estimation is chosen to be the sampled time: $L = T_s = 0.03$ s and the time delayed acceleration $\ddot{q}_i(t-L)$ for $i = 1, \cdots, 7$ is obtained using the approximation based on the joint position as:

$$\ddot{q}_i(t-L) = \frac{1}{T_s^2}\big[q_i(t-L) - 2q_i(t-2L) + q_i(t-3L)\big]$$

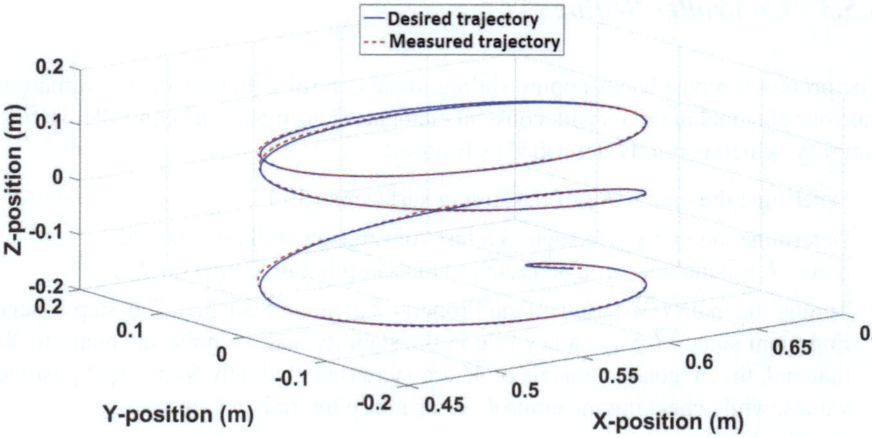

Fig. 5.9 3D Task space tracking trajectory

The controller gains are chosen such as the condition of stability in Eq. (5.25) is verified:

$$K_1 = \text{diag}\,(2, 2, 2, 2, 2, 2, 2)$$

$$K_2 = \text{diag}\,(1.2, 1.2, 1.2, 1.2, 1.2, 1.2, 1.2)$$

$$k_i = 3.8;\ \delta_i = 0.7;\ \gamma_i = 20;\ p_i = 1\ for\ i = 1:7$$

The experimental results are shown in Figs. 5.9, 5.10, 5.11, 5.12, 5.13 and 5.14: 3D task space tracking, workspace tracking position and error, joint space tracking position and error and control torque inputs. In these results, the performances produced by the proposed controller are illustrated for the ANAT robot by neglecting the Coriolis matrix and adding a load of 5.5 kg.

The proposed controller ensures the convergence of the end-effector position to the task space desired position with high accuracy due to a good estimation of uncertainties as shown in Figs. 5.9, 5.10 and 5.11.

Figure 5.12 illustrates the high accuracy tracking position in joint space even in presence of uncertainties and external disturbances confirmed by the small tracking errors as depicted in Fig. 5.13.

Furthermore, it can be seen from Fig. 5.14 that the control torque inputs are smooth and chattering free with small values that are acceptable for our robot, this is thanks to the exponential reaching law that make the controller adapted to the variations of the sliding surface and forcing $k_i/N_i(S_i)$ to vary between k_i and k_i/δ_i for $i = 1, \cdots, n$ (here $n = 7$).

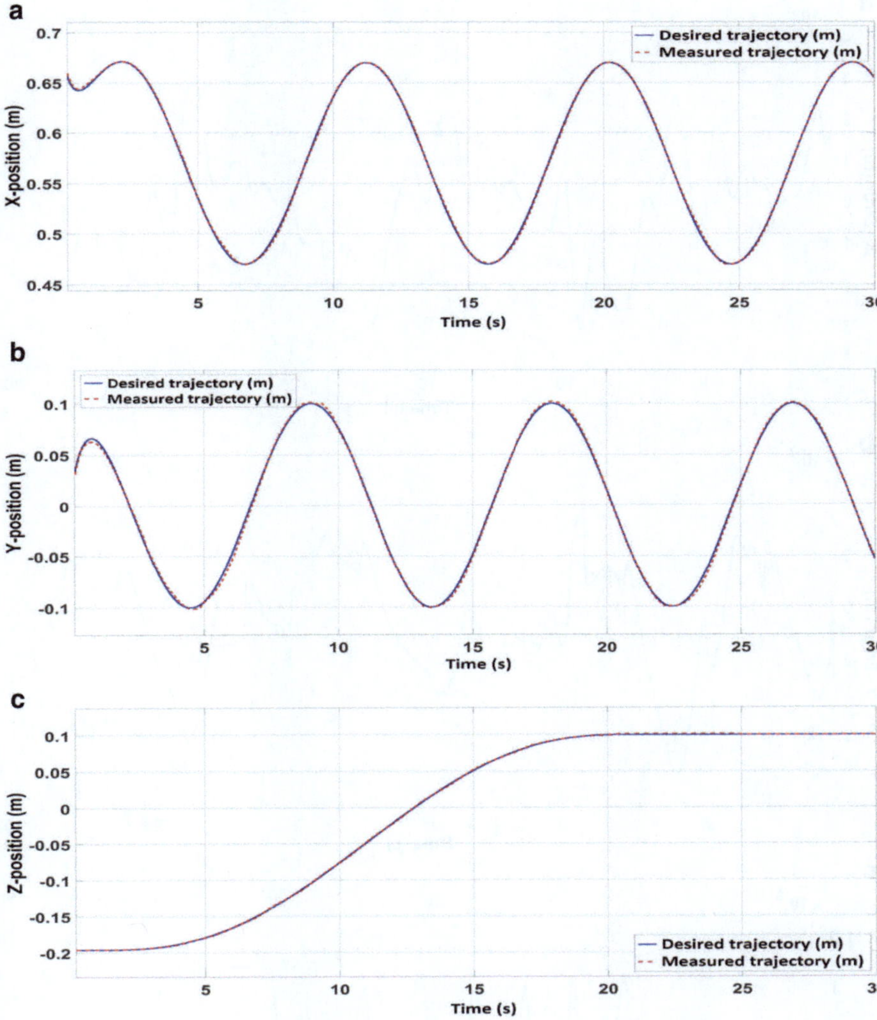

Fig. 5.10 Task space tracking position. (**a**) X-tracking. (**b**) Y-tracking. (**c**) Z-tracking

To evaluate the performance of the proposed modified backstepping sliding mode, it is compared with the conventional SMC and the SMTDC (Kali et al. 2015) for the spiral desired trajectory in presence of uncertainties using two performance indices.The first one, the Root-Mean-Squared (RMS) error $\|e_i\|_{RMS}$ is a numerical measure of average tracking performance while and the second one is the RMS torque $\|\tau_i\|_{RMS}$ which is the average control torque input allowing the evaluation of the control effort. The expressions of $\|e_i\|_{RMS}$ and $\|\tau_i\|_{RMS}$ are given as follows:

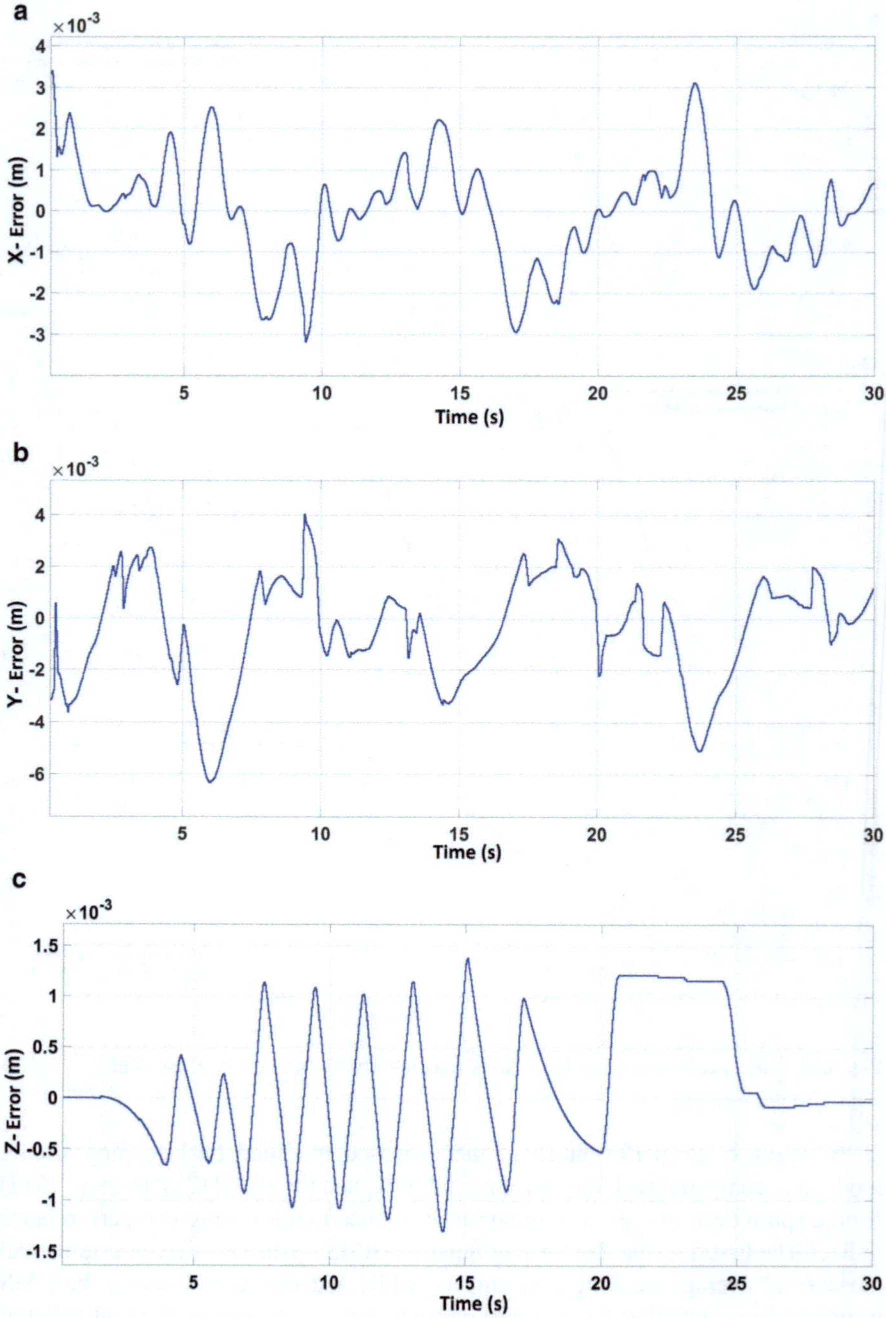

Fig. 5.11 Task space tracking error. (**a**) X-error. (**b**) Y-error. (**c**) Z-error

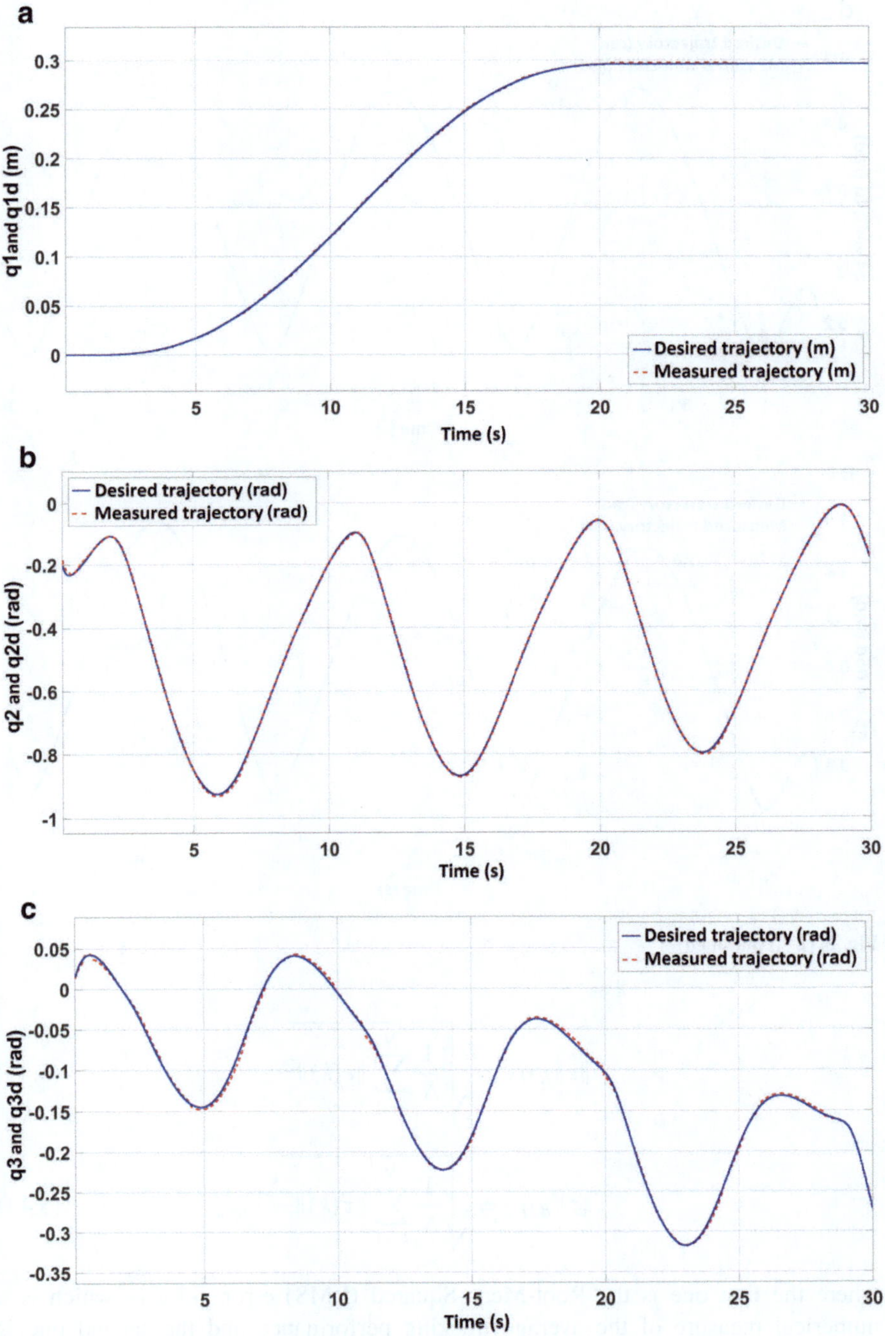

Fig. 5.12 Joint space tracking position. (**a**) Tracking position of joint 1. (**b**) Tracking position of joint 2. (**c**) Tracking position of joint 3. (**d**) Tracking position of joint 4. (**e**) Tracking position of joint 5

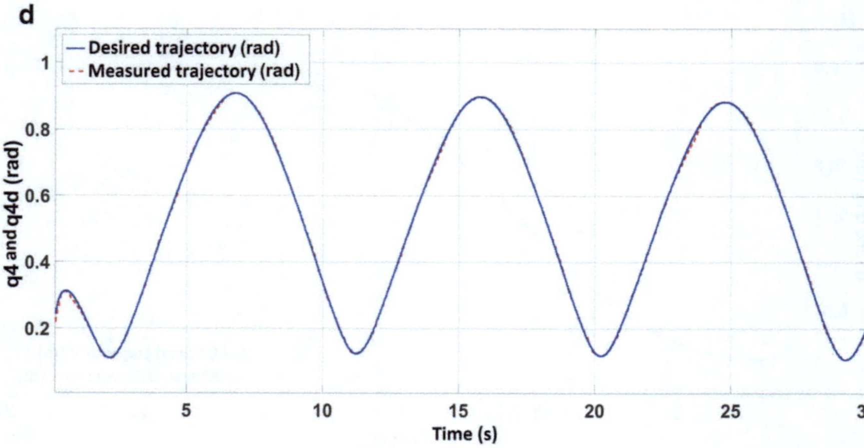

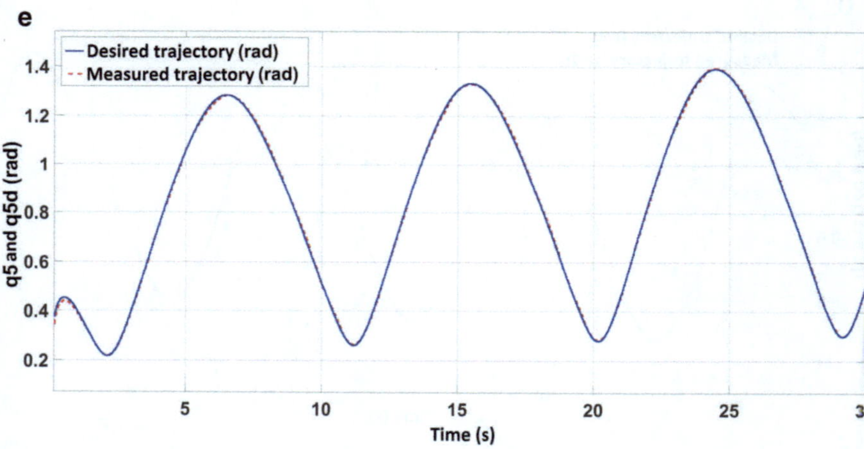

Fig. 5.12 (continued)

$$\|e\|_{RMS} = \sqrt{\frac{1}{N}\sum_{k=1}^{N}\|e(k)\|^2} \tag{5.43}$$

$$\|\tau\|_{RMS} = \sqrt{\frac{1}{N}\sum_{k=1}^{N}\|\tau(k)\|^2} \tag{5.44}$$

where the first one is the Root-Mean-Squared (RMS) error $\|e\|_{RMS}$ which is a numerical measure of the average tracking performance and the second one is the RMS torque $\|\tau\|_{RMS}$ which is the average control torque input allowing the

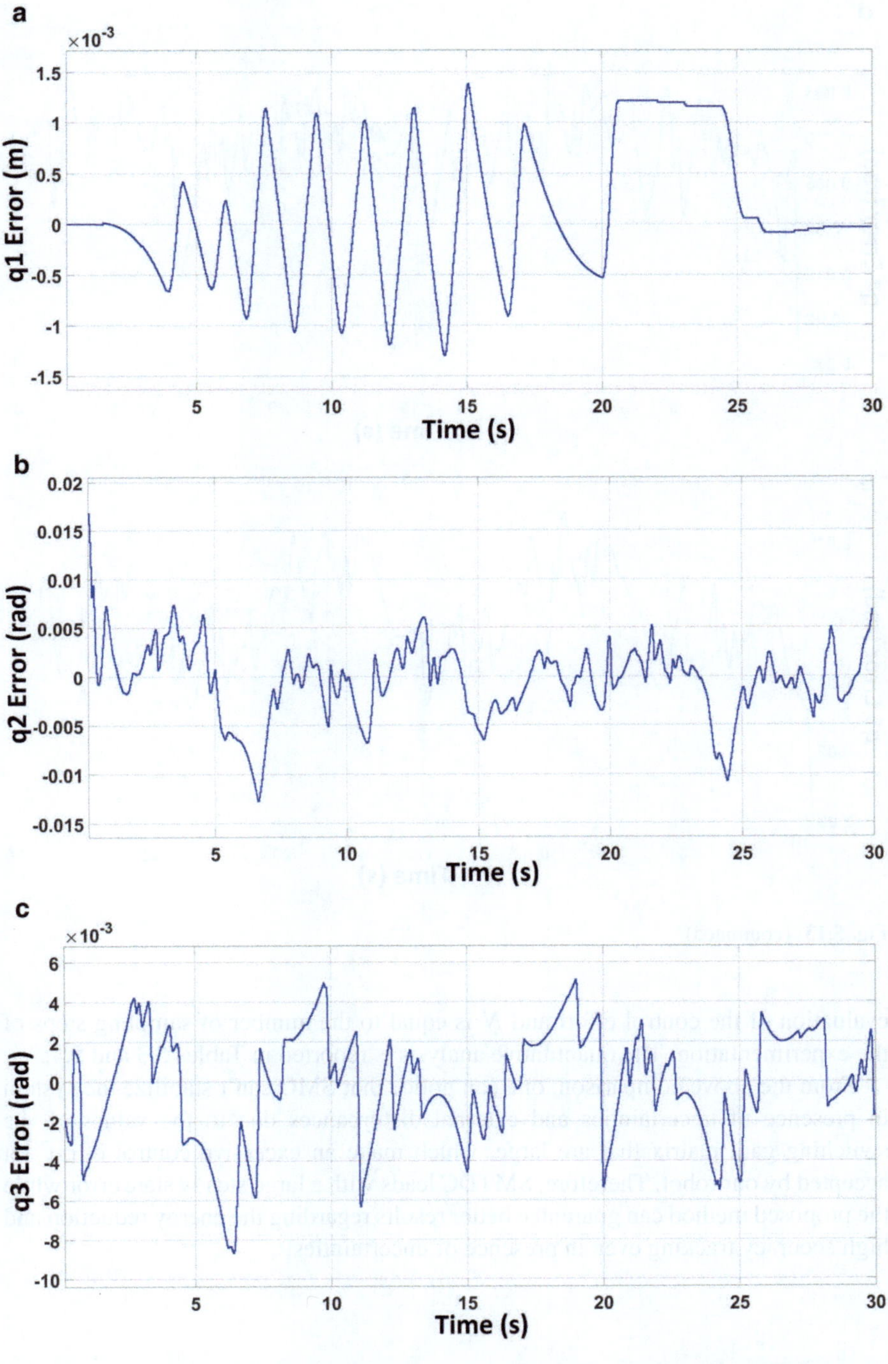

Fig. 5.13 Joint space tracking error. (**a**) Tracking error of joint 1. (**b**) Tracking error of joint 2. (**c**) Tracking error of joint 3. (**d**) Tracking error of joint 4. (**e**) Tracking error of joint 5

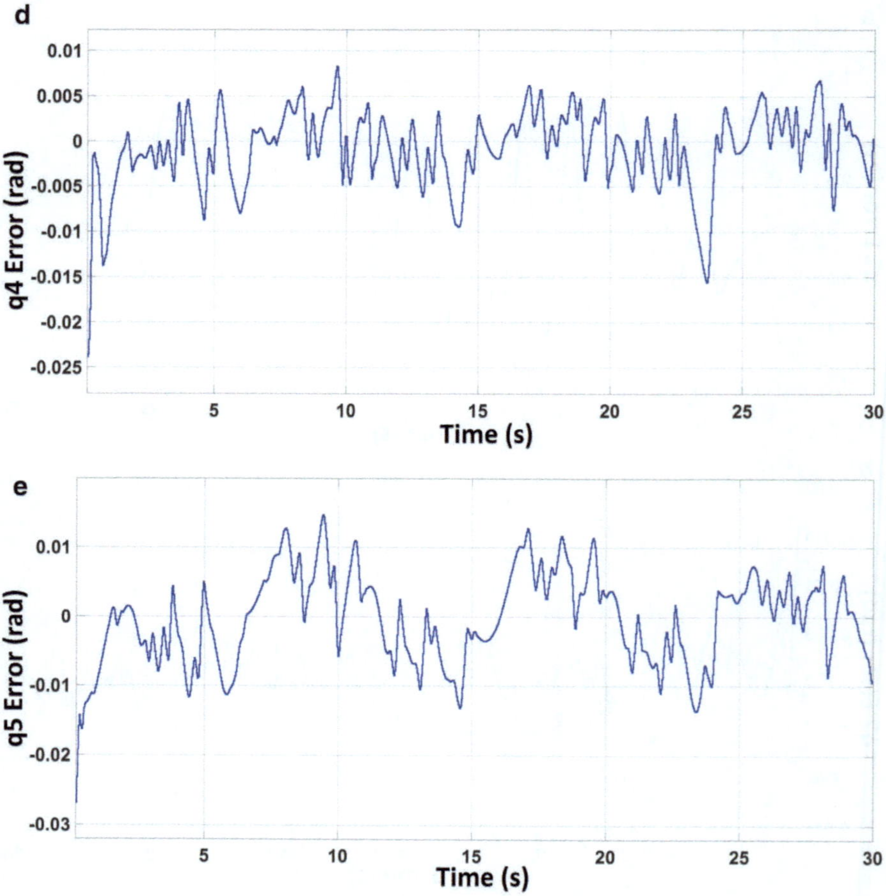

Fig. 5.13 (continued)

evaluation of the control effort and N is equal to the number of sampling steps of the experimentation. The quantitative analysis is reported in Tables 5.3 and 5.4.

From the above comparison, one can notice that SMC can't stabilize the system in presence of uncertainties and external disturbances due to the values of the switching gain matrix that are large, which make an excessive control effort not accepted by our robot. Therefore, SMTDC leads with a large steady state error while the proposed method can guarantee better results regarding the energy reduction and high accuracy tracking even in presence of uncertainties.

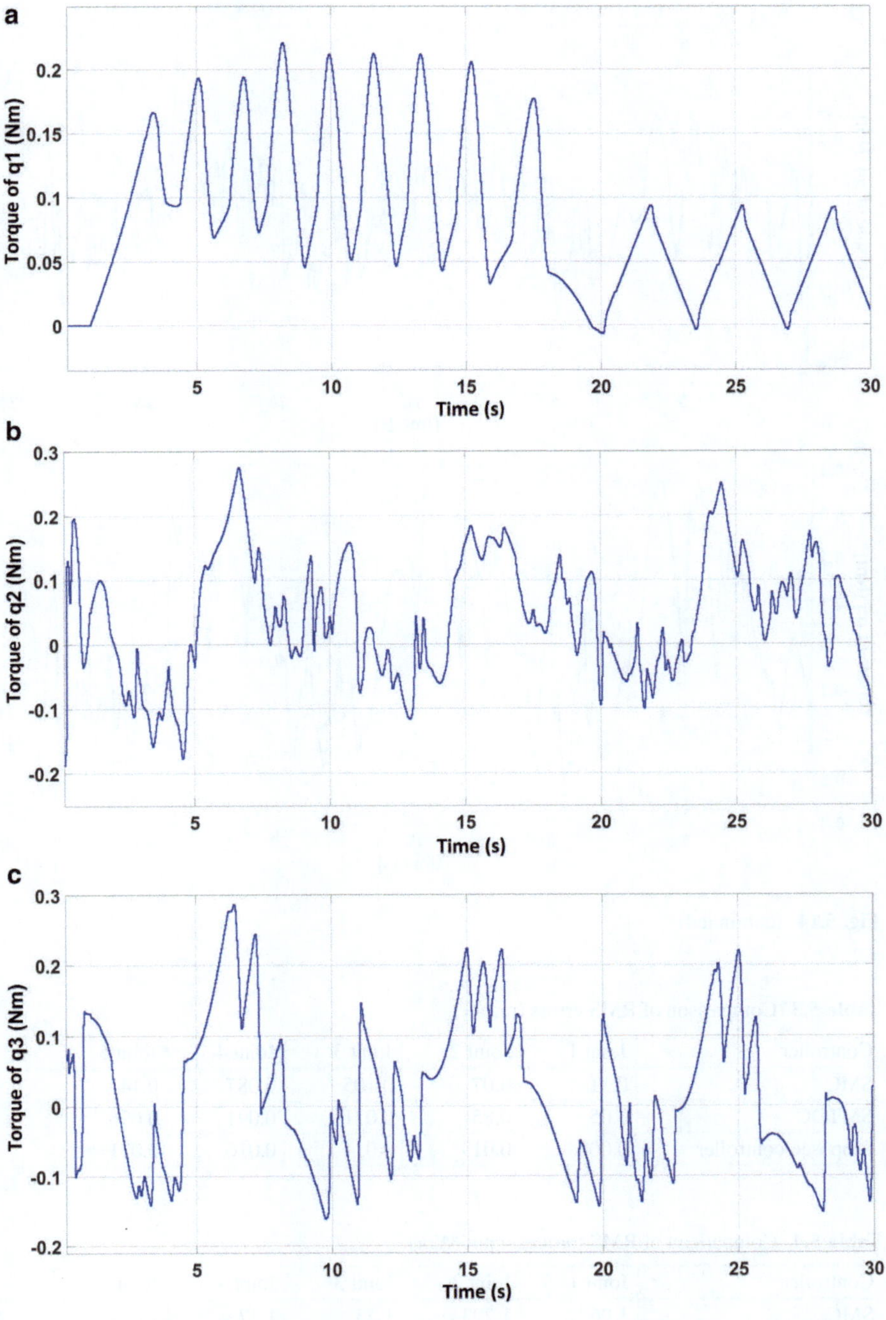

Fig. 5.14 Control torque inputs. (**a**) Torque input of joint 1. (**b**) Torque input of joint 2. (**c**) Torque input of joint 3. (**d**) Torque input of joint 4. (**e**) Torque input of joint 5

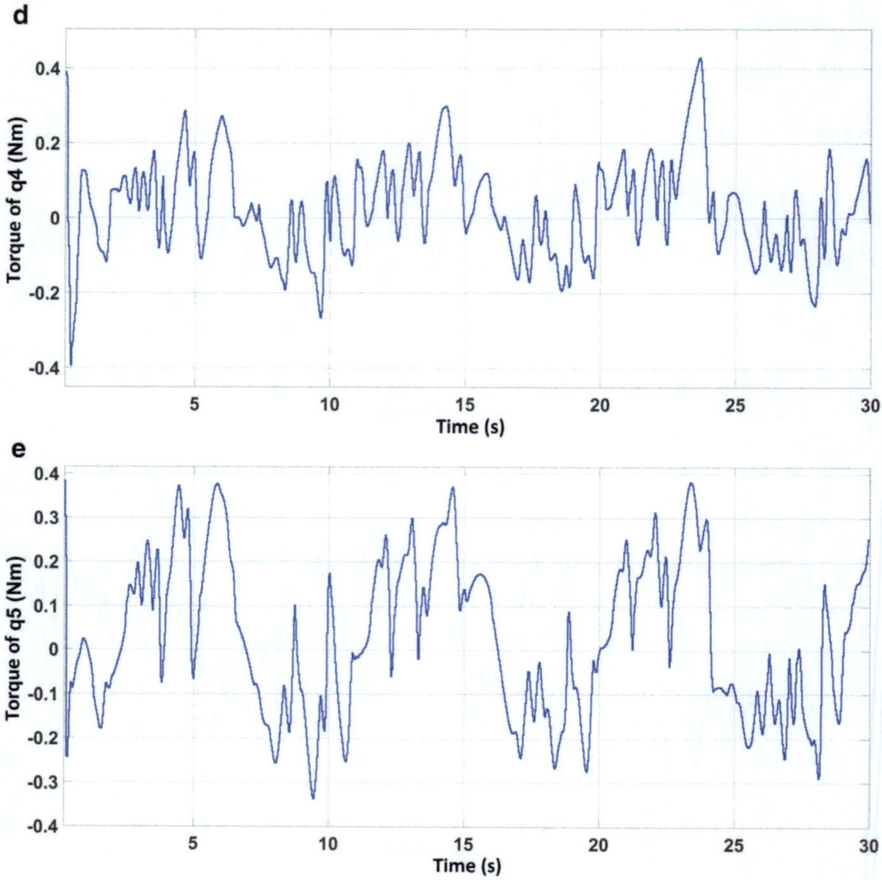

Fig. 5.14 (continued)

Table 5.3 Comparison of RMS errors (rad)

Controller	Joint 1	Joint 2	Joint 3	Joint 4	Joint 5
SMC	0.11	0.07	0.105	0.087	0.14
SMTDC	0.05	0.85	0.071	0.091	0.096
Proposed controller	0.008	0.019	0.014	0.016	0.021

Table 5.4 Comparison of RMS torques input (Nm)

Controller	Joint 1	Joint 2	Joint 3	Joint 4	Joint 5
SMC	1.067	1.293	1.53	1.47	1.628
SMTDC	0.728	0.637	0.94	0.824	0.641
Proposed controller	0.233	0.392	0.381	0.271	0.48

5.6 Conclusion

In this work, a high accuracy tracking control based on backstepping sliding mode with time delay estimation for robot manipulators systems with unknown dynamics and external disturbances has been presented. Sufficient condition is established to guarantee the stability by defining an appropriate Lyapunov function. The experimental results show the merit and the effectiveness of this new control. The results were satisfactory. The tracking trajectory was done with high accuracy and convenable control torque input even in the presence of uncertainties and external disturbance.

Acknowledgements The authors are grateful to Nabil Derbel (University of Sfax, Tunisia), Jawhar Ghommam (University of Tunis, Tunisia) and Quanmin Zhu (University of the West of England) for the opportunity to contribute to the New developments and advances in the field of Robotics.

References

Boiko, I. (2013). Chattering in sliding mode control systems with boundary layer approximation of discontinuous control. *International Journal of Systems Science, 44*, 1126–1133.

Boiko, I., & Fridman, L. (2005). Analysis of chattering in continuous sliding-mode controllers. *IEEE Transactions on Automatic Control, 50*, 1442–1446.

Chang, Y. (2011). Block backstepping control of MIMO systems. *IEEE Transactions on Automation Control, 56*(5), 1191–1197.

Cheah, C., Liu, C., & Slotine, J. (2006). Adaptive Jacobian tracking control of robots with uncertainties in kinematic, dynamic and actuator models. *IEEE Transactions on Automatic Control, 51*, 1024–1029.

Craig, J. (2005). *Introduction to robotics: Mechanics and control* (3rd ed.). Upper Saddle River, NJ: Pearson/Prentice-Hall.

de Wit, C. C., & Slotine, J.-J. (1991). Sliding observers for robot manipulators. *Automatica, 27*(5), 859–864.

Dotoli, M. (2003). Fuzzy sliding mode control with piecewise linear switching manifold. *Asian Journal of Control, 5*(4), 528–542.

Fallaha, C., Saad, M., Kanaan, H., & Al-Haddad, K. (2011). Sliding mode robot control with exponential reaching law. *IEEE Transactions on Industrial Electronics, 58*, 600–610.

Fridman, L. (2001). An averaging approach to chattering. *IEEE Transactions on Automatic Control, 46*, 1260–1265.

Fridman, L., & Levant, A. (2002). Higher order sliding mode. In J. P. Barbot & W. Perruguetti (Eds.), *Sliding mode control in engineering* (Systems and control book series). Marcel Dekker.

Geng, J., Sheng, Y., Liu, X., & Liu, B. (2013). Second order time varying sliding mode control for uncertain systems. In *25th Chinese Control and Decision Conference (CCDC)*, Guiyang (pp. 3880–3885).

Guo, Y., & Chung, M.-J. (2003). An adaptive fuzzy sliding mode controller for robotic manipulators. *IEEE Transactions on Systems, Man, and Cybernetics, 33*(2), 149–159.

Hsia, T., & Jung, S. (1995). A simple alternative to neural network control scheme for robot manipulators. *IEEE Transactions on Industrial Electronics, 42*(4), 414–416.

Jamoussi, K., Ouali, M., Chrifi-Alaoui, L., & Benderradji, H. (2013). Robust sliding mode control using adaptive switching gain for induction motors. *International Journal of Automation and Computing, 10*(4), 303–311.
Jezernik, K., Rodic, M., Safaric, R., & Curk, B. (1997). Neural network sliding mode robot control. *Robotica, 15*, 23–30.
Jin, M., Kang, S. H., & Chang, P. H. (2008). Robust compliant motion control of robot with nonlinear friction using time-delay estimation. *IEEE Transactions on Industrial Electronics, 55*(1), 258–269.
Kali, Y., Saad, M., Benjelloun, K., & Benbrahim, M. (2015). Sliding mode with time delay control for mimo nonlinear systems with unknown dynamics. *International Workshop on Recent Advances in Sliding Modes*, Istanbul, Turkey, April 9–11.
Kali, Y., Saad, M., Benjelloun, K., & Benbrahim, M. (2017a). *Sliding mode with time delay control for robot manipulators* (pp. 135–156). Singapore: Springer.
Kali, Y., Saad, M., Benjelloun, K., & Fatemi, A. (2017b). Discrete-time second order sliding mode with time delay control for uncertain robot manipulators. *Robotics and Autonomous Systems, 94*, 53–60.
Levant, A. (1993). Sliding order and sliding accuracy in sliding mode control. *International Journal of Control, 58*, 1247–1263.
Lewis, F., Abdallah, C., & Dawson, D. (1993). *Control of robot manipulators*. New York: Macmillan Publishing Company.
Lin, C. (2003). Nonsingular terminal sliding mode control of robot manipulators using fuzzy wavelet networks. *IEEE Transactions on Fuzzy Systems, 14*(6), 849–859.
Ling, R., Wu, M., Dong, Y., & Chai, Y. (2012). High order sliding mode control for uncertain nonlinear systems with relative degree three. *Nonlinear Science and Numerical Simulation, 17*, 3406–3416.
Liu, J., & Wang, X. (2012). *Advanced sliding mode control for mechanical systems*. Berlin/Heidelberg: Springer.
Rahman, M., Saad, M., Kenné, J., & Archambault, P. (2013). Control of an exoskeleton robot arm with sliding mode exponential reaching law. *International Journal of Control, Automation and Systems, 11*(1), 92–104.
Seraji, H. (1987). Adaptive control of robotic manipulators. In *26th IEEE Conference on Decision and Control*, Los Angeles (p. 26).
Slotine, J., & Li, W. (1991). *Applied nonlinear control*. Taipei: Printice-Hall International.
Slotine, J. J., & Sastry, S. S. (1983). Tracking control of non-linear systems using sliding surfaces, with application to robot manipulators. *International Journal of Control, 38*(2), 465–492.
Slotine, J. J., Hedrick, J. K., & Misawa, E. A. (1986). On sliding observers for nonlinear systems. In *American Control Conference*, Seattle (pp. 1794–1800).
Spong, M., Hutchinson, S., & Vidyasagar, M. (2005). *Robot dynamics and control*. New York: Wiley.
Toumi, K., & Ito, O. (1990). A time delay controller for systems with unknown dynamics. *ASME Journal of Dynamic System, Measurement and Control, 112*, 133–141.
Toumi, K., & Shortlidge, C. (1991). Control of robot manipulators using time delay. In *IEEE International Conference of Robotics and Automation*.
Utkin, V. (1992). *Sliding mode in control and optimization*. Berlin: Springer.
Utkin, V., Guldner, J., & Shi, J. (1999). *Sliding mode control in electomechanical systems* (2nd ed.). Boca Raton/London: CRC Press.
Utkin, V. (2016). Discussion aspects of high-order sliding mode control. *IEEE Transactions on Automatic Control, 61*(3), 829–833.
Yi, S., & Woo, P.-Y. (1997). A robust fuzzy logic controller for robot manipulators with uncertainties. *IEEE Transactions on Systems, Man, and Cybernetics, 24*(4), 599–602.

Chapter 6
Robot Manipulator Control Using Backstepping with Lagrange's Extrapolation and PI Compensator

Yassine Kali, Maarouf Saad, Jean-Pierre Kenné, and Khalid Benjelloun

Abstract A robust nonlinear backstepping technique with Lagrange's extrapolation and PI compensator is proposed in this chapter for high accuracy trajectory tracking of robot manipulators with uncertain dynamics and unexpeted disturbances. The proposed controller is synthesized by using Lagrangian extrapolation method with PI compensator to estimate the uncertainties and disturbances and to deal with the effect of hard nonlinearities caused by the estimation error while nonlinear backstepping technique is used to ensure good tracking. The stability analysis is accomplished recursively using appropriate Lyapunov functions candidate. As a result, the proposed control technique shows better performances via experimental results on a 7-DOF robot arm in comparison with the classical backstepping and sliding mode control.

Keywords Backstepping · Lagrange's extrapolation · PI compensator · Uncertain robot manipulators · Lyapunov · Trajectory tracking

6.1 Introduction

One of the challenging problems in control engineering practice is to design robust controllers for robot manipulators in the presence of uncertainties and unexpected disturbances. Several controllers are proposed in literature; among them, Predictive control (Wilson et al. 2016), Fuzzy control (Al-Hadithi et al. 2007), Neural Network control (Fierro and Lewis 1998), Backstepping (Khalil 1992; Slotine and Li 1991), Sliding Mode control (Utkin 1992; Utkin et al. 1999). The design of

Y. Kali (✉) · K. Benjelloun
Ecole Mohammadia d'Ingénieurs, University of Mohammed V, Rabat, Morocco
e-mail: yassinekali@research.emi.ac.ma; bkhalid@emi.ac.ma

M. Saad · J.-P. Kenné
École de Technologie Supérieure, University of Quebec, Montreal, QC, Canada
e-mail: maarouf.saad@etsmtl.ca; jean-pierre.kenne@etsmtl.ca

N. Derbel et al. (eds.), *New Developments and Advances in Robot Control*, Studies in Systems, Decision and Control 175,
https://doi.org/10.1007/978-981-13-2212-9_6

those controllers is based on the mathematical model of the robots. Unfortunately, in reality, these models are uncertain due to uncertain parameters, load variations, unmodelled dynamics. Therefore, its difficult to fulfill the control objective.

One of the most important nonlinear control techniques is the Backstepping approach. This control scheme was introduced in the beginning of 1990s by Kokotovic et al. (1995). Due to its design and implementation simplicity, Backstepping method has been used to control many systems in different area. A recursive Lyapunov procedure is the basic idea in the design. This idea consists of designing a controller by considering virtual controls that represent in reality some state variables and then, ensuring them stability by selecting appropriate stabilization functions (Lewis et al. 1993; Slotine and Li 1991). This technique is developed to ensure the stability of the controlled system at any time, in addition to the control objective (tracking and/or stabilization). However, as all nonlinear controllers, this method has many drawbacks. The major one is its sensitivity to uncertainties and external disturbances. Otherwise, the control may easily cause unacceptable practical complications and degrade the control objective.

To cope with this problem, an adaptive backstepping has been proposed that provide an adaptation of the control to ensure robustness towards uncertain nonlinear dynamics and disturbances by an overparametrization that might cause inequality of the number of parameter estimates and the number of unknown parameters (Hu et al. 2012; Shieh and Hsu 2008; Zhou and Wen 2008). In Chen and Fu (2015), Choi and Farrell (2000), and Zhou and Er (2007), by considering the problem of controlling and stabilizing a particular class of systems with uncertainties, an observer-based backstepping control has been considered. Moreover, other researchers proposed a combination of backstepping and intelligent control techniques (neural-network and/or fuzzy logic) (Jagannathan and Lewis 1998; Su et al. 2015; Weisheng et al. 2015; Yoo and Ham 2000). It is known that these techniques have the merit to estimate uncertainties and disturbances. However, they are difficult to implement due to the large number of fuzzy rules introduced (case of fuzzy logic) and the large number of parameters (case of neural-network) that increase when the number of robot's DOF increases.

In this chapter, a robust backstepping controller with Lagrange's extrapolation and PI compensator is designed for high accuracy trajectory tracking for robot manipulators in the presence of uncertainties and external disturbances. The aim of Lagrange's extrapolation method is to provide a good estimation of the unknown dynamics and disturbances using time delayed signals of control torque inputs, joints position and velocity without a prior exact knowledge of robot model (Yaramasu and Wu 2014). However, hard nonlinearities will appear and make the controlled system leads with a large steady state error. To deal with these hard nonlinearities, the PI compensator is used. Moreover, nonlinear backstepping (Skjetne and Fossen 2004; Tan et al. 2000) is used to ensure robustness and high accuracy tracking. The stability of the closed-loop system is proved recursively using Lyapunov theory.

The rest of the chapter is organized as follows: in the next section, the dynamics equation of n-link rigid robot manipulators is presented and the problem formulation is stated. The proposed control scheme and its stability analysis for the resulting closed-loop are provided in Sect. 6.3. Simulation results are presented in Sect. 6.4, to illustrate the superiority of the proposed technique over the classical sliding mode and backstepping technique. In Sect. 6.5, the proposed controller is implemented in real time on the 7-DOF ANAT robot arm and experimental results are presented. The paper is closed Sect. 6.6.

6.2 Preliminaries

Consider the dynamics of n-DOF robot manipulator in the following matrix equation (Spong et al. 2005):

$$M(q(t))\ddot{q}(t) + C(q(t), \dot{q}(t))\dot{q}(t) + G(q(t)) + F(\dot{q}(t)) = \tau(t) + \tau_d(t) \tag{6.1}$$

where:

- $q(t), \dot{q}(t), \ddot{q}(t) \in R^n$ represent the joint space position, velocity and acceleration vectors, respectively.
- $M(q(t)) \in R^{n\times n}$ denotes the inertia matrix which is symmetric definite positive and its inverse always exists.
- $C(q(t), \dot{q}(t)) \in R^{n\times n}$ is the centrifugal and Coriolis matrix.
- $G(q(t)) \in R^n$ represents the gravitational vector.
- $F(\dot{q}(t)) = F_v\dot{q}(t) + F_s sign(\dot{q}(t)) \in R^n$ is the vector of viscous/static friction torque at the joints with $F_v,\ F_s \in R^{n\times n}$ are positive diagonal matrices.
- $\tau(t) \in R^n$ is the torque input vector.
- $\tau_d(t) \in R^n$ denotes the disturbance vector.

Due to load variations, unknown parameters, unmodel dynamics, the dynamic matrices can be divided into two parts as:

$$\begin{cases} M(q(t)) = M_0(q(t)) + \Delta M(q(t)) \\ C(q(t), \dot{q}(t)) = C_0(q(t), \dot{q}(t)) + \Delta C(q(t), \dot{q}(t)) \\ G(q(t)) = G_0(q(t)) + \Delta G(q(t)) \\ F(\dot{q}(t)) = F_0(\dot{q}(t)) + \Delta F(\dot{q}(t)) \end{cases} \tag{6.2}$$

where $M_0(q(t))$, $C_0(q(t), \dot{q}(t))$, $G_0(q(t))$ and $F_0(\dot{q}(t))$ are the nominal parts while $\Delta M(q(t))$, $\Delta C(q(t), \dot{q}(t))$, $\Delta G(q(t))$ and $\Delta F(\dot{q}(t))$ denote the uncertain parts of the inertia matrix $M(q(t))$, the centrifugal and Coriolis matrix $C(q(t), \dot{q}(t))$, the gravitational vector $G(q(t))$ and the vector of viscous/static friction torque at the joints $F(\dot{q}(t))$, respectively.

Then, the dynamic equation of the robot manipulator given in partitioned form in Eq. (6.1) can be rewritten as follows:

$$M_0(q(t))\ddot{q}(t) + N(q(t), \dot{q}(t)) + H(q(t), \dot{q}(t), \ddot{q}(t)) = \tau(t) \tag{6.3}$$

where:

$$N(q(t), \dot{q}(t)) = C_0(q(t), \dot{q}(t))\dot{q}(t) + G_0(q(t)) + F_0(\dot{q}(t)) \tag{6.4}$$

$$\begin{aligned} H(q(t), \dot{q}(t), \ddot{q}(t)) &= \Delta M(q(t))\ddot{q}(t) + \Delta C(q(t), \dot{q}(t))\dot{q}(t) \\ &\quad + \Delta G(q(t)) + \Delta F(\dot{q}(t)) - \tau_d(t) \end{aligned} \tag{6.5}$$

For simplicity, let us denote $M_0(q(t)) = M_0(t)$, $N(t) = N(q(t), \dot{q}(t))$ and $H(t) = H(q(t), \dot{q}(t), \ddot{q}(t))$.

In general, the control objective is to make the joint positions track a desired trajectory with high accuracy. However, since $H(t)$ is unknown, the control performance will be affected. Then, the objective is to design a robust joint space controller which ensures the convergence of the error even in the presence of unknown dynamics and unexpected disturbances. To this end, we will design the controller and carry out its stability analysis based on the following property and assumptions:

Proper The nominal inertia matrix $M_0(t)$ is symmetric definite positive and bounded such as:

$$0 < m_1 \leq \|M_0(t)\| \leq m_2$$

with m_1 and m_2 are two known positive constants (Spong et al. 2005).

Assumption 1 The joint position and velocity states are available for measurements.

Assumption 2 The functions $H_i(t)$ for $i = 1, \cdots, n$ of the uncertain vector $H(t) = [H_1(t), \cdots, H_n(t)]^T$ are continuously differentiable with respect to the time variable and don't vary largely during a small L period of time (Toumi and Ito 1990).

6.3 Controller Design

To achieve the control objective, a combination of nonlinear backstepping and Lagrange's extrapolation and PI compensator will be designed in this section. The architecture of the closed-loop system is represented in Fig. 6.1.

The control design consists on three steps described as follows:

Step 1 This step consists on defining the regulated variable to be the joint position tracking error variable as follows:

$$\begin{aligned} E_1(t) &= e(t) \\ &= q(t) - q_d(t) \end{aligned} \tag{6.6}$$

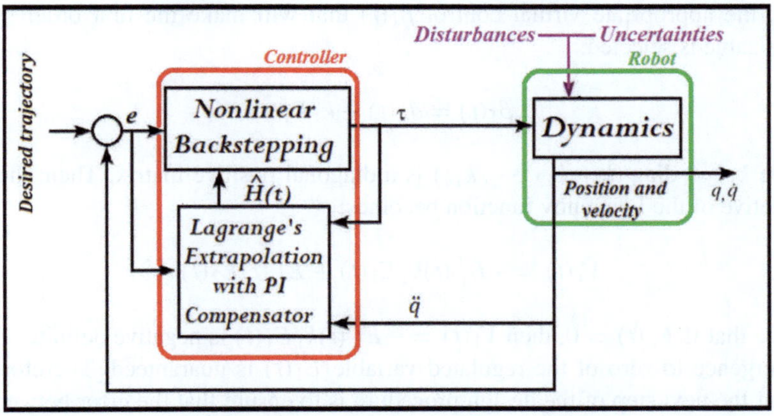

Fig. 6.1 Block diagram of the proposed controller

where $q_d(t) \in R^n$ denotes the desired joint space position trajectory. Therefore, the first time derivative of the dynamics of the new coordinate (i.e. regulated variable) is computed as:

$$\begin{aligned} \dot{E}_1(t) &= \dot{e}(t) \\ &= \dot{q}(t) - \dot{q}_d(t) \end{aligned} \tag{6.7}$$

Considering that $\dot{q}(t)$ is a control variable and defining a virtual control law $\beta_1(t)$ for Eq. (6.7), and let $E_2(t)$ be a new error variable representing the difference between the actual control variable and the virtual control law of (6.7) as follows:

$$E_2(t) = \dot{q}(t) - \beta_1(t) \tag{6.8}$$

Hence, the derivative of the regulated variable $\dot{E}_1(t)$ can be rewritten as:

$$\dot{E}_1(t) = E_2(t) + \beta_1(t) - \dot{q}_d(t) \tag{6.9}$$

In this first step, the control objective is to ensure the convergence of the regulated variable to zero ($E_1(t) \to 0$) by designing an appropriate virtual control law. To that end, let us select the following Lyapunov function:

$$V_1(t) = \frac{1}{2} E_1^T(t) E_1(t) \tag{6.10}$$

Its time derivative is calculated as:

$$\begin{aligned} \dot{V}_1(t) &= E_1^T(t) \dot{E}_1(t) \\ &= E_1^T(t) \left(E_2(t) + \beta_1(t) - \dot{q}_d(t)\right) \\ &= E_1^T(t) \left(\beta_1(t) - \dot{q}_d(t)\right) + E_1^T(t) E_2(t) \end{aligned} \tag{6.11}$$

Now, the appropriate virtual control $\beta_1(t)$ that will make the first order system stabilizable is selected:

$$\beta_1(t) = \dot{q}_d(t) - k_1 E_1(t) \tag{6.12}$$

where $k_1 = \text{diag}\,(k_{11}, k_{12} \cdots, k_{1n})$ is a diagonal positive matrix. Then, the time derivative of the Lyapunov function becomes:

$$\dot{V}_1(t) = -E_1^T(t) k_1 E_1(t) + E_1^T(t) E_2(t) \tag{6.13}$$

Notice that if $E_2(t) = 0$, then $\dot{V}_1(t) = -E_1^T(t) k_1 E_1(t)$ is negative definite and the convergence to zero of the regulated variable $E_1(t)$ is guaranteed. Therefore, the aim of the next step of the design procedure is to ensure that the error between the actual control variable and the virtual control law $E_2(t)$ converges to zero.
Step 2 Now, let us recall the expression of $E_2(t)$:

$$\begin{aligned} E_2(t) &= \dot{q}(t) - \beta_1(t) \\ &= \dot{q}(t) - \dot{q}_d(t) + k_1 E_1(t) \end{aligned} \tag{6.14}$$

Differentiating the above equation with respect to time leads to:

$$\begin{aligned} \dot{E}_2(t) &= \ddot{q}(t) - \dot{\beta}_1(t) \\ &= \ddot{q}(t) - \ddot{q}_d(t) + k_1 \dot{E}_1(t) \\ &= M_0^{-1}(t)\,[\tau(t) - N(t) - H(t)] - \ddot{q}_d(t) + k_1 \dot{E}_1(t) \end{aligned} \tag{6.15}$$

Therefore, using Eqs. (6.6) and (6.14), $\dot{E}_1(t)$ can be rewritten as follows:

$$\dot{E}_1(t) = E_2(t) - k_1 E_1(t) \tag{6.16}$$

Thus, the second Lyapunov function is selected as:

$$V(t) = V_1(t) + \frac{1}{2} E_2^T(t) E_2(t) + \frac{1}{2} \varepsilon^T(t) \Gamma \varepsilon(t) \tag{6.17}$$

where $\varepsilon(t) = M_0^{-1}(t)\left[\hat{H}(t) - H(t)\right]$ denotes the estimation error with $\hat{H}(t)$ is the estimate of the unknown vector $H(t)$ (its expression will be determinate later) and $\Gamma = \text{diag}\,(\Gamma_1, \Gamma_2 \cdots, \Gamma_n)$ is an adaptive gain positive matrix.

Taking the time derivative of the second Lyapunov function (6.17) gives:

$$\begin{aligned} \dot{V}(t) &= \dot{V}_1(t) + E_2^T(t) \dot{E}_2(t) + \varepsilon^T(t) \Gamma \dot{\varepsilon}(t) \\ &= E_1^T(t) \dot{E}_1(t) + E_2^T(t) \dot{E}_2(t) + \varepsilon^T(t) \Gamma \dot{\varepsilon}(t) \end{aligned}$$

$$= E_2^T(t)\left(M_0^{-1}(t)\left[\tau(t) - N(t) - H(t)\right] - \ddot{q}_d(t) + k_1\dot{E}_1(t)\right)$$
$$+ E_1^T(t)(E_2(t) - k_1E_1(t)) + \varepsilon^T(t)\Gamma\dot{\varepsilon}(t)$$
$$= E_2^T(t)\left(M_0^{-1}(t)\left[\tau(t) - N(t) - H(t)\right] - \ddot{q}_d(t) + k_1\dot{E}_1(t)\right)$$
$$+ E_2^T(t)E_1(t) - E_1^T(t)k_1E_1(t) + \varepsilon^T(t)\Gamma\dot{\varepsilon}(t) \quad (6.18)$$

Choosing:

$$M_0^{-1}(t)\left[\tau(t) - N(t) - H(t)\right] - \ddot{q}_d(t) + k_1\dot{E}_1(t)$$
$$+ E_1(t) = -k_2E_2(t) + \varepsilon(t) \quad (6.19)$$

where $k_2 = \text{diag}\,(k_{21}, k_{22}, \cdots, k_{2n})$ is a diagonal positive gain matrix, the expression of the torque control input is obtained as:

$$\tau(t) = M_0(t)u(t) + N(t) + \hat{H}(t) \quad (6.20)$$

where

$$u(t) = \ddot{q}_d(t) - k_1\dot{E}_1(t) - E_1(t) - k_2E_2(t) \quad (6.21)$$

Step 3 In this step, the expression of $\hat{H}(t)$ will be determinate. For that, let us substitute the obtained control law (6.20) in the derivative of the second selected Lyapunov function (6.18) as:

$$\dot{V}(t) = -E_1^T(t)k_1E_1(t) + E_2^T(t)\left(-k_2E_2(t) + \varepsilon(t)\right) + \varepsilon^T(t)\Gamma\dot{\varepsilon}(t)$$
$$= -E_1^T(t)k_1E_1(t) - E_2^T(t)k_2E_2(t) + \varepsilon^T(t)\left(\Gamma\dot{\varepsilon}(t) + E_2(t)\right) \quad (6.22)$$

To guarantee $\dot{V}(t)$ is a negative definite function (i.e., the error variable $E_2(t)$ converges to zero), the following condition is set:

$$\Gamma\dot{\varepsilon}(t) = -E_2(t) \quad (6.23)$$

Integrating the above equation between 0 and t gives:

$$\varepsilon(t) = -\Gamma^{-1}\int_0^t E_2(t)dt$$
$$= -\Gamma^{-1}\int_0^t \left(\dot{E}_1(t) + k_1E_1(t)\right)dt$$
$$= -\Gamma^{-1}\left(E_1(t) - E_1(0) + k_1\int_0^t E_1(t)dt\right) \quad (6.24)$$

where $-\Gamma^{-1}\left(E_1(t) - E_1(0) + k_1\int_0^t E_1(t)dt\right)$ represents the PI compensator with Γ^{-1} and $\Gamma^{-1}k_1$ are the proportional gain matrix and the integral gain matrix, respectively, and $E_1(0)$ is the initial value of the tracking error. Remark that the PI compensator is updated to the direction of decreasing $\varepsilon(t)$. Consequently:

$$M_0^{-1}(t)\left[\hat{H}(t) - H(t)\right] = -\Gamma^{-1}\left(E_1(t) - E_1(0) + k_1\int_0^t E_1(t)dt\right) \tag{6.25}$$

Then, the estimate $\hat{H}(t)$ is obtained as:

$$\hat{H}(t) = H(t) - M_0(t)\Gamma^{-1}\left(E_1(t) - E_1(0) + k_1\int_0^t E_1(t)dt\right) \tag{6.26}$$

where $H(t)$ can be obtained using Lagrange's extrapolation using the last three points:

$$H(t) = 3H(t-L) - 3H(t-2L) + H(t-3L) \tag{6.27}$$

with L is the estimation time delay (in practice, L is chosen to be the sampled time period) and $H(t-aL)$ for $a = 1, 2, 3$ is obtained using Eq. (6.3) as:

$$H(t-aL) = \tau(t-aL) - M_0(t-aL)\ddot{q}(t-aL) - N(t-aL) \tag{6.28}$$

The past acceleration $\ddot{q}(t-aL)$ at time $t-aL$ can be obtained using one of the following approximations:

$$\ddot{q}(t-aL) = \frac{1}{L}\big(\dot{q}(t-aL) - \dot{q}(t-(a+1)L)\big) \tag{6.29}$$

$$\ddot{q}(t-aL) = \frac{1}{L^2}\big(q(t-aL) - 2q(t-(a+1)L) + q(t-(a+2)L)\big) \tag{6.30}$$

Finally, using the obtained expression of $\hat{H}(t)$ in Eq. (6.26), the time derivative of the Lyapunov function $V(t)$ becomes:

$$\begin{aligned}\dot{V}(t) &= -E_1^T(t)k_1E_1(t) - E_2(t)^Tk_2E_2(t)\\ &= -E^T(t)KE(t)\\ &\le -\lambda_{min}(K)\|E(t)\|^2\end{aligned} \tag{6.31}$$

where $E(t) = [E_1(t), E_2(t)]^T$ and $\lambda_{min}(K)$ denotes the minimum eigenvalue of the matrix K defined by:

$$K = \begin{bmatrix} k_1 & 0_{n\times n} \\ 0_{n\times n} & k_2 \end{bmatrix} \tag{6.32}$$

Hence, the stability of the closed loop system is proven.

6.4 Numerical Simulations

In this section, three numerical simulations are presented for trajectory tracking of a two-link rigid robotic manipulator shown in Fig. 6.2 using Matlab/Simulink software with ODE 4 Solver and a sampled time equal to 0.001 s.

For this two-link manipulator, the dynamic equation (6.1) has the following matrices,

$$M(q(t)) = \begin{bmatrix} l_2^2 m_2 + 2 l_1 l_2 m_2 c_2 + l_1^2 (m_1 + m_2) + J_1 & l_2 m_2 (l_1 + l_2) \\ l_2 m_2 (l_1 + l_2) & l_2^2 m_2 + J_2 \end{bmatrix},$$

$$C(q(t), \dot{q}(t))\dot{q}(t) = \begin{bmatrix} -l_1 l_2 m_2 s_2 \dot{q}_2^2(t) - 2 l_1 l_2 m_2 s_2 \dot{q}_1(t)\dot{q}_2(t) \\ l_1 l_2 m_2 s_2 \dot{q}_2^2(t) \end{bmatrix},$$

$$G(q(t)) = \begin{bmatrix} l_2 m_2 g c_{12} + (m_1 + m_2) l_1 g c_1 \\ l_2 m_2 g c_{12} \end{bmatrix},$$

$$\tau_d(t) = \begin{cases} \begin{bmatrix} 2\sin t + 0.5\sin(200\pi t) \\ \cos 2t + 0.5\sin(200\pi t) \end{bmatrix}, & if\ t < 6\,\mathrm{s} \\ \\ \begin{bmatrix} 20\sin t + 5\sin(200\pi t) \\ 10\cos 2t + 5\sin(200\pi t) \end{bmatrix}, & else \end{cases}$$

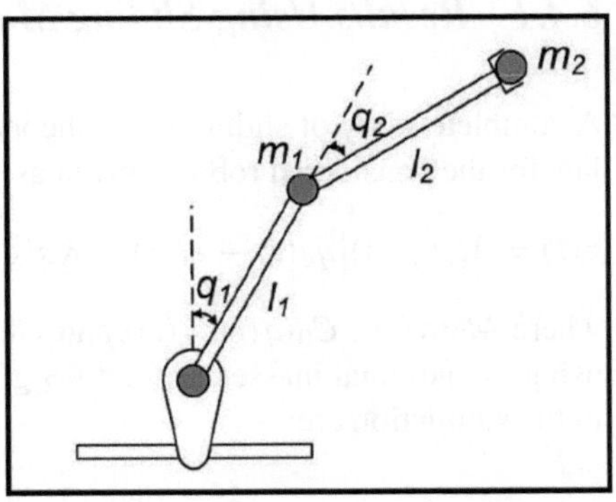

Fig. 6.2 Considered two-link robot manipulator

Table 6.1 Parameters of the two-link robot manipulator

Symbol	Definition	Value	S.I. Unit
l_1	Length of the first link	1	(m)
l_2	Length of the second link	0.85	(m)
J_1	Moment of inertia of the first motor	5	(kg m^2)
J_2	Moment of inertia of the second motor	5	(kg m^2)
m_1	Mass of link 1	0.5	(kg)
m_2	Mass of link 2	1.5	(kg)
$\hat{m}_1$	Nominal mass of link 1	0.4	(kg)
$\hat{m}_2$	Nominal mass of link 2	1.2	(kg)
g	Gravitational constant	9.81	(m/s^2)

where s_i, c_i and c_{ij} are defined by $s_i = \sin(q_i(t))$, $c_i = \cos(q_i(t))$ and $c_{ij} = \cos(q_i(t) + q_j(t))$, respectively. The parameters of the above elements are reported in Table 6.1.

The control objective here is to follow the reference signals defined by:

$$q_{1d}(t) = 1.25 - (7/5)e^{-t} + (7/20)e^{-4t} \tag{6.33}$$

$$\dot{q}_{1d}(t) = (7/5)e^{-t} - (7/5)e^{-4t} \tag{6.34}$$

$$\ddot{q}_{1d}(t) = -(7/5)e^{-t} + (28/5)e^{-4t} \tag{6.35}$$

$$q_{2d}(t) = 1.25 + e^{-t} - (1/4)e^{-4t} \tag{6.36}$$

$$\dot{q}_{2d}(t) = -e^{-t} + e^{-4t} \tag{6.37}$$

$$\ddot{q}_{2d}(t) = e^{-t} - 4e^{-4t} \tag{6.38}$$

The initial states (joint positions and joint velocities) are selected to be $q_1(0) = 0.4$ rad, $q_2(0) = 1.8$ rad and $\dot{q}_1(0) = \dot{q}_2(0) = 0$ rad/s.

6.4.1 Results Using Sliding Mode Control

A complete study of sliding mode theory can be found in Utkin (1992). The control law for the considered robot is given as:

$$\tau(t) = M_0(q(t))\big[\ddot{q}_d(t) - \lambda\dot{e}(t) - K sign(\sigma(t))\big] + C_0(q(t), \dot{q}(t))\dot{q}(t) + G_0(q(t))$$

where $M_0(q(t))$, $C_0(q(t), \dot{q}(t))$ and $G_0(q(t))$ are the nominal matrices calculated using the nominal masses $\hat{m}_1$ and $\hat{m}_2$ given in Table 6.1. The used controller gains in the simulation are:

$$\lambda = \text{diag}\ (2, 2), \quad K = \text{diag}\ (5, 5)$$

The simulation results are given in Fig. 6.3.

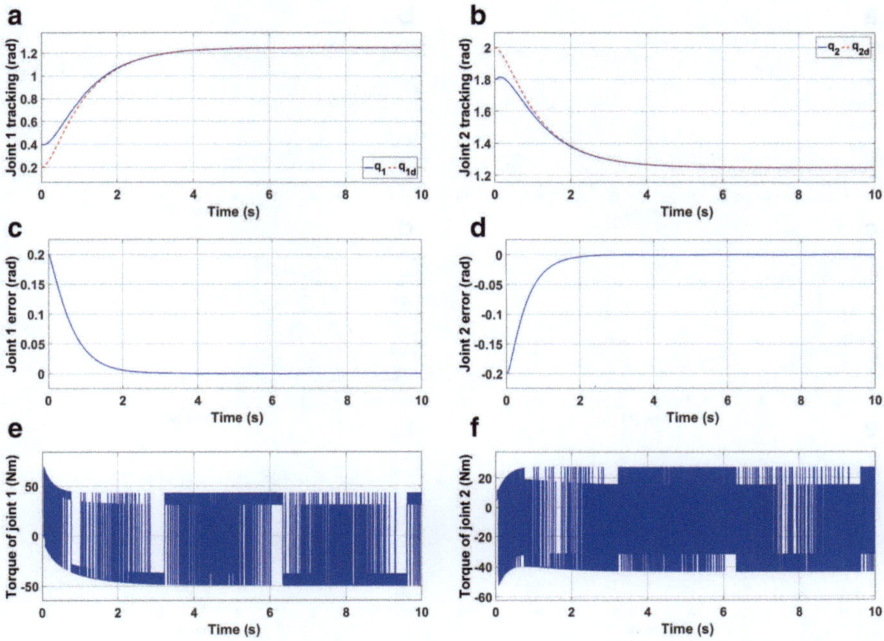

Fig. 6.3 Obtained simulation results via classical sliding mode control. (**a**) Joint 1 tracking. (**b**) Joint 2 tracking. (**c**) Joint 1 error. (**d**) Joint 2 error. (**e**) Torque of joint 1. (**f**) Torque of joint 2

6.4.2 Results Using Backstepping Method

The theoretical development of the nonlinear backstepping scheme can be found in Slotine and Li (1991) and Liu and Wang (2012). The control law for robot is given as:

$$\tau(t) = M_0(q(t))\left[\ddot{q}_d(t) - (\lambda_1 + \lambda_2)\dot{e}(t) - (I + \lambda_1\lambda_2)e(t)\right] \\ +C_0(q(t), \dot{q}(t))\dot{q}(t) + G_0(q(t))$$

In this simulation, the used parameters for sliding mode control using saturation function with time delay estimation are:

$$\lambda_1 = \text{diag}\ (2, 2), \quad \lambda_2 = \text{diag}\ (3, 3)$$

The obtained simulation results are shown in Fig. 6.4.

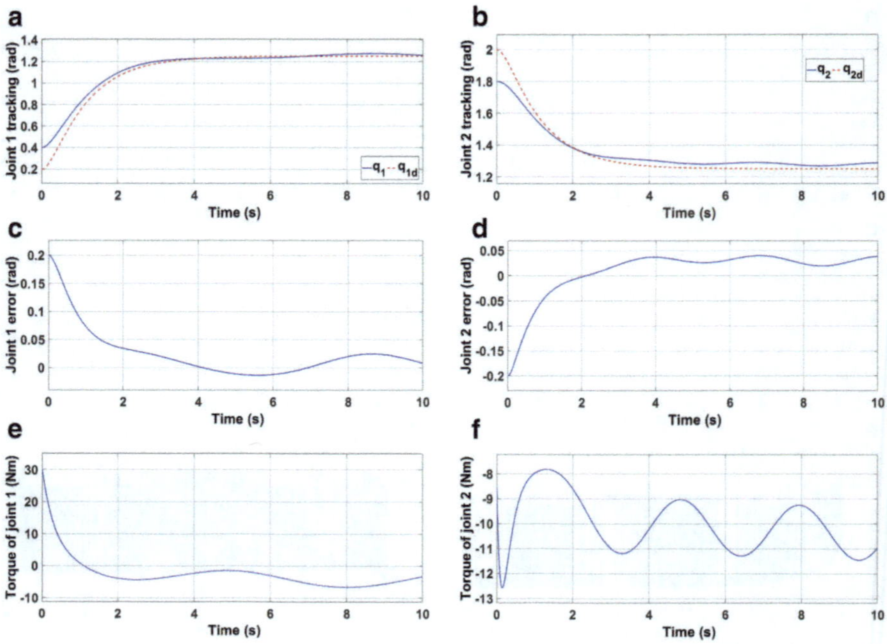

Fig. 6.4 Obtained simulation results via backstepping method. (**a**) Joint 1 tracking. (**b**) Joint 2 tracking. (**c**) Joint 1 error. (**d**) Joint 2 error. (**e**) Torque of joint 1. (**f**) Torque of joint 2

6.4.3 Results Using the Proposed Method

The proposed nonlinear backstepping with Lagrange's extrapolation and PI compensator simulated on the two-link robot is given in Eq. (6.20) where the controller gains are chosen as:

$$K_1 = \text{diag}\,(2, 2), \quad K_2 = \text{diag}\,(5, 5)$$

$$\Gamma^{-1} = \text{diag}\,(0.04, 0.04), \quad L = T_s = 0, 001\,\text{s}$$

Moreover, the time delayed accelerations $\ddot{q}(t - aL)$ for $a = 1, 2, 3$ are calculated using the second approximation given in Eq. (6.30). The simulation results are depicted in Fig. 6.5.

6.4.4 Discussion

6.4.4.1 Classical Sliding Mode Controller

The joint positions q_1 and q_2 shown respectively in Fig. 6.3a, b follow their desired references properly. However, Fig. 6.3e, f show that the chattering phenomenon and

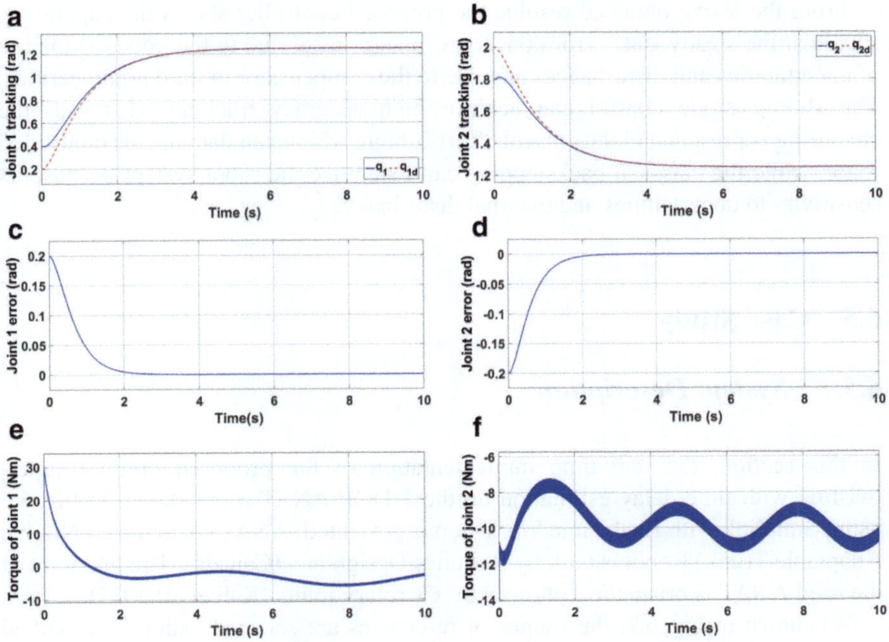

Fig. 6.5 Obtained simulation results via the proposed controller. (**a**) Joint 1 tracking. (**b**) Joint 2 tracking. (**c**) Joint 1 error. (**d**) Joint 2 error. (**e**) Torque of joint 1. (**f**) Torque of joint 2

the control effort are very important due to the large choice of the switching gains which give an unacceptable torque inputs.

6.4.4.2 Nonlinear Backstepping

Figure 6.4e, f show that the control effort is very reduced when it is compared to the obtained results with the robust classical sliding mode. However, the robot system do not converge to zero due to its sensitivity to uncertainties and disturbances as depicted in Fig. 6.4c, d.

6.4.4.3 Proposed Backstepping with Lagrange's Extrapolation and PI Compensator

The obtained simulation results show the effectiveness of the proposed nonlinear backstepping sliding mode method with Lagrange's extrapolation. The joints q_1 and q_2 shown respectively in Fig. 6.5a, b converge with high accuracy to their desired trajectories. This is confirmed in the tracking error figures (Fig. 6.5c, d). In addition, the control torque inputs show acceptable effort and values as depicted in Fig. 6.5e, f.

From the above obtained results, the proposed controller shows the superiority such that the steady state error converges to zero faster due to the good estimation of uncertainties and disturbances and due to the compensator of hard nonlinearities. The sliding mode control can achieve high accuracy tracking. However, the chattering is present and the control effort is high, which can damage the controlled robot, while the classical backstepping can't stabilize the robot system because its sensitivity to uncertainties and external disturbances.

6.5 Case Study

6.5.1 System Description

In this section, the real time implementation of the proposed optimal super-twisting with time delay estimation on the 7-DOF ANAT robot shown in Fig. 6.6 using Simulink with Real-Time Workshop is presented. ANAT (Articulated Nimble Adaptable Trunk) Robot is built by Robotics Design inc. (Canada). The first joint of the used ANAT is prismatic, followed by six rotary joints (Kali et al. 2017).

As shown in Fig. 6.7, the frames of references are selected using the modified Denavit-Hartenberg (Spong et al. 2005) (D-H) convention. The D-H parameters that determine the homogeneous transformation matrices are given in Table 6.2.

The initial task space position of the end-effector that is placed on the last joint (7th joint) is: $x(0) = [0.6764\ 0\ \ -0.19]^T$ while the initial joint positions are 0 rad and joint velocities are 0 rad/s. The equation of motion of the ANAT robot are as in Eq. (6.3). In addition, the ANAT model verifies the assumptions given in Sect. 6.2.

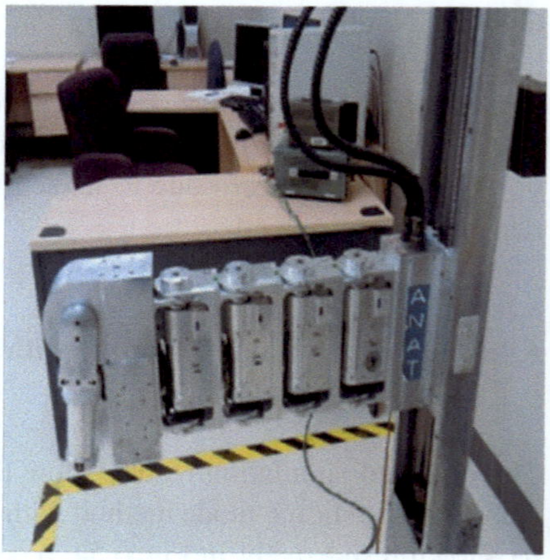

Fig. 6.6 7-DOF ANAT robot arm

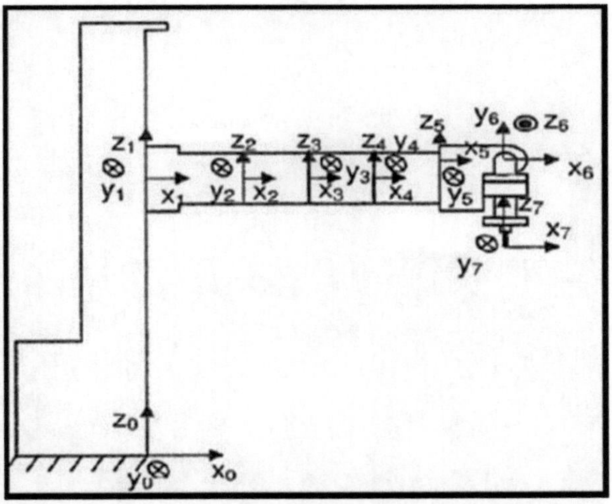

Fig. 6.7 D-H frames of the ANAT robot arm

Table 6.2 D-H parameters of the 7-DOF ANAT robot

Joints	α_{i-1} (rad)	a_{i-1} (m)	d_i (m)	q_i (rad)
1	0	0	q_1	0
2	0	L_1	0	q_2
3	0	L_0	0	q_3
4	0	L_0	0	q_4
5	0	L_0	L_2	q_5
6	$\pi/2$	L_3	0	q_6
7	$-\pi/2$	0	$-L_4$	q_7

6.5.2 *Real-Time Setup*

All robot joints are actuated by DC motors. The actuators encoders provide position measurements. The ATMEGA 16 microcontrollers are used to send the angles positions to simulink, and using the forward kinematics, the joint space trajectory is transformed to the workspace trajectory. The microcontrollers are also used to transform the control input signals into a Pulse Width Modulation (PWM) signal. Then, this PWM signal is applied to the H-bridge drive of the actuators of the 7-DOF ANAT robot. A current sensor located in the H-bridge drive provides the current measurement of each actuator. In the experimentation, the sixth, and seventh joints are chosen to stay at their initial position. A sampling time $T_s = 0.03$ s has been selected for the control loop (Kali et al. 2017). The above real-time setup is illustrated in Fig. 6.8.

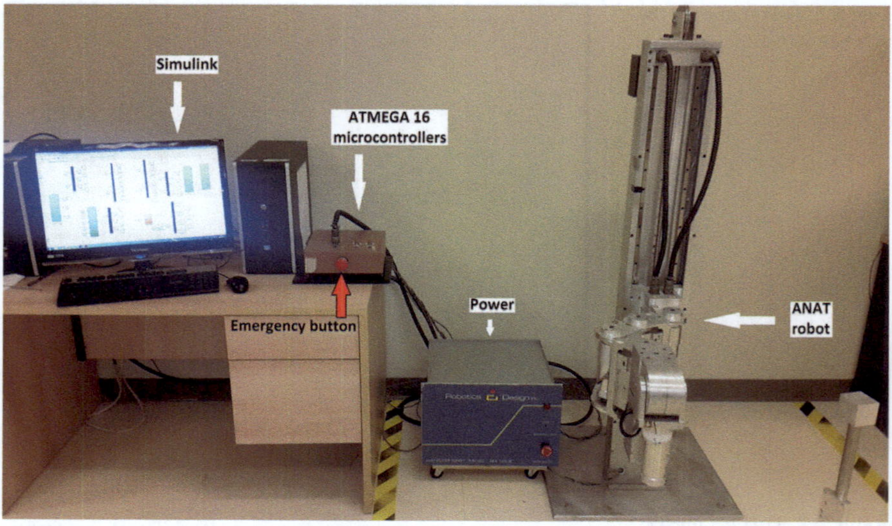

Fig. 6.8 Experimental setup

6.5.3 Controller Setting

The proposed nonlinear backstepping with Lagrange's extrapolation has three diagonal matrices with constant elements. This makes the controller setting simplify, which is briefly described as follows:

- Determine the matrix K_1 of the virtual control β_1 in Eq. (6.12).
- Tuning the matrix K_2 such as a fast convergence will be obtained.
- Select L as small as possible (equal to the sampling time interval T_s).
- Finally the constants Γ_i for $i = 1, \cdots, n$ are chosen such as the condition of stability in Eq. (6.24) is verified, they should be further tuned to reduce the effect of hard nonlinearities caused by Lagrange's extrapolation.

6.5.4 Experimental Results

The control objective here is to track a triangular desired trajectory. For this application the desired trajectory is chosen such as all kinematic singularities are avoided, the relation between the desired accelerations in the task space and the joint space is given by:

$$\ddot{q}_d(t) = J^+\ddot{x}_d(t) - J^+\dot{J}J^+\dot{x}_d(t) \tag{6.39}$$

where $J^+ = J^T(JJ^T)^{-1}$ denotes the generalized inverse of the Jacobian matrix while $\dot{J}$ is its derivative, $\ddot{q}_d(t)$ is the desired joint acceleration vector, $\ddot{x}_d(t)$ and $\dot{x}_d(t)$ are the task space desired acceleration and velocity vectors, respectively. The desired velocity in the joint space $\dot{q}_d(t)$ is obtained from $\ddot{q}_d(t)$ using an integrator while the desired position in the joint space $q_d(t)$ is obtained from $\dot{q}_d(t)$ using another integrator.

In Sect. 6.2, the proposed controller that ensures the convergence of the states trajectories to is given in Eq. (6.20). The time delayed acceleration $\ddot{q}_i(t - aL)$ for $a = 1, 2, 3$ and for $i = 1, \cdots, 7$ are obtained using the approximation based on the joint position as:

$$\ddot{q}_i(t - aL) = \frac{1}{T_s^2}\left[q_i(t - aL) - 2q_i(t - (2 + a)L) + q_i(t - (3 + a)L)\right]$$

while the controller gains are reported in Table 6.3.

The experimental results are shown in Figs. 6.9, 6.10, 6.11, 6.12, and 6.13: 2D task space tracking (The arrows in Fig. 6.9 shows the direction of displacements), X and Y tracking, joint space tracking position, joint space tracking error and control

Table 6.3 Controller parameters

Parameter	Value
λ	diag (0.75; 0.75; 0.75; 0.75; 0.75; 0.75; 0.75)
k_1	diag (1.5; 1.5; 1.5; 1.5; 1.5; 1.5; 1.5)
k_2	diag (2; 2; 2; 2; 2; 2; 2)
Γ^{-1}	diag (0.02; 0.02; 0.02; 0.02; 0.02; 0.02; 0.02)

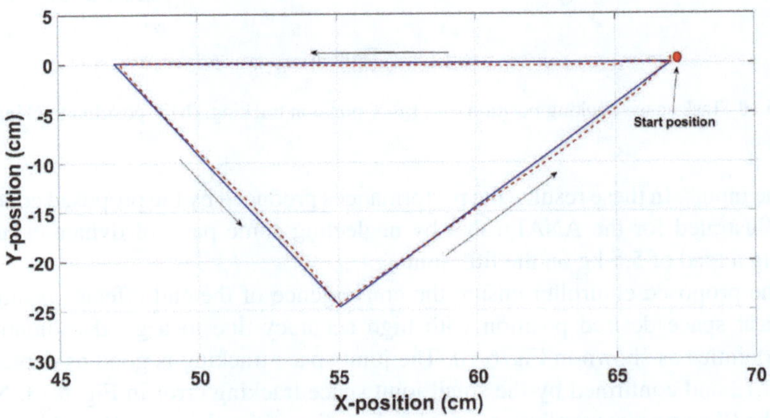

Fig. 6.9 2D Task space tracking

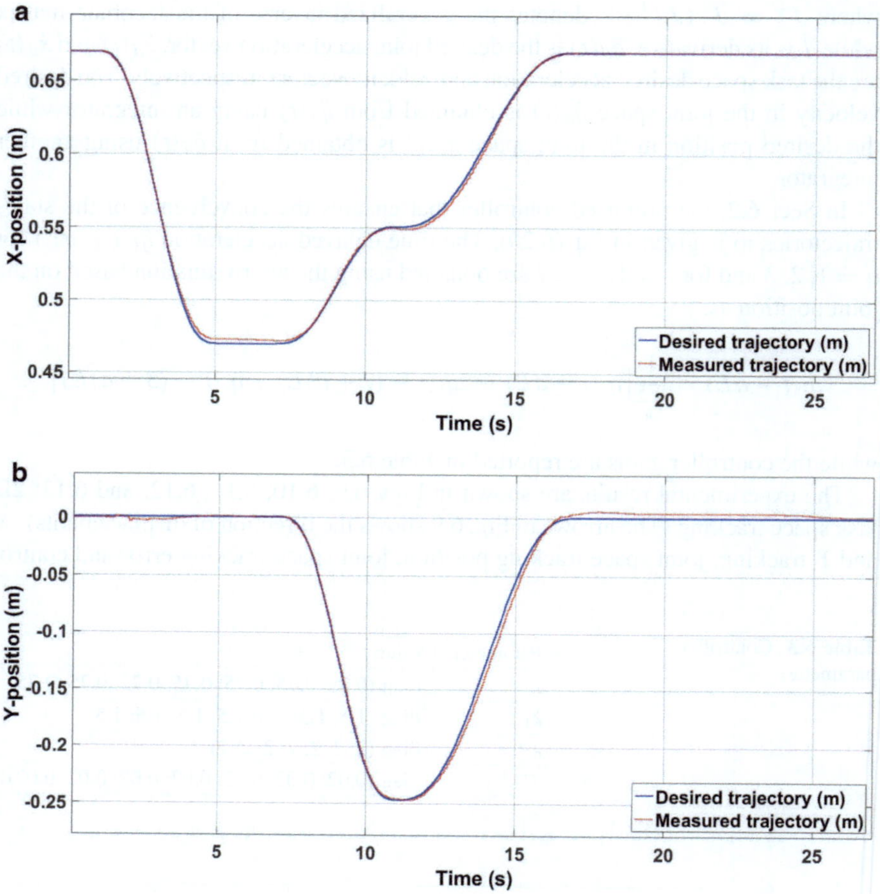

Fig. 6.10 Task space tracking position (m). (**a**) X-position tracking. (**b**) Y-position tracking.

torque inputs. In these results, the performances produced by the proposed controller are illustrated for the ANAT robot by neglecting some parts of dynamics and by adding a load of 5.5 kg on the fifth joint.

The proposed controller ensure the convergence of the end-effector position to the task space desired position with high accuracy due to a good estimation of uncertainties as shown in Fig. 6.10. The joint space tracking is good as depicted in Fig. 6.12 and confirmed by the small joint space tracking error in Fig. 6.13. Notice that the PI compensator eliminate the hard nonlinearities that can affect the tracking accuracy.

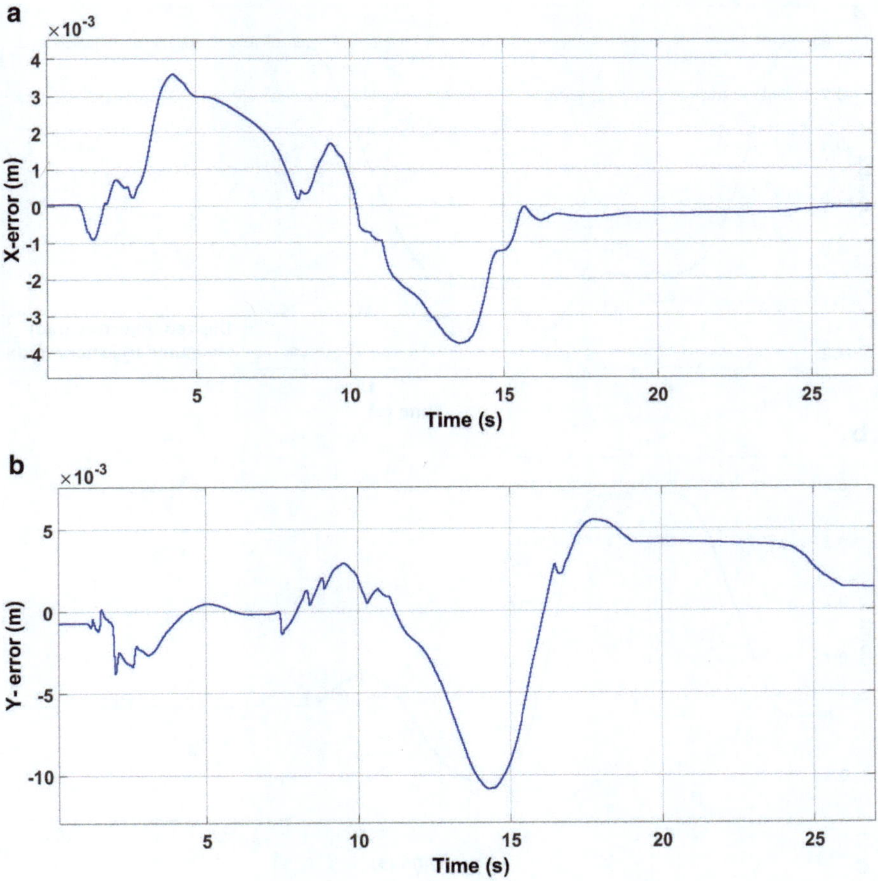

Fig. 6.11 Task space tracking error (m). (**a**) X-position tracking error. (**b**) Y-position tracking error

Furthermore, it can be seen from Fig. 6.14 that the control torque inputs are smooth with small values that are acceptable for our robot, this is thanks to the good estimation.

6.6 Conclusion

For a class n-DOF uncertain robot manipulators in presence of uncertainties and unexpected perturbations, a robust backstepping control with Lagrange's extrapolation and PI compensator is presented in order to achieve ourcontrol objective with

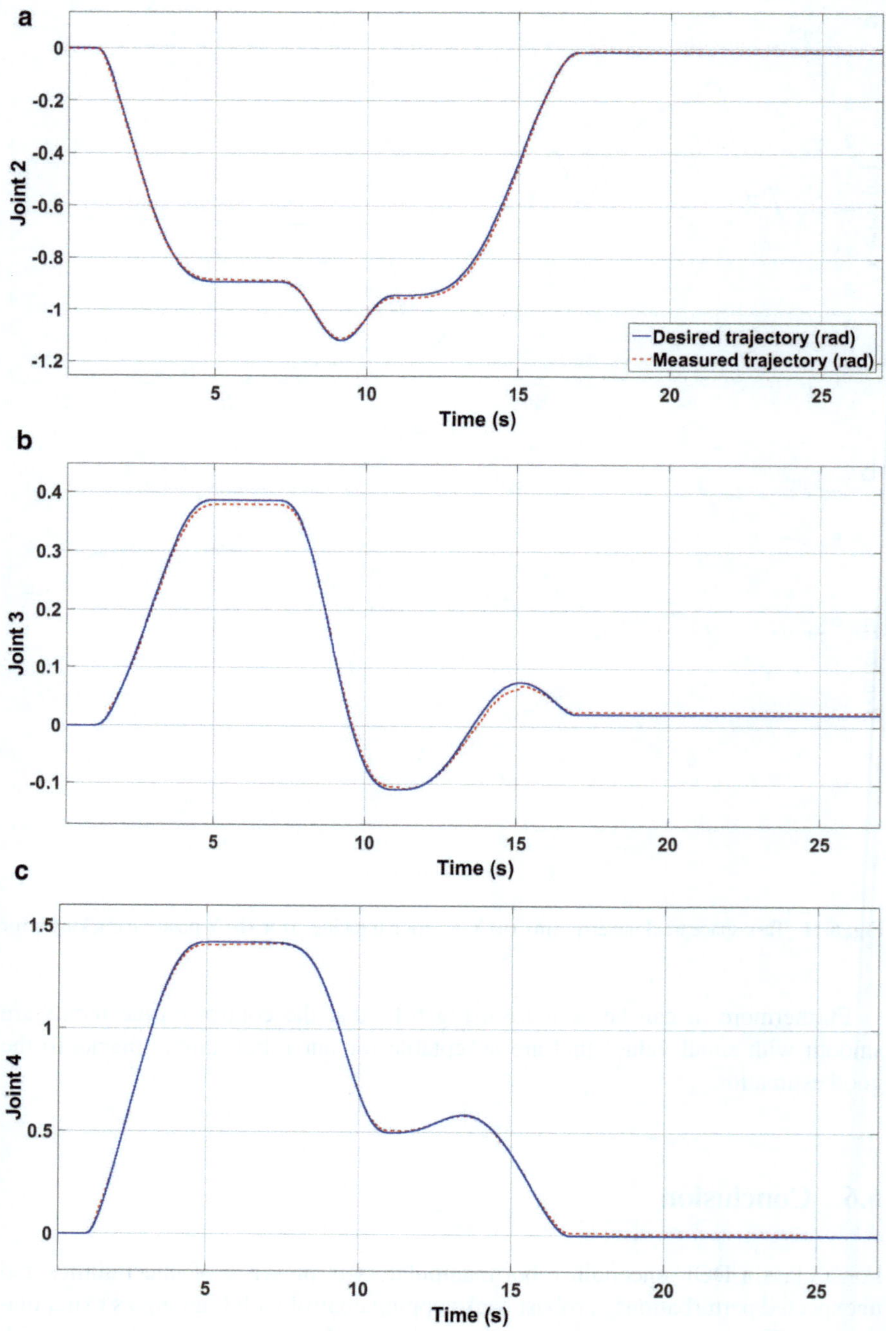

Fig. 6.12 Joint space tracking position (rad). (**a**) Joint 2 tracking. (**b**) Joint 3 tracking. (**c**) Joint 4 tracking. (**d**) Joint 5 tracking

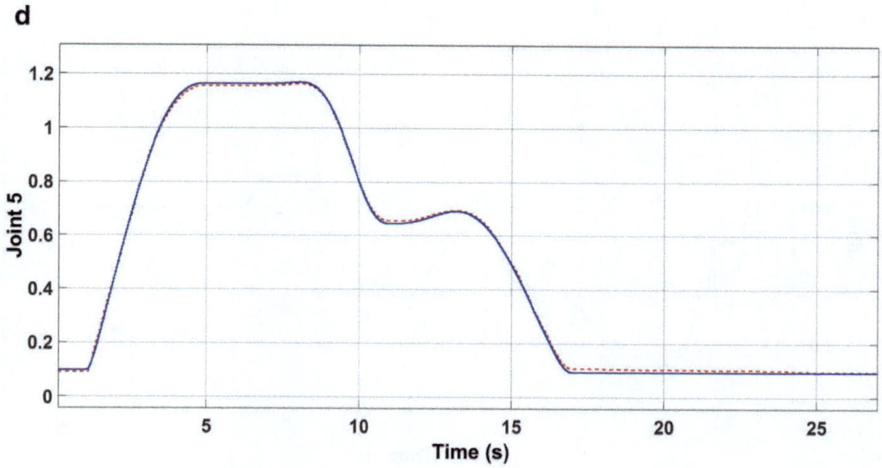

Fig. 6.12 (continued)

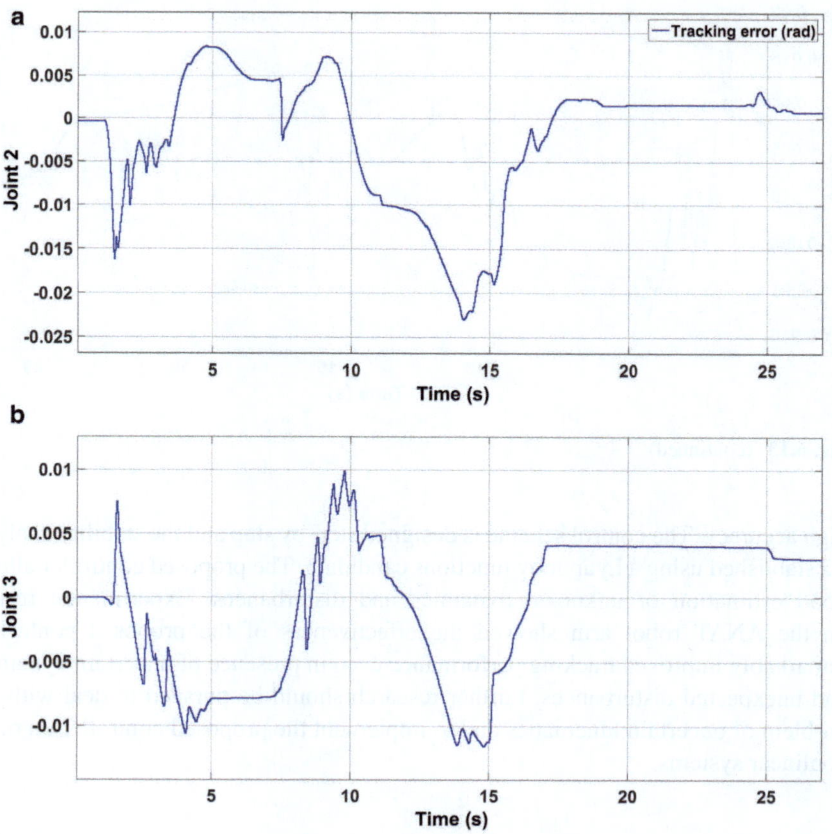

Fig. 6.13 Joint space tracking error (rad). (**a**) Joint 2 tracking error. (**b**) Joint 3 tracking error. (**c**) Joint 4 tracking error. (**d**) Joint 5 tracking error

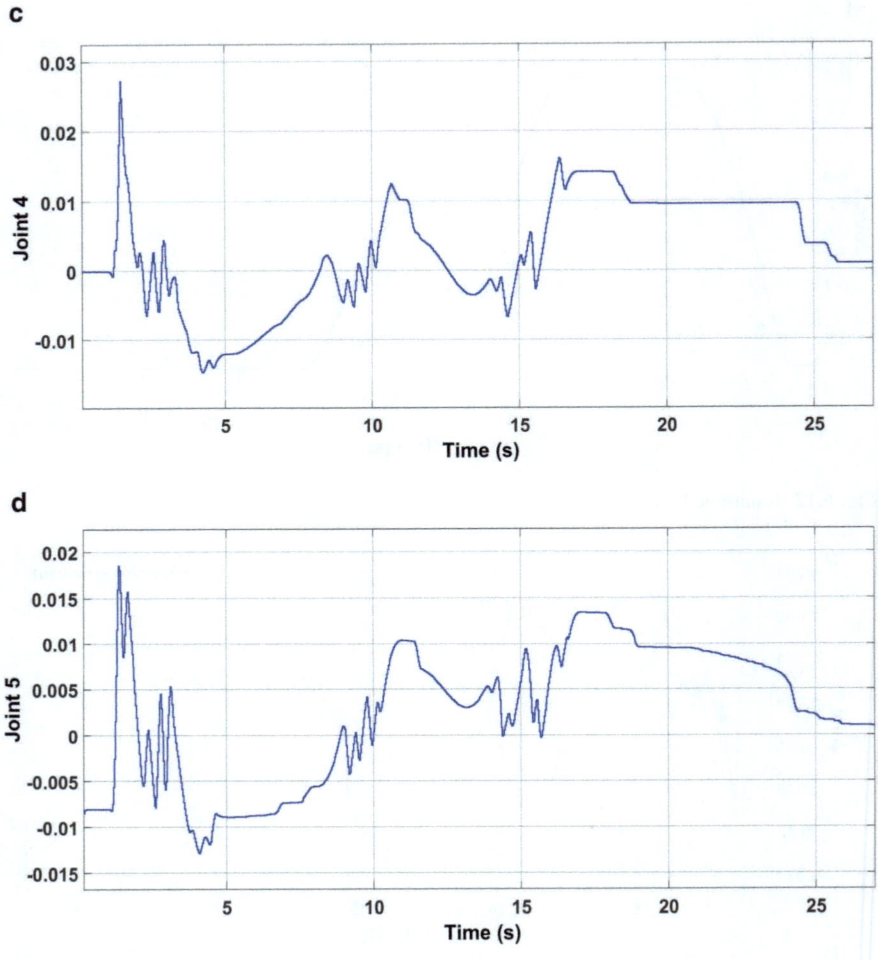

Fig. 6.13 (continued)

high accuracy. The control scheme is designed step-by step and the stability analysis is established using a Lyapunov functions candidate. The proposed controller allows good estimation of unknown dynamics and disturbances. Experimental results on the ANAT robot arm showed the effectiveness of the proposed controller, remarkably improved tracking performance even in presence of uncertain dynamics and unexpected disturbances. Further research should be pursued to deal with the problem of uncertain kinematics and to implement the proposed controller on other nonlinear systems.

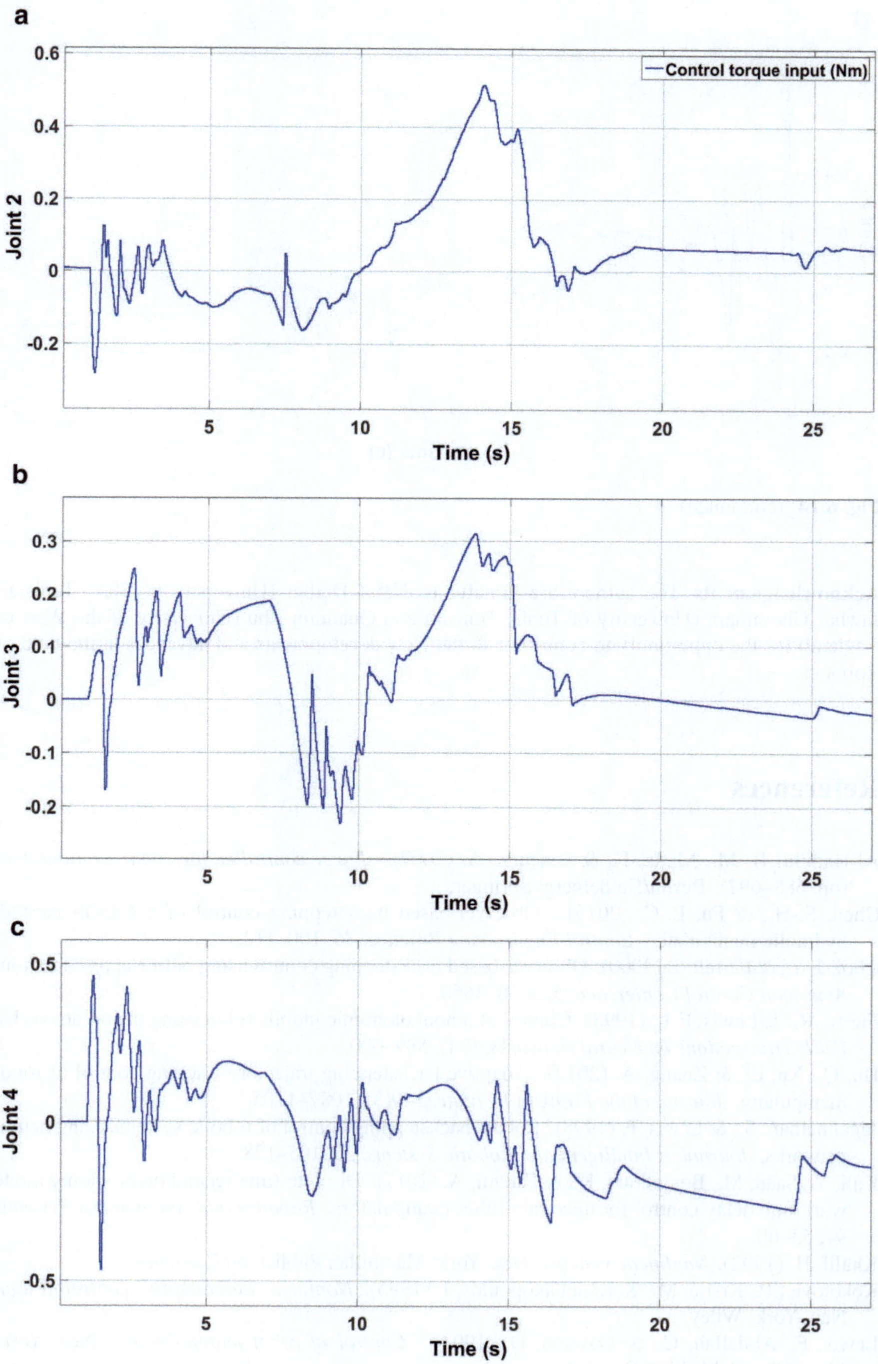

Fig. 6.14 Joint control torques (Nm). (**a**) Torque of joint 2. (**b**) Torque of joint 3. (**c**) Torque of joint 4. (**d**) Torque of joint 5

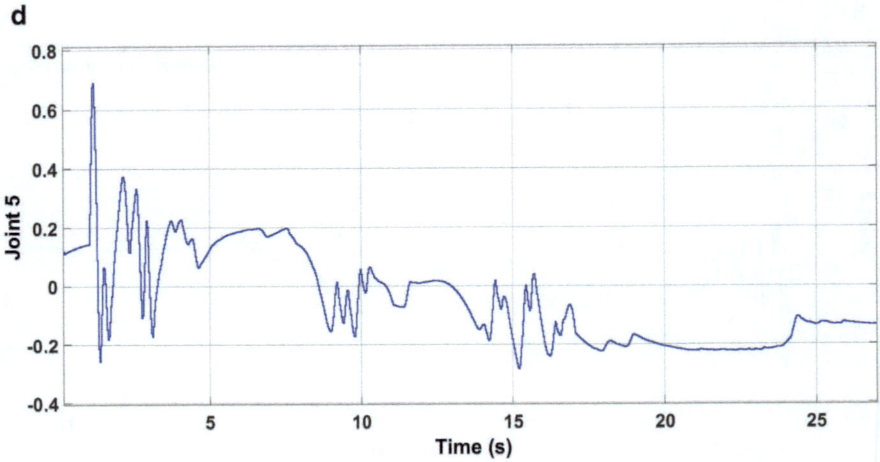

Fig. 6.14 (continued)

Acknowledgements The authors are grateful to Nabil Derbel (University of Sfax, Tunisia), Jawhar Ghommam (University of Tunis, Tunisia) and Quanmin Zhu (University of the West of England) for the opportunity to contribute to the New developments and advances in the field of Robotics.

References

Al-Hadithi, B. M., Matía, F., & Jiménez, A. (2007). *Fuzzy controller for robot manipulators* (pp. 688–697). Berlin/Heidelberg: Springer.

Chen, S.-H., & Fu, L.-C. (2015). Observer-based backstepping control of a 6-DOF parallel hydraulic manipulator. *Control Engineering Practice, 36*, 100–112.

Choi, J. Y., & Farrell, J. (2000). Observer-based backstepping control using online approximation. *American Control Conference, 5*, 3646–3650.

Fierro, R., & Lewis, F. L. (1998). Control of a nonholonomic mobile robot using neural networks. *IEEE Transactions on Neural Networks, 9*(4), 589–600.

Hu, Q., Xu, L., & Zhang, A. (2012). Adaptive backstepping trajectory tracking control of robot manipulator. *Journal of the Franklin Institute, 349*(3), 1087–1105.

Jagannathan, S., & Lewis, F. (1998). Robust backstepping control of robotic systems using neural networks. *Journal of Intelligent and Robotic Systems, 23*, 105–128.

Kali, Y., Saad, M., Benjelloun, K., & Fatemi, A. (2017). Discrete-time second order sliding mode with time delay control for uncertain robot manipulators. *Robotics and Autonomous Systems, 94*, 53–60.

Khalil, H. (1992). *Nonlinear systems*. New York: Macmillan Publishing Company.

Kokotovic, P., Krstic, M., & Kanellakopoulos, I. (1995). *Nonlinear and adaptive control design*. New York: Wiley.

Lewis, F., Abdallah, C., & Dawson, D. (1993). *Control of robot manipulators*. New York: Macmillan Publishing Company.

Liu, J., & Wang, X. (2012). *Advanced sliding mode control for mechanical systems*. Berlin/Heidelberg: Springer.

Shieh, H.-J., & Hsu, C.-H. (2008). An adaptive approximator-based backstepping control approach for piezoactuator-driven stages. *IEEE Transactions On Industrial Electronics, 55*, 1729–1738.

Skjetne, R., & Fossen, T. (2004). On integral control in backstepping: Analysis of different techniques. In *American Control Conference*, Boston, Massachusetts.

Slotine, J., & Li, W. (1991). *Applied nonlinear control.* Taipei: Prentice-Hall International.

Spong, M., Hutchinson, S., & Vidyasagar, M. (2005). *Robot modeling and control.* Wiley.

Su, C.-Y., Li, G., Li, Z., & Su, H. (2015). Fuzzy approximation-based adaptive backstepping control of an exoskeleton for human upper limbs. *IEEE Transactions on Fuzzy Systems, 23*, 555–566.

Tan, Y., Chang, J., Tan, H., & Jun, H. (2000). Integral backstepping control and experimental implementation for motion system. In *IEEE International Conference on Control Applications*, Anchorage, AK.

Toumi, K., & Ito, O. (1990). A time delay controller for systems with unknown dynamics. *ASME Journal of Dynamic System, Measurement and Control, 112*, 133–141.

Utkin, V. (1992). *Sliding mode in control and optimization.* Berlin: Springer.

Utkin, V., Guldner, J., & Shi, J. (1999). *Sliding mode control in electromechanical systems* (2nd ed.). Boca Raton, London.

Weisheng, C., Ge, S., Jian, W., & Maoguo, G. (2015). Globally stable adaptive backstepping neural network control for uncertain strict-feedback systems with tracking accuracy known a priori. *IEEE Transactions on Neural Networks and Learning Systems, 26*, 1842–1854.

Wilson, J., Charest, M., & Dubay, R. (2016). Non-linear model predictive control schemes with application on a 2 link vertical robot manipulator. *Robotics and Computer-Integrated Manufacturing, 41*, 23–30.

Yaramasu, V., & Wu, B. (2014). Model predictive decoupled active and reactive power control for high-power grid-connected four-level diode-clamped inverters. *IEEE Transactions on Industrial Electronics, 61*(7), 3407–3416.

Yoo, B. K., & Ham, W. C. (2000). Adaptive control of robot manipulator using fuzzy compensator. *IEEE Transactions on Fuzzy Systems, 8*, 186–199.

Zhou, J., & Er, M. J. (2007). Adaptive output control of a class of uncertain chaotic systems. *Systems and Control Letters, 56*(6), 452–460.

Zhou, J., & Wen, C. (2008). *Adaptive backstepping control of uncertain systems*. Berlin: Springer.

Chapter 7
Nonlinear Observer-Based Fault Detection and Isolation for a Manipulator Robot

Khaoula Oulidi Omali, M. Nabil Kabbaj, and Mohammed Benbrahim

Abstract Fault Detection and Isolation (FDI) techniques in robot manipulator is becoming one of the most phenomena in robotics in order to ensure higher levels of safety and productivity. Research, has been produced a considerable effort in seeking systematic approaches to fault detection for both linear and nonlinear dynamical systems. In the last decade considerable research efforts have been spent to seek for systematic approaches to Fault Detection (FD) in dynamical systems. Special attention has been addressing for robotic systems, especially for those operating in remote or hazardous environments, where a high degree of safety as well as self-detection capabilities are required. On the other hand, the development of effective strategies of fault detection for robot manipulators operating in an industrial context is a critical research task. Several FD techniques for robot manipulators have been proposed in the literature, although the problem of their application to industrial robots has not been extensively investigated.

In this chapter, we present a high-gain observer based fault detection and isolation scheme for a class of affine nonlinear systems. In order to test the effectiveness and the robustess of the proposed approach, a case study is developed for a special robot manipulator named Articulated Nimble Adaptable Trunk "ANAT" with a five-degree-of-freedom in order to detected and isolated sensor fault.

Keywords Fault detection and isolation · Robot manipulator · High-gain observer · Residuals · Nonlinear systems

---

K. Oulidi Omali (✉) · M. N. Kabbaj · M. Benbrahim
Sidi Mohamed Ben Abdellah University Fez, Fez, Morocco
e-mail: khaoula.oulidiomali@usmba.ac.ma; n.kabbaj@usmba.ac.ma; mohammed.benbrahim@usmba.ac.ma

N. Derbel et al. (eds.), *New Developments and Advances in Robot Control*, Studies in Systems, Decision and Control 175,
https://doi.org/10.1007/978-981-13-2212-9_7

7.1 Introduction

Robotics is a relatively young field of modern technology that crosses traditional engineering boundaries. Understanding the complexity of robots and their applications requires knowledge of electrical engineering, mechanical engineering, systems and industrial engineering, computer science, economics, and mathematics. New disciplines of engineering, such as manufacturing engineering, applications engineering, and knowledge engineering have emerged to deal with the complexity of the field of robotics and factory automation.

Recently, robot manipulators are the most important components of manufacturing and control processes. They have the impact: Improving productivity, increasing the quality of manufactured products, and reducing the cost of labor.

Associated with the increasing industrial demands on safety, reliability, dependability and assuring the normal operation of this robot manipulator, the issue of FDI is important.

The most important steps are fault detection and fault isolation. The methods of fault detection can be separated into methods which don't use plant models and model-based approaches. For the first category are classical limit and trend value checks as well as signal analysis methods, for example, autocorrelation techniques, spectral analysis, etc. Normally, those methods need dedicated sensor systems. Observer-based, parameter estimation and parity relations techniques, on the other hand, the second class of fault detection schemes are the most relevant approaches. Therefore, different methods have been developed for detecting and isolating actuator and sensor faults on robot manipulators, such as parameter estimation, parity relations and observer based approaches (Filaretov et al. 2003; Jollie 2016; Ma and Yang 2016; Schneider and Frank 1996).

In the literature, the parameter estimation method requires knowledge of the model structure of the investigated process and actual process measurements. The actual values of model parameters are then determined from this information. The fault detection is accomplished by a comparison of identified with predetermined parameter values by using thresholds. Note that parameter estimation methods require appropriate excitation of the system. An overview of theory and applications of parameter estimation methods can be found in Isermann (1984) (Fig. 7.1).

The parity space approach for nonlinear systems is studied only in a few papers (Guernez et al. 1997; Krishnaswami and Rissoni 1994; Mironovsky 1989; Shumsky 1998). The papers Mironovsky (1989) and Shumsky (1998) consider the problem of fault detection only; Krishnaswami and Rissoni (1994) gives only the most general recommendations to obtain the parity equations; Guernez et al. (1997) extends the parity space approach to nonlinear polynomial dynamic systems. It is known, however, that the description of manipulation robots includes non polynomial, non linearities such as backlash and Coulomb friction (Filaretov et al. 1999), so the approach (Guernez et al. 1997) cannot be used in this case. There are many papers where the parity space approach was considered for linear systems (see surveys

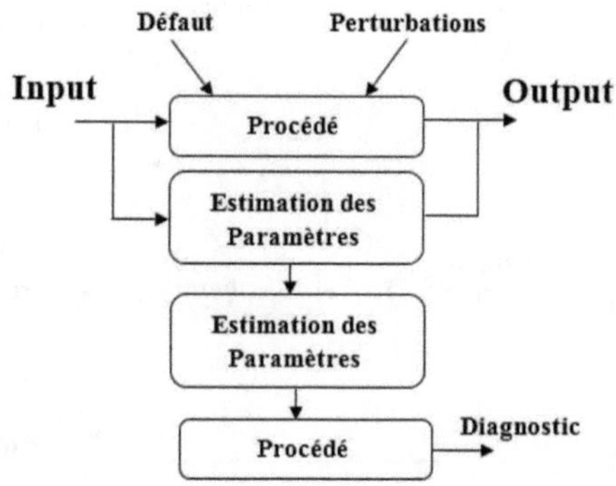

Fig. 7.1 Parameter estimation scheme

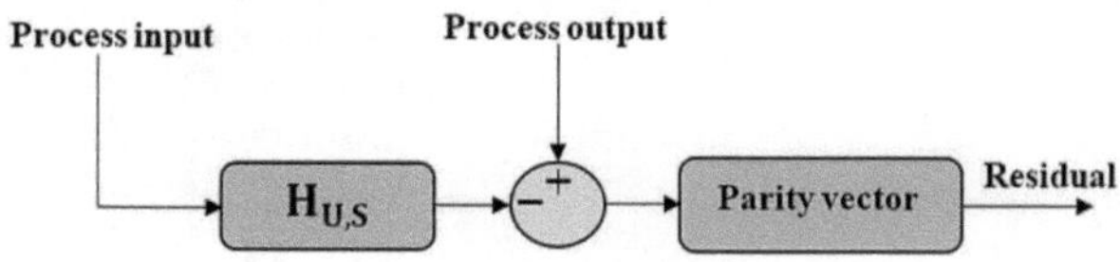

Fig. 7.2 Parity space scheme

Frank 1990; Gertler and Monajemy 1995; Patton 1994). It is well known (Frank 1990; Gertler 1993; Gertler and Monajemy 1995) that the parity space approach is equivalent to the observer-based one in the linear case. However, they aren't equivalent in the nonlinear case (Fig. 7.2).

Observer-based method requires a model of the investigated process also. The model is operated parallel to the real process, the inputs to the model are the same as those to the real process. In contrast to classical state-space observers, e.g., Kalman filters, the diagnostic observers are output observers which are operated in an open-loop configuration. Assuming an exact model of the plant, the difference of measured and calculated process outputs (which is called a residual) will produce a nonzero value when a fault has occurred. A thorough overview of robust observer-based fault detection methods can be found in Frank (1990, 1993). The main advantage of observer-based methods over parameter estimation methods is given by the fact that no special excitation is required. The observer-based method will work in steady-state operation conditions as well.

In any of the systems that were discussed above, in order to have the efficient operation of the process and to increase the reliability and safety, prompt detection of anomalous situations (fault detection) and the fast identification (isolation) of the most probable causes (faults) need to be addressed. FDI can be carried out using analytical or functional information about the system being monitored, i.e., based on a mathematical model of the system. This approach is known as analytical redundancy, which is also known as model based or quantitative FDI. Model based FDI is currently the subject of extensive research and is being used in highly reliable control systems due to the fact that analytical redundancy based techniques are more economical and at times more powerful. These methods have the potential of detecting soft incipient faults even during the system's transient operation. One

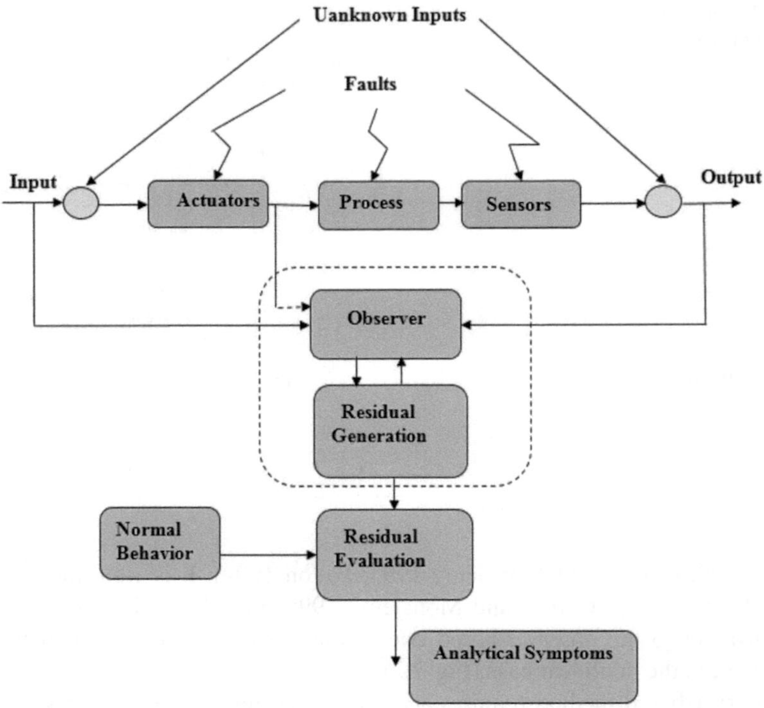

Fig. 7.3 Basic configuration of observer-based fault diagnosis

of the most popular of the model based approaches to FDI is the observers based techniques. The basic idea of observers-based methods consists reconstruct some or all system outputs from accessible process variables. The fault indicators are generated by forming the differences between the estimated outputs and the actual outputs. However, special attention has to be paid when applying observer theory for fault detection and isolation. The basic configuration of observer-based fault diagnosis is shown in Fig. 7.3.

One of the most challenging topics in observer based in the FDI research and application areas is high-gain observers.This type of observers are relatively fast and simple to design as you don't need to solve complex differential equations nor use complicated formulaes, also for large class of nonlinear systems, they can provide global or semi-global stability results for class of systems. This means that their can provide stability guarantees for any arbitrarily initial conditions. For nonlinear systems represented in the normal form, it was shown that when high-gain observers are used in output feedback control they can recover the performance of the state feedback controller. In other words, assuming that all the states are measured, you can design a controller that time. Then, when the states are provided by the high-gain observers, the output feedback controller can provide a very similar performance to the state feedback case, therefore, they can be robust to modeling uncertainty and external disturbances. They were first introduced by Gauthier et al. (1992). They are non-linear observers that take into account the non-stationarity, non-linearity of industrial processes and at the same time ensure a good estimate of the real state with easy adjustment of Gain vector (Farza et al. 2005; Koubaa et al. 2004; Nadri 2001).

High-gain observers have been developed around 30 years ago. In the works (Esfandiari and Khalil 1987; Khalil and Saberi 2007; Saberi and Sannuti 1990; Tornambe 1988), also the recent survey (Khalil and Praly 2014). Choosing the observer gain K large enough, the observer error can be made arbitrarily small, see e.g. Tornambe (1992). The advantages of high-gain observers is that they can be used to estimate the system states without knowing the exact parameters; only some structural assumptions, such as a known relative degree, are necessary. Furthermore, they are robust with respect to input noise. The drawback is that it is not known a priori how large K must be chosen and appropriate values must be identified by offline simulations. If K is chosen unnecessarily large, the sensitivity to measurement noise increases dramatically.

In Gauthier et al. (2001), the authors gave a necessary and sufficient condition that characterizes the class of nonlinear affine systems mono-input, mono-output observable for any input. The authors shows that this class of systems is diffeomorphic to the canonical form which consists of a fixed linear dynamic and a triangular nonlinear dynamic. Using this structure, the authors proposed a high-gain observer synthesized under a Lipschitz hypothesis on nonlinear dynamics. The gain of the proposed observer is derived from the resolution of an algebraic Lyapunov equation which can be explicitly computed. Several generalizations of this result to multi-input systems are proposed in Busawon et al. (1998), Farza et al. (2004), Gauthier and Kupka (2001), Hammouri and Farza (2003), Hou et al. (2000) and Shim et al. (2001). The main characteristic of the high-gain observer proposed in Farza et al. (2005), is its simplicity of implementation since the observer is a copy of the dynamics of the system with a gain whose expression can be explicitly calculated. In addition, the observer setting can be completed by selecting a single synthesis parameter. The effectiveness of this type of observer is proved by several industrial applications (Farza et al. 1999; Hammouri et al. 2001). Indeed, some tests are necessary for a judicious choice of a gain ensuring the compromise between the speed of convergence and the robustness with respect to measurement noises. In this chapter, high-gain observer will be applied to a robot manipulator in objective to detect and isolate the faults.

The chapter is organized as follows. The next section exposes system description and modeling. High-gain observers used for the design of residuals generators are developed in Sect. 7.3. Section 7.4 provides the application for a robot manipulator. The conclusion is given in Sect. 7.5.

7.2 System Description and Modeling

Consider the MIMO nonlinear system with m inputs and p outputs defined by the following state representation (Slotine and Weiping 1991):

$$\begin{cases} \dot{x} = f(x, d, t) + \sum_{i=1}^{m} g_i(x, t)u_i(t) \\ y = Cx \end{cases} \tag{7.1}$$

where $x \in \mathbb{R}^n$ is the state variable vector, $u(t) \in \mathbb{R}^m$ is the input control vector and $y \in \mathbb{R}^p$ is the output vector, $f(x, d, t)$ is the n-dimensional unknown nonlinear dynamics and d is the disturbance, $g(x, t)$ is the $(n \times m)$ nonlinear control dynamics matrix, C is the $(p \times n)$ output distribution matrix.

Such systems must satisfy the following assumptions:

A_1: the integers m, p and n are known such as $m \leq n$ and $p = m$;
A_2: the system is controllable and observable, in limit detectable and stabilizable;
A_3: the unknown nonlinear dynamics and unexpected disturbances are continuously differentiable with respect to the time variable.

7.2.1 The ANAT Robot

A.N.A.T. (Articulated Nimble Adaptable Trunk) technology is a new innovation in robotic architecture that allows the creation of highly robust and intelligent robots with reconfigurable modular architectures. This technology was invented by the founder of Robotics Design Inc., Mr. Charles Kharallah. In use in over seven ground-breaking products worldwide, ANAT technology represents robotic architecture of the future.

ANAT was exclusively designed and patented by Robotics Design Inc. (US patent 6323 615) ANAT has many properties that give it flexibility and dexterity in larger movements than most existing manipulators. This asset comes mainly from the robot's redundant modular trunk, which enables it to perform auxiliary obstacle avoidance tasks, and therefore allows it to work in more complex and difficult-to-access workspaces. In addition, the robot's redundancy ensures the main task continues even in the case where one or more motors of the trunk are defective. Thus, it is possible to implement on ANAT avoidance tasks of obstacle in the cartesian space, provided that they have the necessary resources for the realization.

The invention consists of a series of motorized U and H shaped modules that function much like cells of the human body; they work together to achieve a common goal. These modules connect to each other along four points, allowing them to evenly distribute pressure amongst themselves like the simple yet timeless design of the Roman arches. This allows ANAT robots to carry exceptionally heavy

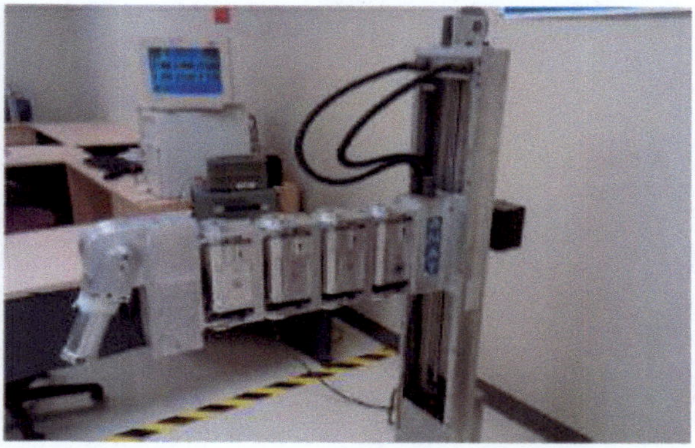

Fig. 7.4 Robot ANAT

payloads, and withstand pressure applied on any point of their modules. ANAT modules can be configured into a diverse spectrum of robot configurations such as fixed manipulators, mobile robots, flexible ergonomic arms (up to 32 D.O.F.), and much more. Motors are contained in the central axis of each module, so regardless of the quantity or impact point of outside pressure placed on the module, no strain is placed on the motor. These modules have one degree of freedom each, bend and fold relatively to each other along their axis, are highly flexible and able to avoid obstacles with ease. Modules can be connected in a myriad of configurations like Lego blocks, and can be reconfigured from their initial form to form another robot. This allows the same modular technology used in the design of an arc-welder or pick and place manipulator to be re-used in forming a mobile robot such as an unmanned mining vessel or vehicle. A robot using this simple and innovative architecture which performs a single specialized application can be formed by modules varying only in size, or a single robot can be formed to specialize in several applications, with the only variable being attached accessories. This also drastically simplifies and reduces costs for maintenance, as a faulty module can be easily replaced with an identical one. In short, nimble, adaptable and durable modular robots are now possible through our technology, and new products are born into the ANAT robot family daily (Fig. 7.4).

Figure 7.5 below shows the schematic representation of the robot. It has seven degrees of freedom composed as follows: a first prismatic joint, a part redundant system consisting of three parallel rotary joints. Finally, an effective effector to orient the tool in a desired position and consisting of three rotary joints whose axes of rotation are perpendicular to one another.

We say redundancy when the number of degrees of freedom of the manipulator is greater than the number of degrees of freedom of the task to be performed. When there is redundancy, there is an infinity of possible positions for a fixed placement of the tool; thus some members of the manipulator may be in motion. Let us call it the internal movement of the robot.

Redundancy allows to optimize the performance of a robot and to avoid constraints imposed by the articular limits or the obstacles.

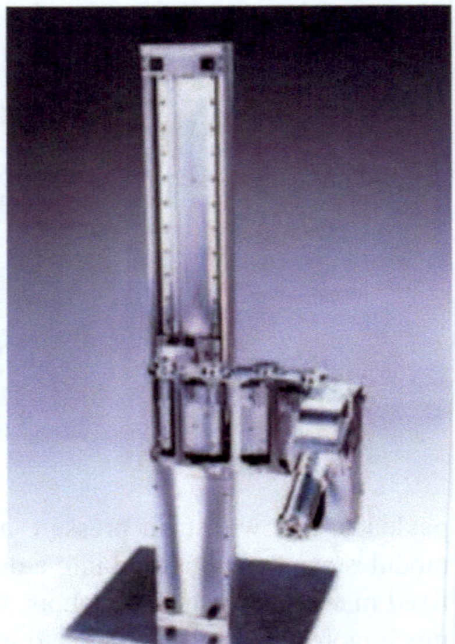

Fig. 7.5 ANAT robot

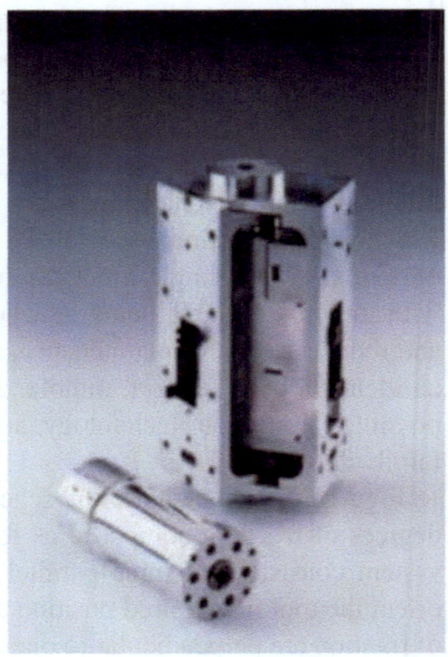

Fig. 7.6 One of the robot modules

An interesting property of ANAT lies in its modularity which facilitates its maintenance and performs "Plug and Play" tasks. Figure 7.6 shows one of the ANAT modules.

Moreover, the mechanical structure of the robot also has interesting advantages over conventional manipulators. First the aluminum structure of ANAT makes it

Table 7.1 Robot workspace

Articulation	Type	Space
1	Prismatique	of L_0=0.57 m at 1.27 m
2	Rotative	of −90° +90°
3	Rotative	of −90° +90°
4	Rotative	of −90° +90°
5	Rotative	of −90° +90°
6	Rotative	of −90° +90°
7	Rotative	of $-n\pi$ at $n\pi$

relatively light. On the other hand, the load distribution on the robot motors is optimized, in contrast to the usual manipulators on which the stresses are unequally distributed. Therefore, for common manipulators, motors may be oversized to accomplish the same task. The industrial applications of such a robot are multiple. Indeed, ANAT can be particularly used in the paint and finishing industries, the automotive industries, as well as the aeronautical and naval industries.

The robot is subjected to some mechanical displacement stresses q_m shown in Table 7.1.

The system (7.1) comes from the dynamic equation of robot (7.2). So the dynamics of n-DOF robot manipulator in the following matrix equation (Craig 1989):

$$M(q)\ddot{q} + C(q, \dot{q})\dot{q} + G(q) + F\dot{q} = \tau + \tau_d \tag{7.2}$$

where $q \in \mathbb{R}^n$ is the vector of the generalized coordinates in the joint space, $\dot{q}$, $\ddot{q}$ $\in \mathbb{R}^n$ the joint velocity and acceleration vectors, respectively, $M(q) \in \mathbb{R}^{n\times n}$ is the inertia matrix, $C(q, \dot{q}) \in \mathbb{R}^{n\times n}$ is the centrifugal and coriolis matrix, $G(q) \in \mathbb{R}^n$ is the gravitational vector, $F\dot{q} \in \mathbb{R}^n$ is the vector of viscous friction torque at the joints, τ, $\tau_d \in \mathbb{R}^n$ denotes the disturbance and torque input vectors, respectively.

The dynamic equation of the robot manipulator given in partitioned form in Eq. (7.2) can be rewritten as follows: with the disturbance and the viscous friction are negligibles. ($\tau_d = 0$), ($F\dot{q} = 0$).

$$\ddot{q} = -M(q)^{-1}F(q, \dot{q}) + M(q)^{-1}\tau \tag{7.3}$$

where M is the inertia matrix, which is symmetric and positive definite. Thus, $M(q)^{-}1$ always exits. F is the centrifugal, coriolis, and gravity vector; q is the joint position vector; τ is the torque input vector of the manipulator. Let $x = [X_1^T, X_2^T]^T$ the state vector with $X_1 = [q_1, q_2, q_3, q_4, q_5]^T$ and $X_2 = [\dot{q}_1, \dot{q}_2, \dot{q}_3, \dot{q}_4, \dot{q}_5]^T$, and $y = X_1$ is the output vector. The description of the system can be given in state representation form as follows:

$$\begin{cases} \dot{x}_1 = x_6 \\ \dot{x}_2 = x_7 \\ \dot{x}_3 = x_8 \\ \dot{x}_4 = x_9 \\ \dot{x}_5 = x_{10} \\ \dot{x}_6 = f_1(x, d, t) + \sum_{i=1}^{5} g_{1i}(x, t)u_i(t) \\ \dot{x}_7 = f_2(x, d, t) + \sum_{i=1}^{5} g_{2i}(x, t)u_i(t) \\ \dot{x}_8 = f_3(x, d, t) + \sum_{i=1}^{5} g_{3i}(x, t)u_i(t) \\ \dot{x}_9 = f_4(x, d, t) + \sum_{i=1}^{5} g_{4i}(x, t)u_i(t) \\ \dot{x}_{10} = f_5(x, d, t) + \sum_{i=1}^{5} g_{5i}(x, t)u_i(t) \end{cases} \tag{7.4}$$

where

$$g(x, t) = M(q)^{-1}$$

$$u_i = \tau_i, \qquad \text{for} \qquad i = 1:5$$

$$f(x, d, t) = -M(q)^{-1} F(q, \dot{q})$$

7.3 High Gain Observer

In general, an observer is a dynamic system that provides estimations of the current state of the system, by using the previous knowledge of the inputs and outputs of the system.

Consider the following class of affine nonlinear system:

$$\begin{cases} \dot{x} = f(x, d, t) + \sum_{i=1}^{m} g_i(x, t)u_i(t) \\ y = Cx \end{cases} \tag{7.5}$$

The system (7.5) has the input $u(t) \in \mathbb{R}^m$ which has a set of admissible values of the input. It is also assumed that there exists a physical domain $\Omega \in \mathbb{R}^n$ (open, bounded) of evolution of the input and that is the domain of interest of the system.

Suppose that system (7.5) is observable in the sense of rank and that $u = 0$ is an universal input, then the jacobian $\{h_1, L_f h_1, \dots, L_f^{n-1} h_1, h_2, L_f^{n-1} h_2, L_f^{n-1} h_p\}$. In the surroundings of a regular point one can select a subset of full rank:

$$\phi = \{z_1, \dots, z_n\}\{h_1, L_f h_1, \dots, L_f^{n-1} h_1, h_2, L_f^{n-1} h_2, L_f^{n-1} h_p\} \tag{7.6}$$

with $\sum_{k=1}^{p} \eta_k = n$ and L is the lie derivative.

The input $h_k(x)$ belongs in the order η_k. This determines a local coordinate system in which the system (7.5) is written as:

$$\begin{cases} \dot{z} = Az + \tilde{\varphi}(z) + \bar{\varphi}(z)u \\ y = Cz \end{cases} \tag{7.7}$$

with $z \in \mathbb{R}^n$, $u \in \mathbb{R}^m$, $y \in \mathbb{R}^p$

$$A = \begin{bmatrix} A_1 & & \\ & \ddots & \\ & & A_p \end{bmatrix}; \qquad B = \begin{bmatrix} C_1 & & \\ & \ddots & \\ & & C_p \end{bmatrix}; \tag{7.8}$$

$$\tilde{\varphi}(z) = \begin{bmatrix} \tilde{\varphi_1}(z) \\ \vdots \\ \tilde{\varphi_p}(z) \end{bmatrix}; \qquad \bar{\varphi}(z) = \begin{bmatrix} \bar{\varphi_1}(z) \\ \vdots \\ \bar{\varphi_p}(z) \end{bmatrix}; \tag{7.9}$$

with

$$A_k = \begin{bmatrix} 0\ 1 \dots 0 \\ \vdots\ \vdots\ \ddots\ \vdots \\ 0\ 0 \dots 1 \\ 0\ 0 \dots 0 \end{bmatrix}; \qquad \tilde{\varphi_k}(z) = \begin{bmatrix} 0 \\ \vdots \\ 0 \\ \tilde{\varphi_k}(z) \end{bmatrix}; \tag{7.10}$$

$$\bar{\varphi_k}(z) = \begin{bmatrix} \bar{\varphi_1} k(z) \\ \vdots \\ \bar{\varphi_{\eta_k}} k(z) \end{bmatrix}; \qquad C_k = \begin{bmatrix} 1\ 0 \dots 0 \end{bmatrix} \tag{7.11}$$

with $\tilde{\varphi_k}(z) = L_f^{\eta k} h_k$, $\bar{\varphi_i} k(z) = L_g L_f^{i-1} h_k$; dim $(A_k) = (\eta_k \times \eta_k)$, dim $(C_k) = (1 \times \eta_k)$ and dim $(\tilde{\varphi_k}(z))$ = dim $(\bar{\varphi_k}(z)) = (\eta_k \times 1)$ for $k = 1 \dots p$, $i = 1 \dots \eta_k$.

For the following theorem, is not used the linearity in u, the following system is then considered:

$$\begin{cases} \dot{z} = Az + \varphi(z, u) \\ y = Cz \end{cases} \tag{7.12}$$

Let K be a matrix $(n \times p, p)$ such that

$$K = \begin{bmatrix} K_1 & & \\ & \ddots & \\ & & K_p \end{bmatrix} \tag{7.13}$$

(with k_k of dimension $n \times 1$), such that for each block k, the matrix $A_k - K_k C$ have negative real parts. Then the system is uniformly locally observable, and it exists $T_0 > 0$, such that, for every T, such $0 < T < T_0$, the following system Constitutes an observer for the system (7.12):

$$\dot{z} = A\hat{z} + \varphi(\hat{\hat{z}}, u) + \Lambda^{-1}(T, \delta)K(y - C\hat{Z}) \tag{7.14}$$

with $\hat{\hat{z}}_{\mu}k = y_k$
$\hat{\hat{z}}_j = \hat{z}_j \neq \mu_k$,

$$\Lambda(T, \delta) = \begin{bmatrix} T^{\delta_1}\Delta_1(T^{\delta_1}) & & \\ & \ddots & \\ & & T^{\delta_p}\Delta_p(T^{\delta_p}) \end{bmatrix} \tag{7.15}$$

with

$$\Delta_k(T) = \begin{bmatrix} T^{\delta_k} & & & & \\ & \ddots & & & \\ & & T^{2\delta_k} & & \\ & & & \ddots & \\ & & & & T^{\eta_k\delta_k} \end{bmatrix} \tag{7.16}$$

Moreover, the standard of the observation error is bounded by an exponential whose decay rate can be chosen arbitrarily large.

Remark 1 The system

$$\dot{\hat{z}} = A\hat{z} + \varphi(\hat{z}, u) + \Lambda^{-1}(T, \delta)K(y - C\hat{Z}) \tag{7.17}$$

is also an observer for system (7.12). If a change of variable $z = \phi(x)$ has been necessary, it is necessary to return to the old database by$\widehat{x} = \phi^{-1}(\widehat{z})$. By applying this change of coordinates to the previous system, we obtain the observer in the old coordinates.

$$\dot{\widehat{x}}(t) = f(\widehat{x}(t)) + \sum_{i=1}^{m} g_i(\widehat{x}(t))u_i(t) + \left[\frac{\partial\phi(x)}{\partial x}\right]_{\widehat{x}}^{-1} \Lambda^{-1}K[y(t) - C\widehat{x}(t)] \tag{7.18}$$

Remark 2 The observer is written in the new coordinates as a system copy plus a non-linear correction. Implementation requires writing the observer into the original frame.

Therefore the observer becomes:

$$\begin{cases} \dot{\widehat{x}}(t) = f(\widehat{x}(t)) + \sum_{i=1}^{m} g_i[\widehat{x}(t)]u_i(t) + \left[\frac{\partial\phi(x)}{\partial x}\right]_{\widehat{x}}^{-1} \Lambda^{-1}K[y(t) - C\widehat{x}(t)] \\ y = Cx \end{cases} \tag{7.19}$$

The correction term $\left[\frac{\partial\phi(x)}{\partial x}\right]_{\widehat{x}}^{-1} \Lambda^{-1}K[y(t) - C\widehat{x}(t)]$ is then explicited as follows:

$$\left[\frac{\partial\phi(x)}{\partial x}\right]_{\widehat{x}}^{-1} = \begin{bmatrix} 1\,0\,0\,0\,0 \\ 0\,1\,0\,0\,0 \\ 0\,0\,1\,0\,0 \\ 0\,0\,0\,1\,0 \\ 0\,0\,0\,0\,1 \end{bmatrix} \tag{7.20}$$

$$\Lambda^{-1} = \begin{bmatrix} T^{-\delta_1} & 0 & 0 & 0 & 0 \\ 0 & T^{-2\delta_1} & 0 & 0 & 0 \\ 0 & 0 & T^{-\delta_2} & 0 & 0 \\ 0 & 0 & 0 & T^{-2\delta_2} & 0 \\ 0 & 0 & 0 & 0 & T^{-\delta_3} \end{bmatrix} \tag{7.21}$$

The gain T^{-1} must be selected as $0 < T \leq T_0 < 1$. T_0 is defined according to different parameters (η^2, the constant Lipschitz of the function $\varphi(z, u)$ defined by variable change,...). And the gain K is given by:

$$K = \begin{bmatrix} K_1 & & & & & & & & \\ & \ddots & & & & & & & \\ & & K_2 & & & & & & \\ & & & \ddots & & & & & \\ & & & & K_3 & & & & \\ & & & & & \ddots & & & \\ & & & & & & K_4 & & \\ & & & & & & & \ddots & \\ & & & & & & & & K_5 \end{bmatrix} \tag{7.22}$$

The pair (A, C) is observable, and it is an easy matter to assign eigenvalues to the matrix $(A - KC)$, that has the companion structure:

$$A - KC = \begin{bmatrix} -K_1 & 1 \dots 0 \\ \vdots & \vdots \ddots \vdots \\ -K_{n-1} & 0 \dots 1 \\ -K_n & 0 \dots 0 \end{bmatrix} \tag{7.23}$$

If a n-pla $\lambda = (\lambda_1,\dots,\lambda_n)$ of eigenvalues has to be assigned, the vector $K(\lambda)$ is the vector that contains the coefficients of the monic polynomial that has λ as roots. If the assigned eigenvalues are distinct, matrix $(A - KC)$ can be diagonalized by a vandermonde matrix:

$$V \equiv V(\lambda) \begin{bmatrix} \lambda_1^{n-1} & \dots & \lambda_1 & 1 \\ \vdots & \ddots & \vdots & \vdots \\ \lambda_n^{n-1} & \dots & \lambda_n & 1 \end{bmatrix} \tag{7.24}$$

So that

$$V(\lambda)(A - K(\lambda)C)V(\lambda)^{-1} = \text{diag}\,\{\lambda\} = \Lambda \tag{7.25}$$

Remark 3 Given a set λ of n eigenvalues to be assigned to A-KC, the gain $K(\lambda)$ is readily computed through the formula:

$$K(\lambda) = -V^{-1}(\lambda)[\lambda_1{}^n \dots \lambda_n{}^n]^T \tag{7.26}$$

which is not difficult to check. It is well known that a vandermonde matrix $V(\lambda)$ is singular if and only if two (or more) eigenvalues in the set λ coincide. It is also well known that the smaller is the minimum difference between eigenvalues in λ, the larger is the norm of $V^1(\lambda)$. For reasons that will be made clear in the following section it is important to choose eigenvalues for matrix $(A - K(\lambda)C)$ keeping bounded the norm of the inverse of the vandermonde matrix $V(\lambda)$. In Saberi and Sannuti (1990) it is shown that if the n eigenvalues are chosen as $\lambda_j = \lambda_j(w) = -w^j$, for $j = 1, \ldots, n$, with $w > 0$, then

$$\lim_{w \to \infty} ||V^{-1}(\lambda(w))|| = 1 \tag{7.27}$$

So, the term of correction of the system is:

$$\left[\frac{\partial \phi(x)}{\partial x}\right]^{-1}_{\widehat{x}} \Lambda^{-1} K[y(t) - C\widehat{x}(t)] = \begin{bmatrix} T^{-\delta_1} K_1(q_1 - \widehat{q_1}) \\ T^{-2\delta_1} K_2(q_2 - \widehat{q_2}) \\ T^{-\delta_2} K_3(q_3 - \widehat{q_3}) \\ T^{-2\delta_2} K_4(q_4 - \widehat{q_4}) \\ T^{-\delta_3} K_5(q_5 - \widehat{q_5}) \end{bmatrix} \tag{7.28}$$

7.4 High Gain Observer Applied to the ANAT Robot

In the observer based scheme, reconstruct some or all system outputs from accessible process variables. The fault indicators are generated by forming the differences between the estimated outputs and the actual outputs. In the absence of faults, the differences converge to zero as soon as a fault occurs, these differences are no longer equal to zero (Dixon et al. 2000; Farza et al. 2004; Filaretov et al. 2003).

Using the system model (7.5), a high-gain observer is developed as explained in Sect. 7.3:

$$\begin{cases} \dot{\widehat{x}}(t) = f(\widehat{x}(t)) + \sum_{i=1}^{m} g_i(\widehat{x}(t))u_i(t) + \left[\frac{\partial \phi(x)}{\partial x}\right]^{-1}_{\widehat{x}} \Lambda^{-1} K[y(t) - C\widehat{x}(t)] \\ y = Cx \end{cases} \tag{7.29}$$

The gain K and T, show the effectiveness of this observer. Thus, the choice of the gain of high-gain observer is based on a compromise between the speed of convergence of the observer and insensitivity to measurement noise. The gain K are chosen according to Remark 3, and T between 0 and 1.

Therefore, the system model and the observer are shown in Fig. 7.7 in order to clarified the structure of residuals generation.

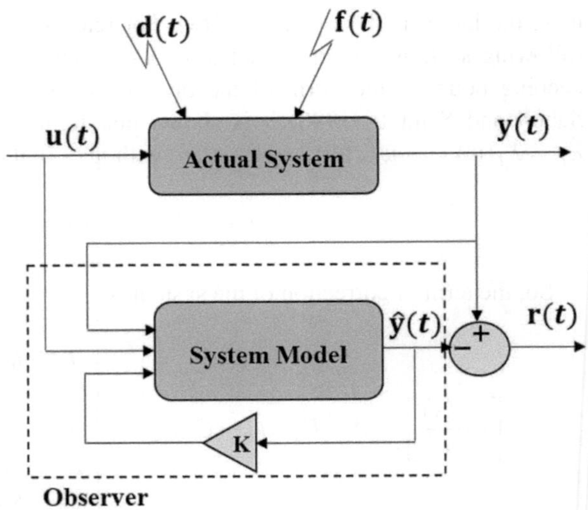

Fig. 7.7 Residual generation

7.4.1 Simulation Results

The state estimation error $r(t)$ can be calculated as:

$$r(t) = y(t) - \widehat{y}(t) \tag{7.30}$$

The residuals are supposed to differ from zero in case of faults $r(t) \neq 0$ and to be zero when there are no faults on the sensors $r(t) = 0$ (Fig. 7.8).

So the residuals are evaluated as:

$$r_1 = |q_1 - \widehat{q}_1| \tag{7.31}$$

$$r_2 = |q_2 - \widehat{q}_2| \tag{7.32}$$

$$r_3 = |q_3 - \widehat{q}_3| \tag{7.33}$$

$$r_4 = |q_4 - \widehat{q}_4| \tag{7.34}$$

$$r_5 = |q_5 - \widehat{q}_5| \tag{7.35}$$

Table 7.2 represents the fault signatures matrix for these residuals. Assuming that simultaneous faults cannot occur, we find that the signatures for each of the failures are quite different.

We simulate the system during $T = 30$ s. It is also noted that all faulty signals are additive.

- a fault d_1 is injected on the first joint (q_1) at the time $t = 2$ s. Figure 7.9 shows that the residual r_1 is different from zero during the presence of fault, and the residuals (r_2; r_3; r_4; r_5) are equal to zero.

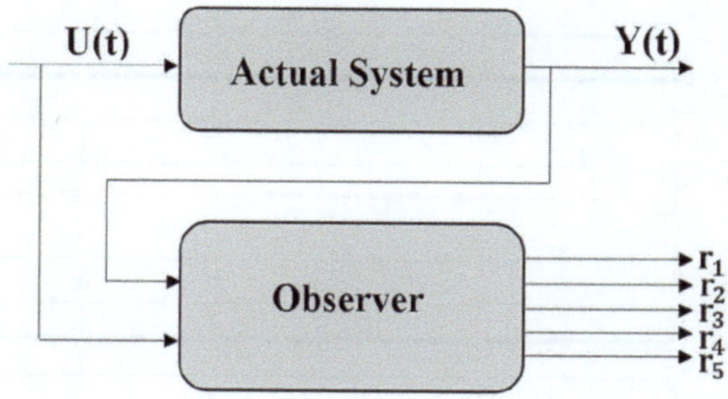

Fig. 7.8 Observer based residual generation

Table 7.2 Fault signatures matrix

d_i/r_i	r_1	r_2	r_3	r_4	r_5
d_1	1	0	0	0	0
d_2	0	1	0	0	0
d_3	0	0	1	0	0
d_4	0	0	0	1	0
d_5	0	0	0	0	1

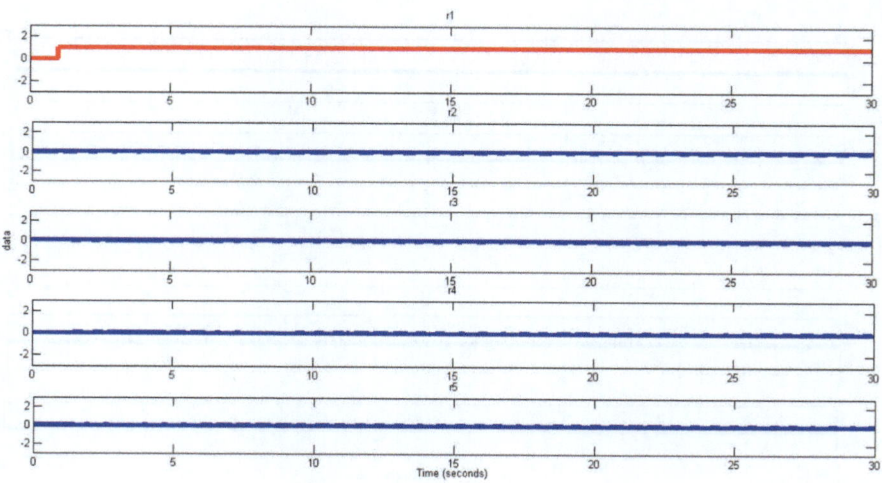

Fig. 7.9 Residuals evolution of the system with fault d_1

- at $t = 2$ s, Fig. 7.10 shows that the residual r_2 is sensitive to this fault, and the residuals (r_1; r_3; r_4; r_5) are equal to zero.
- a fault d_3 is injected on the joint (q_3) at the instant $t = 2$ s. Figure 7.11 shows the evolution of the different residuals. It can be seen that the residuals (r_1; r_2; r_4; r_5) are equal to zero and the residuals r_3 is sensitive to the fault.
- a fault d_4 injected on the fourth articulation (q_4) at the instant $t = 2$ s. Figure 7.12 shows that the residu r_4 differ from zero during the failure time.
- a fault d_5 is injected on the articulation (q_5) at the time $t = 2$ s. Figure 7.13 shows that the residual r_5 is sensitive to this fault.

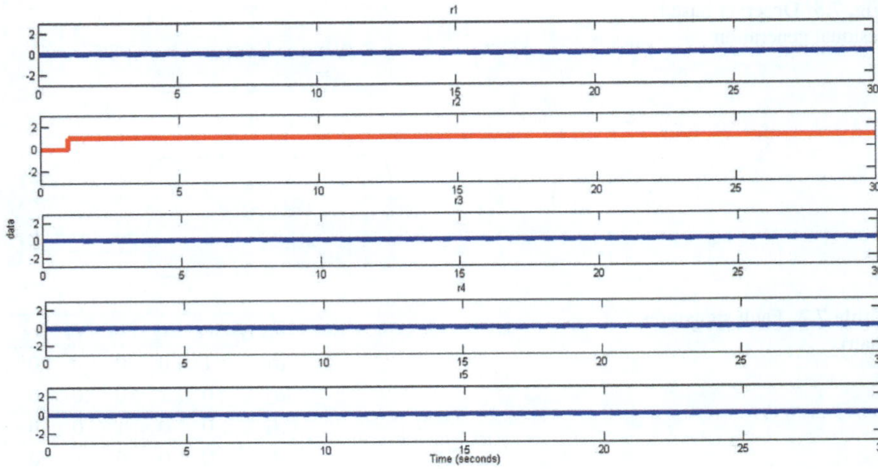

Fig. 7.10 Residuals evolution of the system with fault d_2

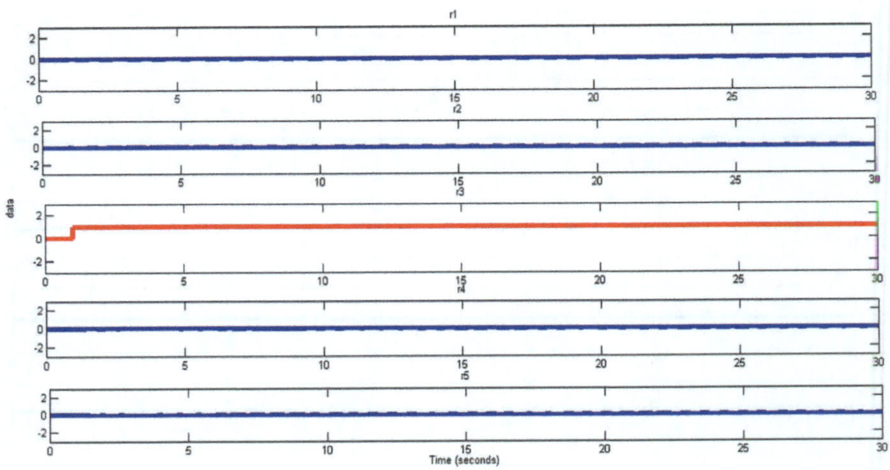

Fig. 7.11 Residuals evolution of the system with fault d_3

Therefore, we simulate the system at the time of $T = 10$ s. It is also noted that all faulty signals are additive.

Those faults are injected in different times of appear and disappear.

The initial condition of the system chosen as $q_1 = 0.25$ rad/s; $q_2 = 0.75$ rad/s; $q_3 = 0.5$ rad/s; $q_4 = 1.25$ rad/s; $q_5 = 0.45$ rad/s, and for the observer are $q_1 = 0.75$ rad/s; $q_2 = 1.25$ rad/s; $q_3 = 0.25$ rad/s; $q_4 = 0.25$ rad/s; $q_5 = 0.75$ rad/s.

- The first Fig. 7.14 represents the residual r_1 when fault d_1 is injected on the articulation q_1 between the time $t_i = 2$ s and $t_f = 4$ s. The residual r_1 is sensitive to this fault changed from zero during the failure time.

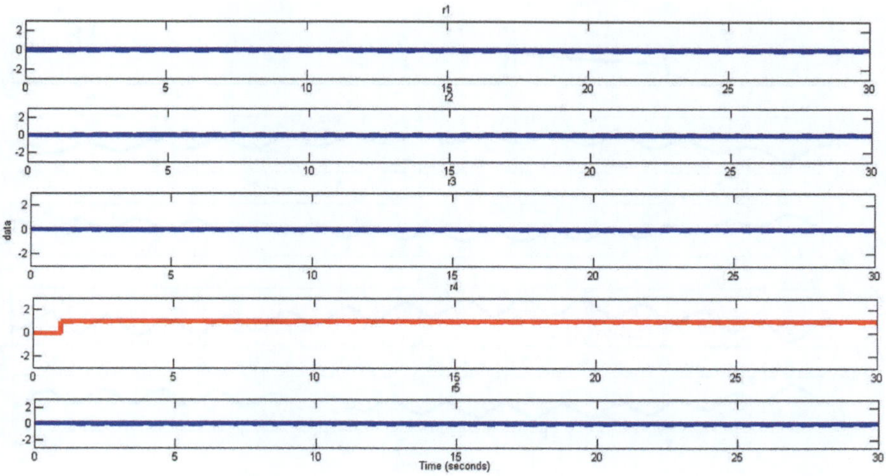

Fig. 7.12 Residuals evolution of the system with fault d_4

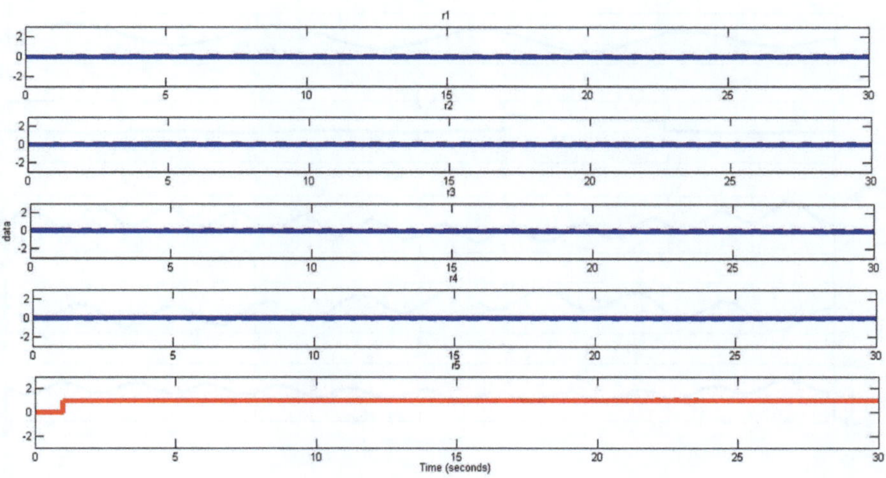

Fig. 7.13 Residuals evolution of the system with fault d_5

- Figure 7.15 represents the residual r_2 when a fault d_2 is injected on the articulation q_2 between the time $t_i = 4$ s and $t_f = 6$ s, presents that the residual r_2 is sensitive to this fault changed from zero during the failure time.
- Figure 7.16 represents the residual r_3 when a fault d_3 is injected on the articulation q_3 between the time $t_i = 2$ s and $t_f = 3$ s. The residual r_3 is sensitive to this fault changed from zero during the failure time.
- Figure 7.17 represents the residual r_4 when a fault d_4 is injected on the articulation q_4 between the time $t_i = 5$ s and $t_f = 6$ s. The residual r_4 is sensitive to this fault changed from zero during the failure time.

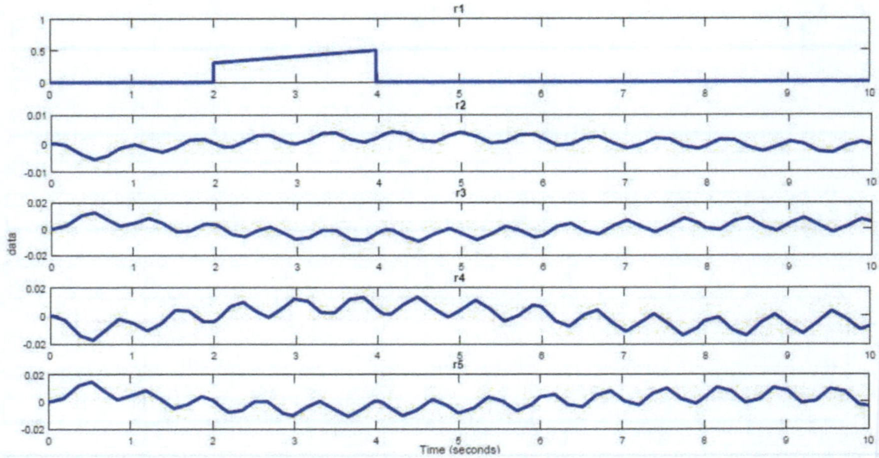

Fig. 7.14 Residuals evolution of the system with fault d_1

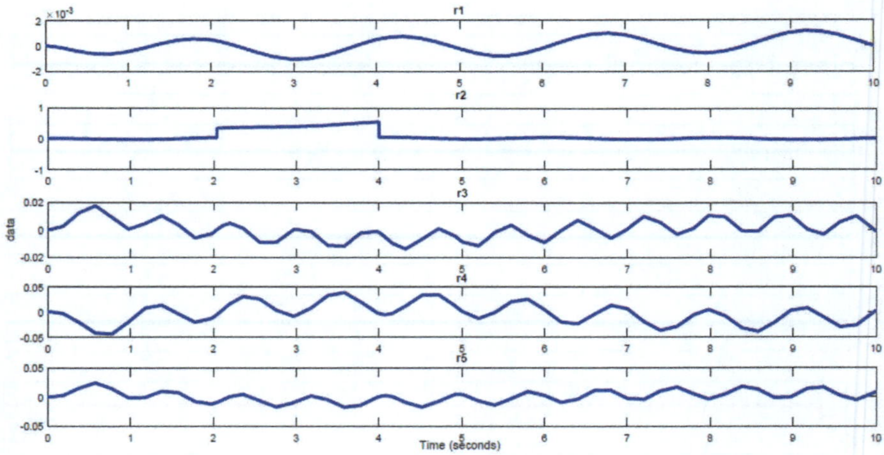

Fig. 7.15 Residuals evolution of the system with fault d_2

– Figure 7.18 represents the residual r_5 when a fault d_5 is injected on the articulation q_5 between the time $t_i = 4$ s and $t_f = 8$ s. The residual r_5 is sensitive to this fault changed from zero during the failure time.

A comparison of the results obtained by the first part of simulation and the second shows that: the first gives a good results, the second part show that when the initials conditions are differents to zeros gives also all residuals near of zero.

So, we can say that the proposed methods can give a good results and it can detect and isolate the sensor faults in a robot manipulator.

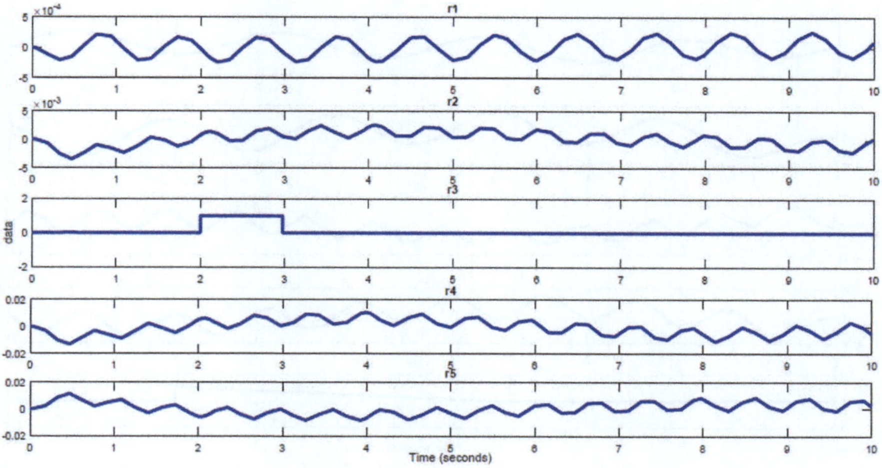

Fig. 7.16 Residuals evolution of the system with fault d_3

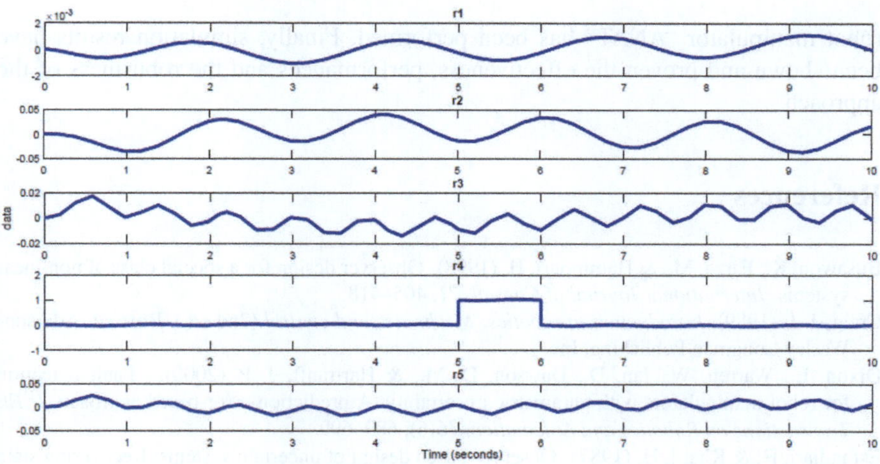

Fig. 7.17 Residuals evolution of the system with fault d_4

7.5 Conclusion

In this chapter, an approach for fault detection and isolation based on high-gain observer, for a class of affine nonlinear systems has been proposed. The approach is applied to a robot manipulator. Through the simulation results, it is shown that all sensors faults defined in specifications are detected and isolated. First, high-gain observer approach has been presented. Then, application for this methods in

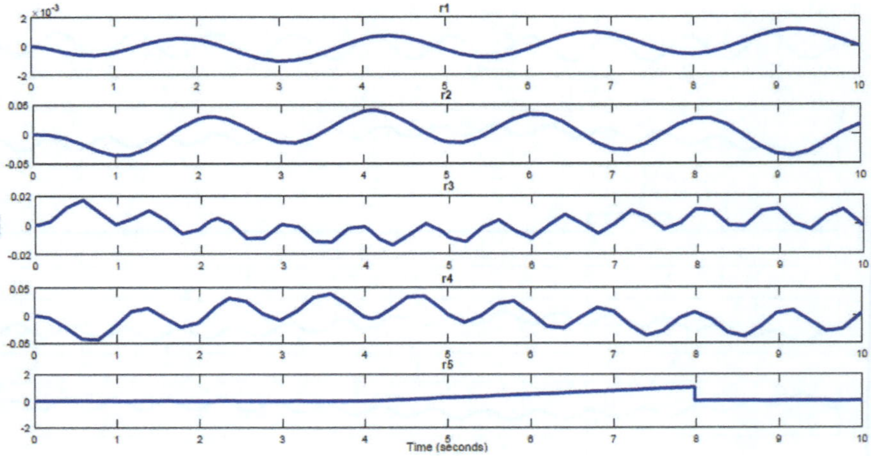

Fig. 7.18 Residuals evolution of the system with fault d_5

robot manipulator "ANAT" has been performed. Finally, simulation results have been shown and proven the effectiveness, performances and the robustness of the approach.

References

Busawon, K., Farza, M., & Hammouri, H. (1998). Observer design for a special class of nonlinear systems. *International Journal of Control, 71*, 405–418.

Craig, J. J. (1989). *Introduction to robotics: Mechanics and control* (2nd ed.). Boston: Addison-Wesley Longman Publishing, Inc.

Dixon, E., Warren, W., Ian, D., Dawson, D. M., & Hartranft, J. P. (2000). Fault detection for robot manipulators with parametric uncertainty: A prediction-error-based approach. *IEEE Transactions on Robotics and Automation, 16*(6), 689–699.

Esfandiari, F., & Khalil, H. (1987). Observer-based design of uncertain systems: Recovering state feedback robustness under matching conditions. In *Allerton Conference*, Monticello (pp. 97–106).

Farza, M., Hammouri, H., Jallut, C., & Lieto, J. (1999). State observation of a nonlinear system: Application to (bio) chemical processes. *AICHE Journal, 45*, 93–106.

Farza, M., M'Saad, M., & Rossignol, L. (2004). Observer design for a class of MIMO nonlinear systems. *Automatica, 40*, 135–143.

Farza, M., M'saad, M., & Sekher, M. (2005). A set of observers for a class of nonlinear systems. In *International Federation of Automation Control-IFAC'05*, Prague, 37.

Filaretov, V., Vukobratovic, M., & Zhirabok, A. (1999). Observer-based fault diagnosis in manipulation robots. *Mechatronics, 9*, 929–939.

Filaretov, V., Vukobratovic, M., & Zhirabok, A. (2003). Parity relation approach to fault diagnosis in manipulation robots. *Mechatronics, 13*, 141–152.

Frank, P. (1990). Fault diagnosis in dynamic systems using analytical and knowledge-based redundancy? A survey and some new results. *Automatica, 26*, 459–474.

Frank, P. (1993). Advances in observer-based fault diagnosis. In *TOOLDIAG'93, International Conference on Fault Diagnosis*, Toulouse.

Gauthier, J., & Kupka, I. (2001). Observability and observers for nonlinear systems. *SIAM Journal Control Optimisation, 32*, 181–188.
Gauthier, J. P., Hammouri, H., & Othman, S. (1992). A simple observer for nonlinear systems – application to bioreactors. *IEEE Transactions on Automatic Control, 37*, 875–880.
Gauthier, J. P., Hammouri, H., & Othman, S. (2001). A simple observer for nonlinear systems – application to bioreactors. *IEEE Transactions on Automatic Control, 37*, 875–880.
Gertler, J. (1993). Residual generation in model-based fault diagnosis. *Control Theory Advanced Technology, 9*, 259–285.
Gertler, J., & Monajemy, R. (1995). The state of the art. *Automatica, 31*, 627–635.
Guernez, C., Cassar, J., & Staroswiecki, M. (1997). Process fault diagnosis based on modeling and estimation methods-a survey automatica. In *IFAC Symposium SAFEPROCESS'97*, Kingston upon Hull.
Hammouri, H., & Farza, M. (2003). Nonlinear observers for locally uniformly observable systems. *ESAIM, 9*, 353–370.
Hammouri, H., Busawon, K., Yahoui, A., & Grellet, G. (2001). A nonlinear observer for induction motors. *European Physical Journal-Applied Physics, 15*, 181–188.
Hou, M., Busawon, K., & Saif, M. (2000). Observer design for a class of MIMO nonlinear systems. *IEEE Transactions on Automatic Control, 45*, 1350–1355.
Isermann, R. (1984). Process fault diagnosis based on modeling and estimation methods-a survey automatica. *International Federation of Automatic Control, 20*, 387–404.
Jollie, I. T. (2016). Simultaneous fault diagnosis for robot manipulators with actuator and sensor faults. *Information Sciences, 366*, 12–30.
Khalil, H., & Praly, L. (2014). High-gain observers in nonlinear feedback control. *International Journal of Robust and Nonlinear Control, 24*, 993–1015.
Khalil, H., & Saberi, A. (2007). Adaptive stabilization of a class on nonlinear systems using high-gain feedback. *IEEE Transactions on Automatic Control, 32*, 1031–1035.
Koubaa, Y., Farza, M., & M'saad, M. (2004). Obsevateur adaptatif pour une classe des systèmes non linéaires. In *5th Conférence Internationale des Sciences et Techniques de l'Automatique, STA'04*.
Krishnaswami, V., & Rissoni, G. (1994). Nonlinear parity equation residual generation for fault detection and isolation. In *IFAC Symposium SAFEPROCESS'94*, Espoo (Vol. 1, pp. 317–322).
Ma, H. J., & Yang, G. H. (2016). Simultaneous fault diagnosis for robot manipulators with actuator and sensor faults. *Information Sciences, 366*, 12–30.
Mironovsky, L. (1989). Functional diagnosis of nonlinear discrete-time processes. *Autom Remote Control, 6*, 150–157.
Nadri, M. (2001). Observation et commande des systèmes non linéaires et application aux bioprocédés. *Thèse de doctorat, Université Claude Bernard Lyon-1*.
Patton, R. (1994). The state of the art. In *IFAC Symposium SAFEPROCESS'94*, Espoo (Vol. 1, pp. 1–24).
Saberi, A., & Sannuti, P. (1990). Observer design for loop transfer recovery and for uncertain dynamical systems. *IEEE Transactions on Automatic Control, 35*, 878–897.
Schneider, H., & Frank, P. M. (1996). Observed- based supervision and fault detection in robots using nonlinear and fuzzy logic residual evaluation. *IEEE Transactions on Control Systems Technology, 4*(3), 274–282.
Shim, H., Son, Y., & Seo, J. (2001). Semi-global observer for multi-output nonlinear systems. *Systems and Control Letters, 42*, 233–244.
Shumsky, A. (1998). Parity relation method and its application to fault detection in nonlinear dynamic systems. *Automation and Remote Control, 9*, 155–165.
Slotine, J. J., & Weiping, L. (1991). *Applied nonlinear control*. Englewood Cliffs: Printice-Hall International.
Tornambe, A. (1988). Use of asymptotic observers having high gains in the state and parameter estimation. In *27th IEEE Conference on Decision and Control*, Austin (pp. 1791–1794).
Tornambe, A. (1992). High-gain observers for non-linear systems. *International Journal of Systems Science, 23*, 1475–1489.

Chapter 8
Crone Controller Design for a Robot Arm

Ahmed Abid, Rim Jallouli–Khlif, Nabil Derbel, and Pierre Melchior

Abstract The CRONE strategy (French acronym of Commande Robuste d'Ordre Non Entier) is assumed to give rise to a great number of applications in different industrial fields. This paper introduces the design of a second generation controller intended to control in position a two degrees of freedom robot arm. Great performances and robust behaviour are obtained and validated especially by a comparative study.

Keywords CRONE control · Second generation · Robustness · Robot arm

8.1 Introduction

In the two last decades in the field of automatic control, fractional systems have been of growing interest in analysis, system identification and control domains. Concerning this last domain, several authors have suggested to introduce fractional controllers (Machado 1997; Manabe 1961; Petras and Dorcak 1999; Podlubny 1999). This approach leads indeed to a better flexibility in the structure of the controller. In CRONE control, fractional differentiation is also used in the design procedure of the controller to ensure the robustness of the feedback system. Although robustness is a very wide notion, it always means insensitivity or very nearly so. In the fractional CRONE approach (Oustaloup 1991a,b; Oustaloup and Mathieu 1999) the severest robustness of the stability degree (stability margins)

A. Abid (✉) · R. Jallouli–Khlif · N. Derbel
Control and Energy Management Lab (CEM-Lab LR11ES47), National Engineering School of Sfax, Sfax, Tunisia
e-mail: rim.jalloulikhlif@isims.usf.tn; n.derbel@enis.rnu.tn

P. Melchior
Integration from Materials to Systems Lab (IMS-UMR 5218 CNRS), Bordeaux INP, University of Bordeaux, Bordeaux, France
e-mail: pierre.melchior@ims-bordeaux.fr

N. Derbel et al. (eds.), *New Developments and Advances in Robot Control*, Studies in Systems, Decision and Control 175,
https://doi.org/10.1007/978-981-13-2212-9_8

is sought, in order to maintain the frequency or time dynamic performance which measures this degree (performance robustness) (Oustaloup et al. 2013). Hence, the CRONE controller is the first fractional order controller developed. It was launched in 1975, and it was introduced using three generations (Abid et al. 2015; Jallouli-Khlif et al. 2010; Moreau et al. 2018).

CRONE control methodology treats with different domains as automobile (Nouillant et al. 2002; Oustaloup et al. 1996), aerospace, thermal applications (Moreau et al. 2018), industrial systems (Oustaloup et al. 2013), robotics (Delavari et al. 2013)...

This work is dealing with the design of a CRONE controller for a robotic application, notably the position control of a two degrees of freedom robot arm. As a first step, the work deals with a servo $MX - 28AT$, used as an actuator in a humanoid research robot Darwin OP2 (Maximo et al. 2017). The controller is designed using the CRONE toolbox on the base of an experimentally validated plant model (Maximo et al. 2017). Simulation results prove the efficiency of such second generation controller. Moreover, a comparison study with a PID controller arguments the advantages of the fractional approach especially in settling time and robustness against parametric uncertainties. In a second step, the designed fractional controller is introduced to control an XY plane robot arm. Simulations run upon nominal case and two modified ones validate the approach notably by comparing with the classical PID approach.

The remainder of the paper is divided as follows. An overview on the second CRONE generation is given in Sect. 8.2. The robot servo model subject to control is briefly presented in Sect. 8.3. Section 8.4 explains the control strategy based on the CRONE controller and gives simulation results. A comparison study versus a PID controller is achieved in Sect. 8.5. Section 8.6 deals with the use of the designed fractional controller to command in position a robot arm with two degrees of freedom. Efficiency and robustness of the approach are mentioned. Finally, some concluding remarks are drawn in the last Section.

8.2 Second CRONE Generation

For control effort-level problems, it can be difficult to have a crossover frequency ω_{cg} in a frequency interval where the plant phase is constant. For the second generation CRONE approach, as the transfer function of the open loop system is synthesized a posteriori, it is defined as the transfer function of a fractional order integrator:

$$\beta(s) = \left(\frac{\omega_{cg}}{s}\right)^n, \tag{8.1}$$

where ω_{cg} is the frequency for which the uncertainties do not lead to any phase variation, $n \in \mathbb{R}$ and $n \in [1, 2]$. The closed loop transfer function $T(s)$ is defined as follows:

$$T(s) = \frac{1}{1 + \left(\frac{s}{\omega_{cg}}\right)^n}, \tag{8.2}$$

Around the crossover frequency ω_{cg}, the *Black Nichols* plot of the open-loop transfer function $\beta(s)$ is a vertical asymptote with a constant phase. This asymptote allows having (Moreau et al. 2018):

- a robust phase margin M_ϕ equal to $(2 - n) \times \pi/2$;
- a robust resonance factor;
- a robust gain module;
- a robust damping ratio of the closed loop system;

In order to manage accuracy, robustness and control effort problems, Eq. (8.1) of the controller is truncated in frequency, and a low pass filter as well as an integrator are added. Therefore, the new description of the open-loop transfer function will be written as follows:

$$\beta(s) = \beta_0 \left(\frac{1 + \frac{s}{\omega_l}}{\frac{s}{\omega_l}}\right)^{n_l} \left(\frac{1 + \frac{s}{\omega_h}}{1 + \frac{s}{\omega_l}}\right)^n \left(1 + \frac{s}{\omega_h}\right)^{-n_h} \tag{8.3}$$

where ω_l and ω_h are respectively the low and the high transitional frequencies, n is the fractional order around the frequency ω_{cg}, n_l and n_h define the system behavior at low and high frequencies, and β_0 is a constant that guarantees a crossover frequency ω_{cg}. It is expressed in Eq. (8.4) (Oustaloup 1991a):

$$\beta_0 = \left(\frac{\omega_{cg}}{\omega_l}\right)^{n_l} \left[1 + \left(\frac{\omega_{cg}}{\omega_l}\right)^2\right]^{\frac{n-n_l}{2}} \left[1 + \left(\frac{\omega_{cg}}{\omega_h}\right)^2\right]^{\frac{n_h-n}{2}} \tag{8.4}$$

The frequency fractional order behavior of $\beta(s)$ is defined over an interval around the frequency ω_{cg} of a depth of r. Moreover, for the nominal plant, previous studies have shown that it is practical to satisfy such frequency constraints:

$$\begin{cases} \omega_l = \frac{\omega_{cgnom}}{10\sqrt{r}} \\ \omega_h = 10\omega_{cgnom}\sqrt{r} \end{cases} \tag{8.5}$$

Once the open-loop transfer function is calculated, the CRONE controller transfer function is defined as the ratio of the open-loop transfer function $\beta(s)$ over the nominal plant transfer function $H(s)$:

$$C_F(s) = \frac{\beta(s)}{H(s)} \tag{8.6}$$

The rational form $C_R(s)$ is deduced using the recursive poles and zeros distribution of the function. For more details, please refer to Oustaloup (1991a).

8.3 Robot Servo Model

A model for the behavior of the Dynamixel $MX - 28AT$ position servo is developed in Maximo et al. (2017), including phenomena such as viscous friction and saturation. Such a model allows the correct simulation of the servo behavior and the adequate design of the controller. For that purpose, constants for both servo and DC motor are directly obtained from corresponding data sheets or from some standard tests. The servo is built using an armature controlled Maxon DC motor (Maximo et al. 2017). The equivalent electric circuit is depicted in Fig. 8.1, where u is the input voltage at the terminals, e is the back electromotive force, i is the current, R is the resistance and L is the inductance.

Further, the motor is characterised in its mechanical part by a torque constant K_t and a speed constant K_ω. The rotation angle of the motor shaft is θ_m and its moment of inertia is J_m. Alike, the load axis angle of rotation is θ_l and its moment of inertia is J_l.

Usually, models of the servo motor are nonlinear, that means at least one of the states (i: current, θ: motor position) is an argument of a nonlinear function. In order to present such a model as a transfer function, it has to be linearized. However for the servo motor model, nonlinearities are small, so they can be neglected.

The open loop transmittance of the Dynamixel $MX - 28AT$ is expressed by the following transfer Eq. (8.7):

$$H(s) = \frac{\theta(s)}{U(s)} \tag{8.7}$$

Using plant parameters, the transfer function is expressed as follows:

$$H(s) = \frac{N\eta K_t}{s\left[L(J_l+J_m N^2\eta)s^2+[R(J_l+J_m N^2\eta)+Lb_m N^2\eta]s+b_m N^2\eta R+\frac{K_t N^2\eta}{K_\omega}\right]} \tag{8.8}$$

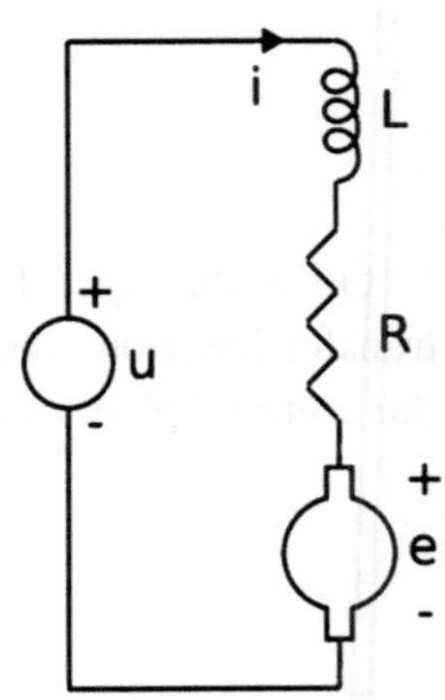

Fig. 8.1 Electrical architecture

Table 8.1 presents the values of the motor parameters supplied by the manufacturer.

8.4 Control Strategy

The manufacturer proposed a PID controller for the $MX-28AT$. Unfortunately, the servomotor electrical parameters L and R (Table 8.1) may vary with the temperature, the saturation of magnetic circuits and frequency variations. To encounter parameter uncertainties effects, this work proposes a robust fractional order controller belonging to the second CRONE generation. The proposed control schema is given in Fig. 8.2.

We assume that the variation of the inductance L can be neglected because the angular velocity of the servo motor and thereby the frequencies are small. So, three parametric states are considered:

- $R = R_{nom} = 8.3\,\Omega$
- $R = R_1 = 150\%$ of R_{nom}
- $R = R_2 = 250\%$ of R_{nom}

Hence, the corresponding identified models are given by:

$$H(s) = \frac{1.298\;10^{-2}s + 1.326\;10^{-2}}{1.123\;10^{-1}s^3 + 1.371s^2 + 1.284s}$$

Table 8.1 Parameters of the $MX-28AT$

Parameters	Notations	Values
Resistance	R	$8.3\,\Omega$
Inductance	L	2.03×10^{-3} H
Gear ratio	N	193
Gear efficiency	η	0.836
Speed constant	K_ω	93.1 rad/V
Torque constant	K_t	0.0107 Nm/A
Inertia	J_m	8.68×10^{-8} kg m^2
Friction	b_m	8.87×10^{-8} Nms

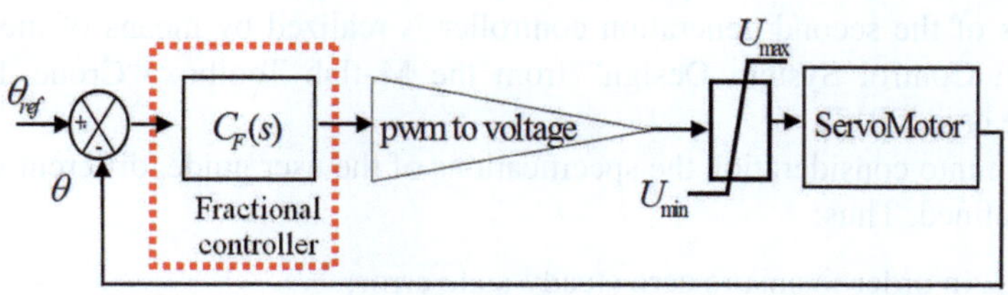

Fig. 8.2 Control schema

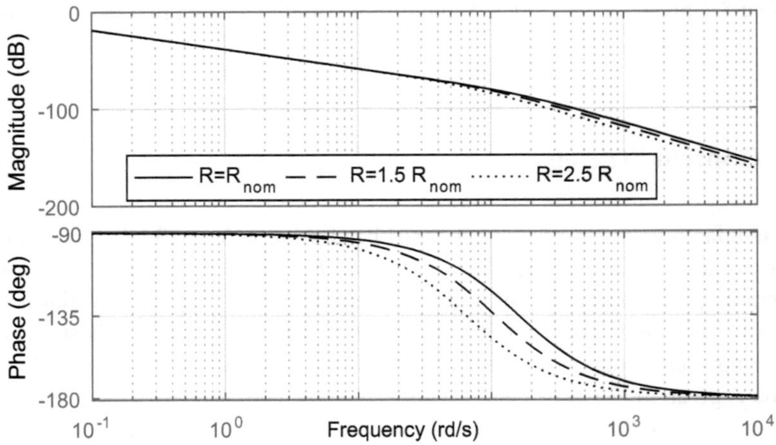

Fig. 8.3 Bode shapes of the plant transfer functions

$$H_1(s) = \frac{8.655\ 10^{-2}s + 8.843\ 10^{-2}}{1.123s^3 + 9.905s^2 + 8.948s} \tag{8.9}$$

$$H_2(s) = \frac{5.193\ 10^{-2}s + 5.306\ 10^{-2}}{1.123s^3 + 6.861s^2 + 5.838s}$$

where $H(s)$ is the nominal model corresponding to $R = R_{nom}$, $H_1(s)$ is the model corresponding to $R_1 = 150\%$ of R_{nom} and $H_2(s)$ corresponds to $R_2 = 250\%$ of R_{nom}.

Figure 8.3 shows the Bode diagrams for the three plants. It is noticed that plant uncertainties are reduced to variations in the gain. Since, the use of the second generation controller is justified.

8.4.1 Controller Design

The purpose is to command the servo in position, so a time response of 0.07 s is adequate. The cross over frequency is fixed to 32 rad/s (Maximo et al. 2017). The synthesis of the second generation controller is realized by means of the module "CRONE Control System Design" from the Matlab Toolbox "Crone Toolbox" (Lanusse et al. 2013).

Taking into consideration the specifications of the user guide, different variables can be defined. Thus:

- $n_l = 2$, in order to ensure zero steady-state error;
- $n_h = 2$, in order to limit the input sensitivity;
- $M_\phi = 90°$ (phase margin);
- $\omega_{cgnom} = 32$ rad/s.

Fig. 8.4 Rational CRONE controller

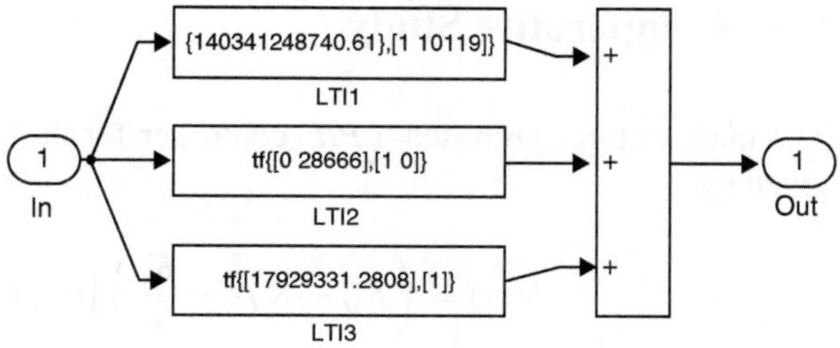

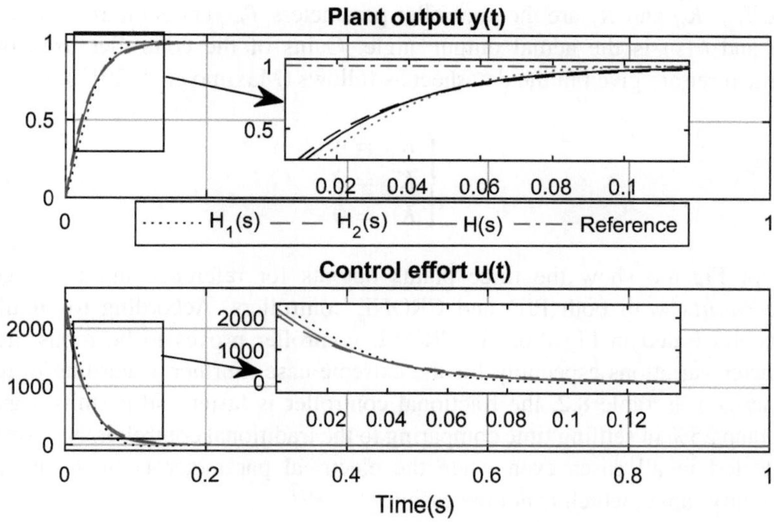

Fig. 8.5 Plant output and control effort versus time

For practical reasons, the second generation CRONE controller is approximated by a rational transmittance, obtained by fitting its frequency response. By export on Simulink, and in order to counter numerical problems, the synthesized CRONE controller is decomposed into parallel transmittances according to Fig. 8.4.

8.4.2 Simulation Results

By introducing the rational CRONE controller in the open loop plant, the following plots resume the outcome (Fig. 8.5).

In fact, for the closed loop step responses, all three outputs have the same behavior respecting the closed loop order: no overshoot. So this proves the robustness of the fractional controller.

8.5 Comparative Study

The manufacturer proposed a *PID* controller for the servo, so the input voltage is given by:

$$U(s) = \left(K_p + K_d s + \frac{K_i}{s}\right)\left[\theta_{ref}(s) - \theta(s)\right] \tag{8.10}$$

where K_p, K_d and K_i are the controller parameters, $\theta_{ref}(s)$ is the reference input angle and $\theta(s)$ is the actual output angle. Gains of the controller used by the manufacturer are given in the datasheet as follows (Maximo et al. 2017):

$$\begin{cases} K_p = 4 \\ K_d = 0 \\ K_i = 0 \end{cases} \tag{8.11}$$

Plots of Fig. 8.6 show the three plants outputs for reference input changes of ± 200 *counts* with both PID and CRONE controllers. According to simulation results illustrated in Fig. 8.6, the CRONE controller proves to be robust against parameter variations especially for the extreme case. Further, according to results summarized in Table 8.2, the fractional controller is faster and permits a gain of more than 25% in settling time comparing to the traditional controller. No overshoot is detected in all cases even when the electrical parameter is in its maximum uncertainty range, which is not the case for the *PID*.

8.6 Robust Control of a Robot Arm

The crone controller designed in Sect. 8.4 is introduced to control two servo-motors coupled to design a robot arm used to move on an XY plane (Fig. 8.7). The arm has two degrees of freedom and is characterized by the parameters given in Table 8.3.

The control strategy is presented in Fig. 8.8.

The controlled arm is modeled by the matrix form expressed in Eq. (8.12).

$$M(\theta)\dot{\theta} + C(\theta, \dot{\theta})\dot{\theta} + G(\theta) = \tau \tag{8.12}$$

where

- $\theta = [\theta_1\ \theta_2]^T \in \mathbb{R}^2$ is the position vector,
- $\dot{\theta} = [\dot{\theta_1}\ \dot{\theta_2}]^T \in \mathbb{R}^2$ is the speed vector,
- $\ddot{\theta} = [\ddot{\theta_1}\ \ddot{\theta_2}]^T \in \mathbb{R}^2$ is the acceleration vector,
- $M(\theta) \in \mathbb{R}^{2\times 2}$ is the inertia matrix, which is symmetric, uniformly bounded and positive definite
- $C(\theta, \dot{\theta}) \in \mathbb{R}^{2\times 2}$, with $C(\theta, \dot{\theta})\dot{\theta}$ represents the Coriolis and centrifugal forces,

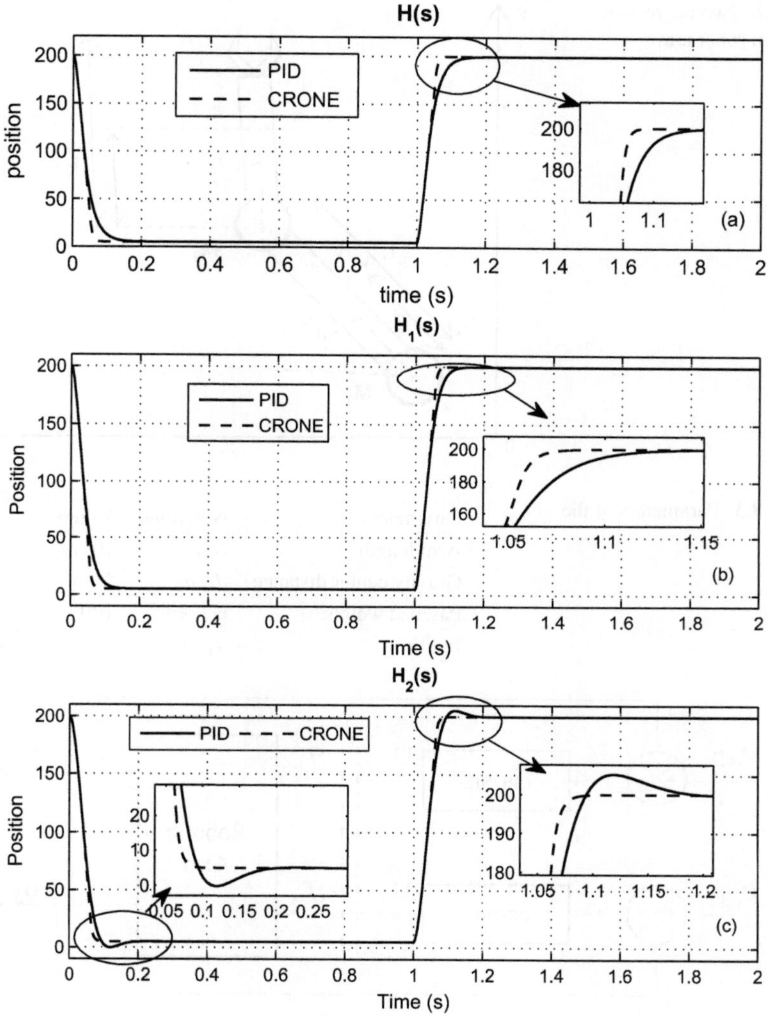

Fig. 8.6 Servo position response. (**a**): $R = R_{nom}$ nominal case. (**b**): $R = 1.5R_{nom}$. (**c**): $R = 2.5R_{nom}$.

Table 8.2 Simulation results of both controllers

	CRONE		PID	
	Overshoot (%)	Settling time $t_{5\%}$[s]	Overshoot (%)	Settling time $t_{5\%}$[s]
$H(s)$	0	0.061	0	0.081
$H_1(s)$	0	0.062	0	0.086
$H_2(s)$	0	0.066	2.56	0.085

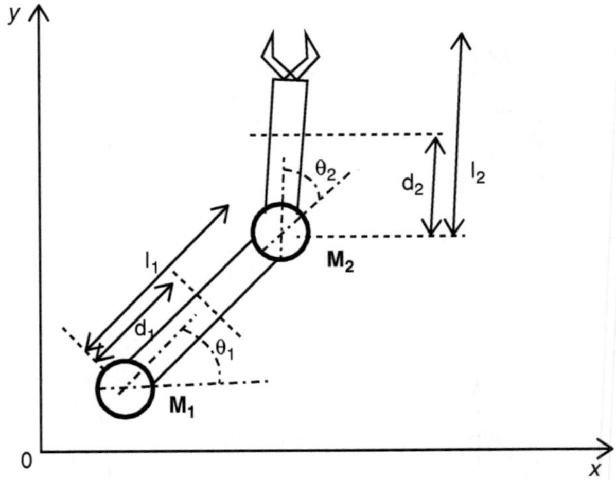

Fig. 8.7 Two degrees of freedom robot arm

Table 8.3 Parameters of the arm

Parameters	Notations	Values
Arm length	l_1, l_2	10 cm
Gravity center distance	d_1, d_2	5 cm
Payload weight	m_1, m_2	100 g
Inertia	J_1, J_2	0.27×10^{-2} kg m^2

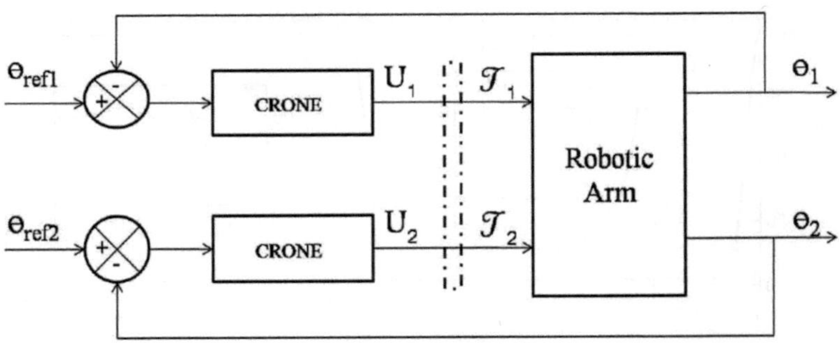

Fig. 8.8 Two degrees of freedom robot arm control strategy

- $G(\theta) \in \mathbb{R}^2$ denotes the gravity vector, is supposed null in our case study,
- $\tau \in \mathbb{R}^2$ is the vector of actuator torques.

The inertia matrix coefficients are expressed as follows:

$$\begin{aligned} M_{11} &= J_1 + J_2 + m_2\left(l_1^2 + d_2^2 + 2l_1 d_2 \cos\theta_2\right) + m_1 d_1^2 \\ M_{12} &= M_{21} = J_2 + m_2\left(d_2^2 + l_1 d_2 \cos\theta_2\right) \\ M_{22} &= J_2 + m_2 d_2^2 \end{aligned} \tag{8.13}$$

Table 8.4 Arm simulation results for position angle θ_1

	CRONE		PID	
	Overshoot (%)	Settling time $t_{5\%}[s]$	Overshoot (%)	Settling time $t_{5\%}[s]$
R_{nom} case	0	0.143	2.3	0.148
R_1 case	1.7	0.151	5.8	0.217
R_2 case	6.4	0.207	12	0.277

Table 8.5 Arm simulation results for position angle θ_2

	CRONE		PID	
	Overshoot (%)	Settling time $t_{5\%}$[s]	Overshoot (%)	Settling time $t_{5\%}$[s]
R_{nom} case	0	0.136	0.8	0.148
R_1 case	0	0.142	2.9	0.151
R_2 case	0	0.153	6.6	0.253

$C(\theta)$ is expressed as:

$$C(\theta, \dot{\theta}) = m_2 l_1 d_2 \sin\theta_2 \begin{pmatrix} -\dot{\theta}_2 & -\dot{\theta}_1 - \dot{\theta}_2 \\ \dot{\theta}_1 & 0 \end{pmatrix} \tag{8.14}$$

and $G(\theta)$ is given by:

$$G(q) = \begin{bmatrix} (m_1 d_1 + m_2 l_1) g \; \cos q_1 + m_2 g d_2 \cos(q_1 + q_2) \\ m_2 d_2 g \cos(q_1 + q_2) \end{bmatrix} \tag{8.15}$$

Simulations are run for the nominal case, and for two cases of uncertainty ($R = R_1 = 150\%$ of R_{nom} and $R = R_2 = 250\%$ of R_{nom}).

Arm responses controlled by the fractional controller are given in Fig. 8.9. As in Fig. 8.5, to prove the efficiency and notably the robustness of the fractional controller, PID-controller results are drawn simultaneously in Fig. 8.10.

Tables 8.4 and 8.5 compare accurately both controllers performances in position control.

According to Figs. 8.9a and 8.10a (the nominal case), both controllers give good results: both position angles θ_1 and θ_2 go towards desired positions perfectly without any overshoot for the fractional case, and with likely small overshoots with PID controller. Response times are likely equivalent for both controllers.

However, if the resistance value is changed, the fractional controller does extremely well notably in cancelling the overshoot for the final desired position as shown in Fig. 8.9b, c. Comparing Figs. 8.9c and 8.10c, the CRONE controller proves its robustness even in the extreme case of uncertainty. Even if PID performances are acceptable (a maximum overshoot of 12%), the CRONE controller makes proof of accurate path tracking and robustness.

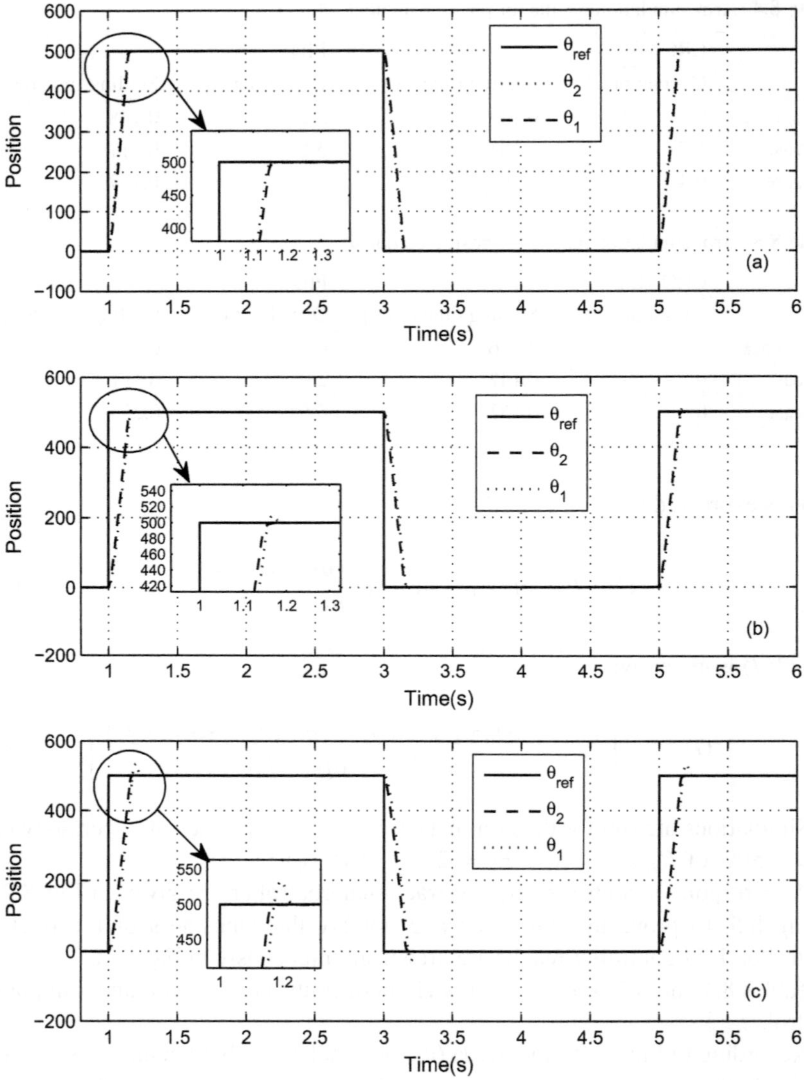

Fig. 8.9 Arm responses with the CRONE controller: position angles θ_1 and θ_2. (**a**): $R = R_{nom}$ nominal case. (**b**): $R = 1.5R_{nom}$. (**c**): $R = 2.5R_{nom}$

8.7 Conclusion

In this work, a fractional order controller is employed in a robotic application. In fact, a second generation controller, based on CRONE approach, is designed first for the position control of a servo $MX - 28AT$, used as actuator in a humanoid research robot Darwin OP2, then to control a robot arm with two degrees of

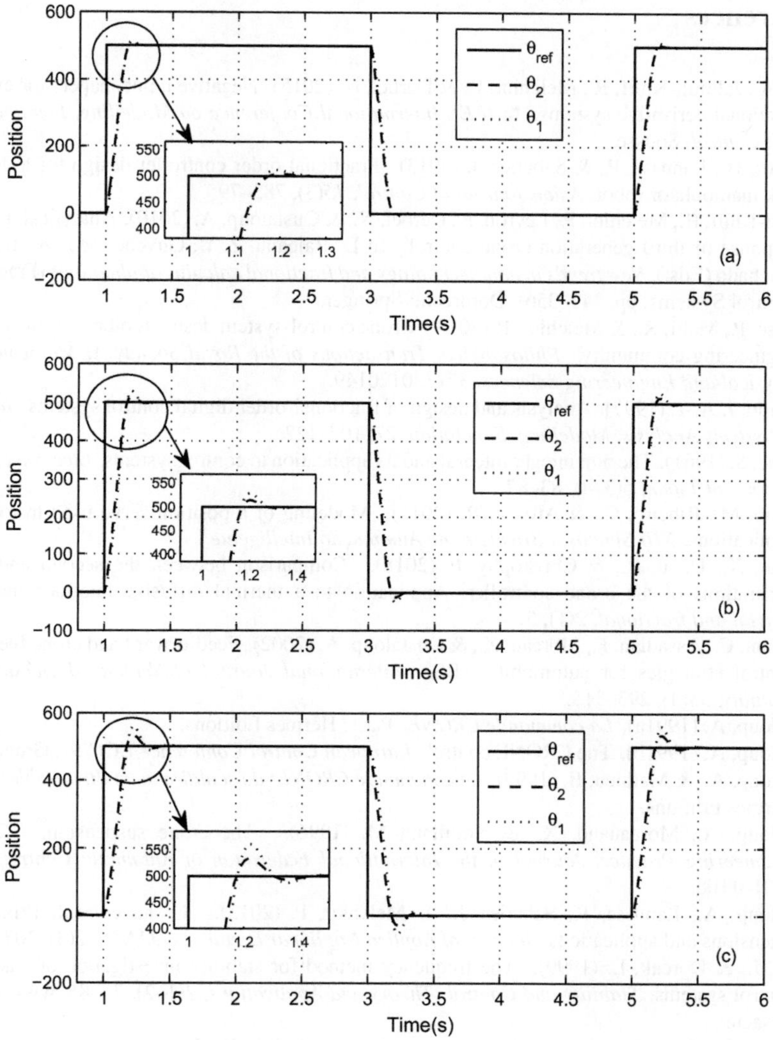

Fig. 8.10 Arm responses with the PID controller: position angles θ_1 and θ_2. (**a**): $R = R_{nom}$ nominal case. (**b**): $R = 1.5R_{nom}$. (**c**): $R = 2.5R_{nom}$

freedom. Simulation results validate the proposed solution in both application cases. Further, a comparison study with the PID controller, traditionally used, proves better temporal performances and distinguished advantage in term of robustness for the fractional controller.

References

Abid, A., Jallouli-Khlif, R., Melchior, P., & Derbel, N. (2015). Negative input shapers for explicit fractional derivative systems. In *IEEE International Conference on Modelling, Identification and Control*, Sousse.

Delavari, H., Lanusse, P., & Sabatier, J. (2013). Fractional order controller design for a flexible link manipulator robot. *Asian Journal of Control, 15*(3), 783–795.

Jallouli-Khlif, R., Melchior, P., Levron, F., Derbel, N., & Oustaloup, A. (2010). Analytical impulse response of third generation crone control. In D. Baleanu, Z. B. Guvenc, & J. A. Tenreiro Machado (Eds.), *New trends in nanotechnology and fractional calculus applications* (Fractional Control Systems, pp. 343–356). Dordrecht: Springer.

Lanusse, P., Malti, R., & Melchior, P. (2013). Crone control-system design toolbox for the control engineering community. *Philosophical Transactions of the Royal Society A: Mathematical, Physical and Engineering Sciences, 371*, 20120149.

Machado, J. A. T. (1997). Analysis and design of fractional-order digital control systems. *Journal of Systems Analysis, Modelling, Simulation, 27*, 107–122.

Manabe, S. (1961). The non integer integral and its application to control systems. *Electrotechnical Journal of Japan, 6*(3–4), 83–87.

Maximo, M., Ribeiro, C., & Afonso, R. (2017). Modeling of a position servo used in robotics applications. *XIII Simposoi Brasileiro de Automaçao Intelligente.*

Moreau, X., Daou, R., & Christophy, F. (2018). Comparison between the second and third generations of the crone controller: Application to a thermal diffusive interface medium. *Fractal and fractional, 2*(1), 5.

Nouillant, C., Assadian, F., Moreau, X., & Oustaloup, A. (2002). Feedforward and crone feedback control strategies for automobile ABS. *International Journal of Vehicle Mechanics and Mobility, 38*(4), 293–315.

Oustaloup, A. (1991a). *La commande CRONE.* Paris: Hermes Editions.

Oustaloup, A. (1991b). The CRONE control. *European Control Conference ECC'91*, Grenoble.

Oustaloup, A., & Mathieu, B. (1999). *La commande CRONE: du scalaire au multivariable*. Paris: Hermes Editions.

Oustaloup, A., Moreauand, X., & Nouillant, M. (1996). The crone suspension. *Control Engineering Practice, Journal of the International Federation of Automatic Control, 4*(8), 1101–1108.

Oustaloup, A., Lanusse, P., Sabatier, J., & Melchior, P. (2013). Crone control: Principles, extensions and applications. *Journal of Applied Nonlinear Dynamics (JAND), 2*(3), 207–223.

Petras, I., & Dorcak, L. (1999). The frequency method for stability investigation of fractional control systems. *Stability and Control: Theory and Applications, 2*(1–2), 75–85. www.cer.co.za/sacta/.

Podlubny, I. (1999). Fractional-order systems and pid-controller. *IEEE Transactions on Automatic Control, 44*(1), 208–214.

Chapter 9
Cartesian Sliding Mode Control of an Upper Extremity Exoskeleton Robot for Rehabilitation

Brahim Brahmi, Maarouf Saad, Cristobal Ochoa-Luna, Mohammad H. Rahman, and Abdelkrim Brahmi

Abstract Rehabilitation robots play an important role in rehabilitation treatment. Unlike conventional rehabilitation approach, the rehabilitation robotics provides an intensive rehabilitation motion with different modes (passive, active and active-assisted) based on the ability of the exoskeleton robot to perform assistive motion for a long period. However, this technology is still an emerging field. In this chapter, we present a Cartesian adaptive control based on a robust proportional sliding mode combined with time delay estimation for controlling a redundant exoskeleton robot called ETS-MARSE subject to uncertain nonlinear dynamics and external forces. The main objective of this research is to allow the exoskeleton robot to perform both rehabilitation modes, passive and active assistive motions with real subjects. The stability of the closed loop system is solved systematically, ensuring asymptotic convergence of the output tracking errors. Experimental results confirm the efficiency of the proposed control to provide an excellent performance despite the presence of dynamic uncertainties and external disturbances.

Keywords Rehabilitation robots · Adaptive control · Time delay estimation · Active and passive motion

---

B. Brahmi (✉) · M. Saad · A. Brahmi
Electrical Engineering Department, École de Technologie Supérieure, Montreal, QC, Canada
e-mail: brahim.brahmi.1@ens.etsmtl.ca; maarouf.saad@etsmtl.ca;
abdelkrim.brahmi.1@ens.etsmtl.ca

C. Ochoa-Luna
School of Physical & Occupational Therapy, Centre for Interdisciplinary Research in Rehabilitation of Greater Montreal, McGill University, Montreal, QC, Canada
e-mail: cristobal.ochoa.luna@mail.mcgill.ca

M. H. Rahman
Mechanical Engineering Department, University of Wisconsin-Milwaukee, Milwaukee, WI, USA
e-mail: rahmanmh@uwm.edu

N. Derbel et al. (eds.), *New Developments and Advances in Robot Control*, Studies in Systems, Decision and Control 175,
https://doi.org/10.1007/978-981-13-2212-9_9

9.1 Introduction

Habitually, survivors of strokes, trauma and/or neurological disorders suffer paralysis and loss of physical strength, often on one side of the body, such as the superior extremity (De Morand 2014). The disability of the upper-limb can create difficulties conducting daily live activities such as eating, dressing and cleanliness tasks.Such disabilities can have a large impaction the victim's behavior (Xie 2016). Currently, effective treatment that helps stroke survivals regain their missing functional capacity, acquire new skills, and improve their quality of life is a rehabilitation treatment (Xie 2016). Physical therapy is a set of medical exercises performed by a physiotherapist in order to increase the range of movement, functional capacity and quality of subject's life (De Morand 2014). Latterly, a new method of rehabilitation has emerged. This approach is named robotics rehabilitation that has attracted a lot of attention from the scientific community since robots are able to supplement the treatments provided by conventional physical therapy.The importance of the rehabilitation robots lies in their ability to provide intensive physiotherapy over a long period of time and not be influenced by factors such as the therapist's unavailability or fatigue (Xie 2016).

Rehabilitation robots are new devices designed to overcome the limitations of conventional physiotherapy approaches and to create new methods of rehabilitation treatment (Xie 2016). Typically, the development of this type of robots is conducted in order to adjust or to be identical to the human arm configuration. To rehabilitate patients with upper-limb impairment, these exoskeletons habitually are worn on the lateral side of the patients' upper-limb. The degree of freedom (DOFs) of the exoskeleton makes it a capable to reach several arm configurations in its workspace. When the degrees of freedom of the robot are increased, it is impossible to obtain the dynamic model of the robot accurately due to the complex mechanical structure and parameters such as nonlinear friction forces, backlash and the complexity of the actuators of the robot (Brahim et al. 2016a,b). Furthermore, the dynamic parameters change due to different physiological conditions of the subjects, such as an external disturbance caused by subject's muscular activity (Brahim et al. 2016a,b). That means the manipulator can be subjected to both parametric uncertainties and unknown nonlinear functions. The uncertain nonlinear dynamic model and external forces can turn into an unknown function that can affect the performance of the exoskeleton In this case, we need a solid adaptive control to better achieve the therapeutic activity by estimating the dynamic model uncertainties and the external forces that negatively affect the overall system performance.

In fact, control of uncertain nonlinear dynamics is one of the new topics of the nonlinear control area. Many control algorithms have been developed to overcome the effect of these uncertainties which can be characterized into two categories. The first one is adaptive control (Slotine and Li 1991), e.g., feedback linearization (Park and Cho 2007), backstepping control (Shieh and Hsu 2008) and H_∞ control (Yim and Park 1999). However, the above controllers assume that the dynamic model of the robot is linear in a set of physical parameters and derive a control law capable of ensuring the stability of the linear system. Actually, the manipulator is highly nonlinear, which means the integration of this law may affect the stability of the system in the presence of even small disturbances. The second one is the deterministic robust control (Khalil and Grizzle 1996; Slotine and Li 1991). One of the successful robust methods that have been used to control the uncertain dynamics is the nonlinear sliding mode control based on high gains switching controller, which forces the system to converge to a selected sliding surface. The robustness of the controller is ensured when the switching gain is larger than the bounded uncertainties (Fridman 1999; Young et al. 1999). To compensate for the chattering problem caused by the high switching gains that provoke damage in the actuators of the robot (Rahman et al. 2013), there are numerous approaches combined with sliding mode control such fuzzy logic (Chen et al. 2009), neural network or/and both (Wang et al. 2009). However, these approaches need heavy computations which make the implementation very difficult. Time-delay control estimation (TDE) (Brahmi et al. 2018a; Youcef-Toumi and Ito 1990) is a strong approach used to limit the effect of the uncertainties. The TDE technique can be combined with sliding mode to control the uncertainties and external forces of the robot (Baek et al. 2016). The time delay estimation technique basically employs time-delayed information from the previous state response of the system and the previous control input to estimate the dynamic uncertainties of the robot. However, the TDE method suffers from time delay error (TDR). This TDR can affect the robustness and the accuracy of the robot performance.

It is interesting to remark that in most rehabilitation applications, the therapeutic tasks are expressed in Cartesian space (Xie 2016). In some cases, the dynamics of the robot is modeled in Cartesian space (Craig 2005). Then, a Cartesian sliding surface is selected to control the uncertain nonlinear dynamics of the robot (Slotine and Li 1991). On the other hand, the dynamics of the manipulator is developed in joint space (Craig 2005). Then, a sliding joint surface is selected to achieve the control. In this case, the transformation between the joint space and Cartesian space is done by the Jacobian matrix (Craig 2005). Unfortunately, it is impossible to determine the dynamic and kinematic parameters of the robot exactly, due to the uncertainty to which the system is subjected. Therefore, the above mentioned methods suffer from the lack of accuracy on tracking the desired trajectory, since the error of the transformation between joint space and Cartesian space is not taken into consideration. This error may affect the accurate performance of the exoskeleton robot.

To address the above problems, we propose time delay estimation based on Cartesian sliding mode controller that allows an exoskeleton robot dealing with dynamic uncertainties. The adaptation of dynamic uncertainties is done by a new time delay approach, where the time delay error (TDR) is taken into consideration. The proposed controller provides a high level of robustness and accuracy by designing a control law consisting of both sliding Cartesian surface and sliding joint surface. This approach may limit the effect of the transformation kinematics error on the performance of the system and reduces the chattering effect.The proposed strategy is not limited to perform a passive rehabilitation but also qualified to perform an active rehabilitation. To complete this protocol, we rely necessarily on the estimation of the Desired Movement Intention (DMI) of the subject using indirect force control. This latter is done by Damped Least Square method (DLS), which has been successfully used in multiple applications (Nakamura and Hanafusa 1986; Wampler 1986). The DLS approach aims to provide a compromise between robustness of the solution and accuracy of the robot's performance. Besides, the proposed control is evaluated with healthy subjects.Anappropriate Lyapunov function is given to solve the adaptation problem systematically, proving the closed-loop stability and ensuring the asymptotic convergence of the output errors. The principal benefit of the designed control is an accurate knowing of the dynamic model, the external disturbances of the robot system are not needed, and afford exceptional tracking performance regardless the existence of uncertainties.

The rest of the chapter is organized as follows. In the Sect. 9.2, the dynamics of the ETS-MARSE is presented. The control approach is detailed in Sect. 9.3. Simulation results are exhibited in Sect. 9.4; finally, the conclusion is presented in Sect. 9.5.

9.2 Dynamics of the ETS-MARSE

9.2.1 Modeling of the Upper Limb ETS-MARSE Robot Definition

The developed exoskeleton robot ETS-MARSE[1] is a redundant robot consisted of 7 degrees of freedom (DOFs), as shown in Fig. 9.1. It is created to provide an assistive physical therapy motion to the injured upper limbs. The idea of the designed exoskeleton is basically extracted from the anatomy of the upper limb of the human to be ergonomic for their wearer along the physical therapy session.The shoulder part consists of three joints, the elbow part comprises by one joint and the wrist part consists of three joints. Each part from the previous parts is responsible for performing a variety of upper limb motions as we show in Table 9.2. The design of the ETS-MARSE has special features compared with the existing exoskeleton robots. All special characteristics of the ETS-MARSE and comparison with similar existing exoskeleton robots are summarized in (Rahman et al. 2013). The modified Denavit-Hartenberg (MDH) parameters are given in Table 9.1. These parameters are obtained from frames reference as shown in Fig. 9.1, and are used to obtain the homogeneous transformation matrices.The workspace of the designed robot is given in Table 9.2 and the details of the parameters and design of ETS- MARSE are given in Rahman et al. (2013, 2015).

[1]École de Technologie Supérieure – Motion Assistive Robotic-exoskeleton for Superior Extremity.

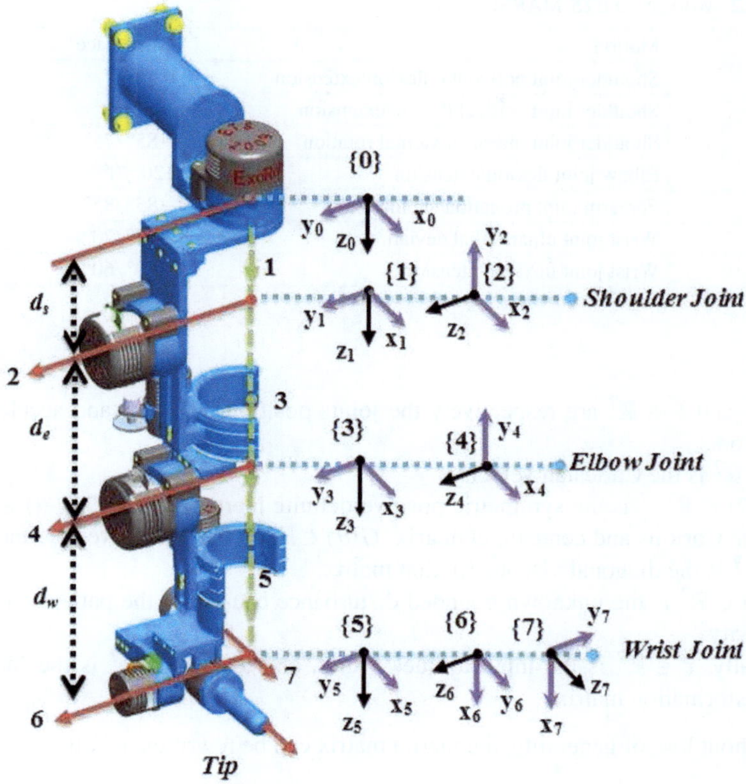

Fig. 9.1 ETS-MARSE robot system (Rahman et al. 2013)

Table 9.1 Modified Denavit-Hartenberg parameters

Joint (i)	a_{i-1}	a_i	d_i	θ_i
1	0	0	d_s	θ_1
2	$-\frac{\pi}{2}$	0	0	θ_2
3	$\frac{\pi}{2}$	0	d_e	θ_3
4	$-\frac{\pi}{2}$	0	0	θ_4
5	$\frac{\pi}{2}$	0	d_w	θ_5
6	$-\frac{\pi}{2}$	0	0	$\theta_6 - \frac{\pi}{2}$
7	$-\frac{\pi}{2}$	0	0	θ_7

9.2.2 *Dynamics of the ETS-MARSE Robot*

The dynamic equation of the ETS-MARSE robot can be shown in joint space, using the Lagrangian method as follows:

$$M(\theta)\ddot{\theta} + C(\theta, \dot{\theta})\dot{\theta} + B_0\dot{\theta} + G(\theta) + d(t) = \tau \tag{9.1}$$

$$x = J(\theta)\dot{\theta} \tag{9.2}$$

Table 9.2 Workspace ETS-MARSE

Joints	Motion	Workspace
1	Shoulder joint horizontal flexion/extension	$0°/140°$
2	Shoulder joint vertical flexion/extension	$140°/0°$
3	Shoulder joint internal/external rotation	$-85°/75°$
4	Elbow joint flexion/extension	$120°/0°$
5	Forearm joint pronation/supination	$-85°/85°$
6	Wrist joint ulnar/radial deviation	$-30°/20°$
7	Wrist joint flexion/extension	$-50°/60°$

where:

- θ, $\dot{\theta}$ and $\ddot{\theta} \in \mathbb{R}^7$ are respectively the joints position, velocity, and acceleration vectors.
- $\dot{x} \in \mathbb{R}^7$ is the Cartesian velocity.
- $M(\theta) \in \mathbb{R}^{7\times7}$ is the symmetric positive-definite inertia matrix, $C(\theta, \dot{\theta}) \in \mathbb{R}^{7\times7}$ is the Coriolis and centrifugal matrix, $G(\theta) \in \mathbb{R}^7$ is the gravity vector, and $B_0 \in \mathbb{R}^{7\times7}$ is the diagonal viscous friction matrix.
- $d(t) \in \mathbb{R}^7$ is the unknown bounded disturbance caused by the patient muscular activity.
- Finally, $\tau \in \mathbb{R}^7$ is the input torques vector, and $J(\theta) \in \mathbb{R}^{6\times7}$ is the Jacobian transformation matrix.

Without loss of generality, the inertia matrix can be rewritten as follows:

$$M(\theta) = M_0(\theta) + \Delta M(\theta) \tag{9.3}$$

where $M_0(\theta)$ is the known part of inertia matrix, and $\Delta M(\theta)$ is the uncertain part. Hence, the dynamic model expressed in (9.1) is rewritten as follows:

$$M_0(\theta)\ddot{\theta} + U(\delta, \ddot{\theta}, \dot{\theta}, \theta, t) = \tau \tag{9.4}$$

where,

- $U(\delta, \ddot{\theta}, \dot{\theta}, \theta, t) = \Delta M(\theta)\ddot{\theta} + C(\theta, \dot{\theta})\dot{\theta} + F(\theta, \dot{\theta})$ is the total unknown nonlinear dynamic uncertainties of the robot.
- δ is the uncertain parameters of the dynamic model
- $F(\theta, \dot{\theta}) = B_0\dot{\theta} + G(\theta) + d(t)$

Let us denote $M_0(\theta) = M_0$ and $U(\delta, \ddot{\theta}, \dot{\theta}, \theta, t) = U(t)$ for an easier handling of the notation of the control design.

Several properties of system exoskeleton robot are given in (9.4) to prove the stability such that:

Property 1 The inertia matrix $M_0(\theta)$ is symmetric, bounded and positive definite for all $\theta \in \mathbb{R}^n$ (Lewis et al. 2003).

Assumption 1 The joint position and joint velocity are measured.

Assumption 2 The pseudo-Jacobian matrix is non-singular.

Assumption 3 The desired trajectory is bounded.

Assumption 4 The function $U(t)$ is a locally Lipschitz function.

9.3 Control Design

The objective of the suggested control is to ensure that the measured Cartesian position x of the robot follows the desired trajectory x_d along the designed therapeutic task. The proposed controller is aimed to provide a high level of accuracy even if the dynamic model of the robot is not completely known and in the existence of external disturbances. All input and output signals in the closed-loop system stay bounded. Along the tracking trajectory, an accurate adaptation of the dynamic model was achieved. The closed-loop of the overall system is presented in Fig. 9.2.

We can start now by the determination of the Cartesian position error and the Cartesian velocity error as follows:

$$\begin{cases} e = x - x_d \\ \dot{e} = \dot{x} - \dot{x}_d \end{cases} \tag{9.5}$$

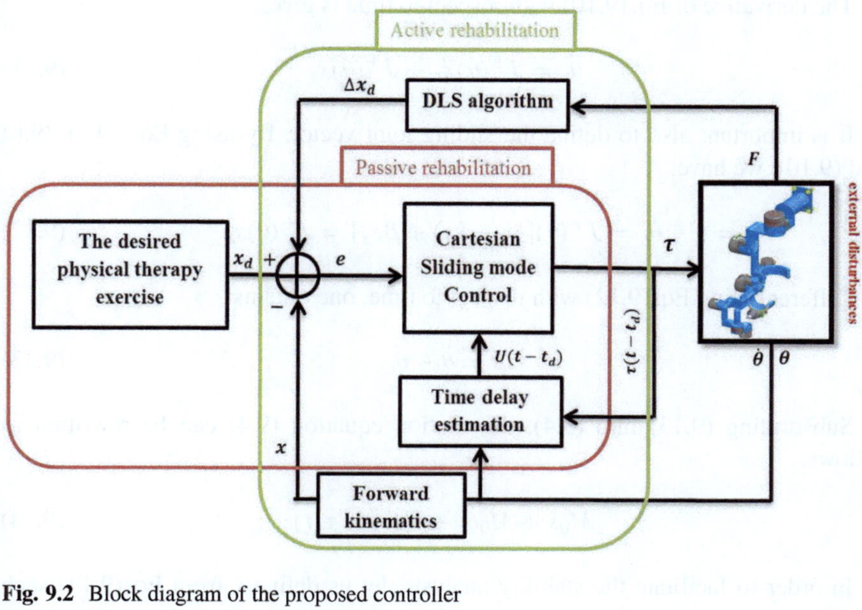

Fig. 9.2 Block diagram of the proposed controller

where, $x_d \in \mathbb{R}^6$ is the desired position and rotation of Cartesian trajectory and $x \in \mathbb{R}^6$ is the measured position and rotation of the end-effector of the exoskeleton robot.

Let us define the required Cartesian velocity as:

$$\dot{x}_r = \dot{x}_d - \beta e_x \tag{9.6}$$

where x $\dot{x}_r \in \mathbb{R}^6$ and β is a scalar positive constant. The derivative of Eq. (9.6) with respect to time yields:

$$\ddot{x}_r = \ddot{x}_d - \beta \ddot{e}_x \tag{9.7}$$

Using Eq. (9.2), we define the sliding Cartesian vector as follows:

$$s_x = \dot{x} - \dot{x}_r = J(\theta)\dot{\theta} - \dot{x}_r \tag{9.8}$$

The derivative of Eq. (9.8) with respect to time yields:

$$\dot{s}_x = \ddot{x} - \ddot{x}_r = J(\theta)\ddot{\theta} + \dot{J}(\theta)\dot{\theta} - \ddot{x}_r \tag{9.9}$$

We can now determine the required joint velocity vector as follows:

$$\dot{\theta}_r = J^+(\theta)\dot{x}_r \tag{9.10}$$

where $J^+(\theta) = J^T(\theta)[J(\theta)J^T(\theta)]^{-1}$ is the pseudo-jacobian matrix.

The derivative of Eq. (9.10) with respect to time is gives:

$$\ddot{\theta}_r = J^+(\theta)\ddot{x}_r + \dot{J}^+(\theta)\dot{x}_r \tag{9.11}$$

It is important also to define the sliding joint vector. By using Eqs. (9.2), (9.6) and (9.10), we have:

$$s = \dot{\theta} - \dot{\theta}_r = J^+(\theta)[(\dot{x} - \dot{x}_d) + \beta e_x] = J^+(\theta)s_x \tag{9.12}$$

Differentiating Eq. (9.12) with respect to time, one obtains:

$$\dot{s} = \ddot{\theta} - \ddot{\theta}_r \tag{9.13}$$

Substituting (9.13) into (9.4), the motion equation (9.4) can be rewritten as follows:

$$M_0\dot{s} + M_0\ddot{\theta}_r + U(t) = \tau(t) \tag{9.14}$$

In order to facilitate the stability analysis, let us define $\dot{s}$ from Eq. (9.14) such that:

$$\dot{s} = M_0^{-1}[\tau(t) - U(t)] - \ddot{\theta}_r \tag{9.15}$$

In this chapter, we proposed a new Cartesian adaptive tracking control tracking by using the Jacobian transpose matrix such that:

$$\tau = M_0[-J^T(\theta)(k_1 e_x + k_2 \dot{e}_x) - J^T(\theta)k_4 s_x - k_3 s + \ddot{\theta}_r] + \widehat{U}(t) + \varepsilon(t) \tag{9.16}$$

with $k_1 \in \mathbb{R}^{6\times6}$, $k_2 \in \mathbb{R}^{6\times6}$, $k_4 \in \mathbb{R}^{6\times6}$, and $k_1 \in \mathbb{R}^{7\times7}$ being diagonal positive matrices. $\widehat{U}(t)$ is the estimation of the uncertain part. $\varepsilon(t)$ is the time delay error that is defined such that:

$$\varepsilon(t) = L\,|t - (t - t_d)| \tag{9.17}$$

where L is the Lipschitz constant; t_d is the sampling time. Let us start by the determination of the uncertain part $\widehat{U}(t)$. If we have a sampling time very small and Assumption 4 is verified, we can estimate the variation of the uncertain dynamics by conventional time delay control using Eq. (9.4):

$$\widehat{U}(t) \simeq U(t - t_d) = \tau(t - t_d) - M_0\ddot{\theta}(t - t_d) \tag{9.18}$$

where $\ddot{\theta}(t - t_d)$, the delayed acceleration input computed using the following approximation:

$$\ddot{\theta}(t - t_d) = \frac{\theta(t) - 2\theta(t - t_d) + \theta(t - 2t_d)}{t_d^2} \tag{9.19}$$

Let us define the term of uncertainties error using Lipschitz condition (Assumption 4) as follows:

$$\widetilde{U}(t) = U(t) - \widehat{U}(t) = U(t) - U(t - t_d) \leq L|t - (t - t_d)| \leq \varepsilon(t) \tag{9.20}$$

In what follows, we will choose appropriate L to always achieve $\widetilde{U}(t) = \varepsilon(t)$, by helping an approximation technique (Reyes-Sierra and Coello 2005). Consider the Lyapunov function candidate that stabilizes the overall system as follows:

$$V = \frac{1}{2}s^T s + \frac{1}{2}e_x^T(k_1 + \beta k_2)e_x \tag{9.21}$$

The derivative of the above Lyapunov function with respect to time is obtained as:

$$\dot{V} = s^T\dot{s} + e_x^T(k_1 + \beta k_2)\dot{e}_x \tag{9.22}$$

Substituting $\dot{s}$ from Eq. (9.15) and control input from Eq. (9.16) into Eq. (9.22), using Eqs. (9.17) and (9.20), we find:

$$\dot{V} = -s^T J^T(\theta)(k_1 e_x + k_2 \dot{e}_x) - s^T J^T(\theta) k_4 s_x - s^T k_3 s$$
$$+ s^T(-M_0^{-1}(U(t) - \widehat{U}(t)) + M_0^{-1}\varepsilon(t)) + e_x^T(k_1 + \beta k_2)\dot{e}_x \quad (9.23)$$

Substituting (9.12) into (9.23), we have:

$$\dot{V} \leq -s_x^T k_1 e_x - s_x^T k_2 \dot{e}_x - s_x^T k_4 s_x - s^T k_3 s + e_x^T(k_1 + \beta k_2)\dot{e}_x \quad (9.24)$$

We have from (9.5), (9.6) and (9.8):

$$s_x = \dot{e}_x + \beta e_x \quad (9.25)$$

Substituting Eq. (9.25) into Eq. (9.24), we find:

$$\begin{aligned}\dot{V} &\leq -(\dot{e}_x + \beta e_x)^T k_1 e_x - (\dot{e}_x + \beta e_x)^T k_2 \dot{e}_x - s_x^T k_4 s_x \\ &\quad - s^T k_3 s + e_x^T(k_1 + \beta k_2)\dot{e}_x \\ &\leq -\dot{e}_x^T k_1 e_x - \beta e_x^T k_1 e_x - \dot{e}_x^T k_2 \dot{e}_x - \beta e_x^T k_2 \dot{e}_x - s_x^T k_4 s_x \\ &\quad - s^T k_3 s + e_x^T k_1 \dot{e}_x + \beta e_x^T k_2 \dot{e}_x \end{aligned} \quad (9.26)$$

Finally, we obtain:

$$\dot{V} \leq -\beta e_x^T k_1 e_x - \dot{e}_x^T k_2 \dot{e}_x - s_x^T k_4 s_x - s^T k_3 s \quad (9.27)$$

It is clear from the $\dot{V}$, which implies that the exoskeleton system is stable.

9.3.1 Active Assistive Motion

This section provides a summary of the estimation method of the Desired Movement Intention (DMI) of the subject. Let us start with the definition of the desired trajectory in the active assistive motion. In this protocol, the desired trajectory is updated as follows (Ochoa-Luna et al. 2015):

$$x_d = x + \Delta x_d \quad (9.28)$$

where $x \in \mathbb{R}^6$ is the measured joint position, and $x_d \in \mathbb{R}^6$ is the estimated of DMI.This quantity of movement is estimated from the measured user's force. If $\Delta x_d \longrightarrow 0$, this means that the exoskeleton's wearer stops exercising forces on the force sensor, making the exoskeleton to decrease its motion and progressively. and whenever $x_d = x$, the exoskeleton rests in its most recent position, Δx_d. Through such a mode, the exoskeleton robot is enabled to track the user's movement intention, while the trajectory tracking adaptive control here guarantees the compensation of the uncertainties dynamics of the robot including the arm weight of the subject. Based on jacobian matrix, it is possible to write the Cartesian displacement (Δx_d) as follows (Craig 2005):

$$\Delta x_d = J(\theta)\Delta\theta_d \tag{9.29}$$

where $\theta_d \in \mathbb{R}^7$ is the joint angular displacement. To estimate the desired movement intention Δx_d from the user's force, we can use the following equation (Khan et al. 2016):

$$F_m = J(\eta_1)\Delta\theta_d \tag{9.30}$$

where $F \in \mathbb{R}^6$ is measured user's force is the measured user's force. To solve the Eq. (9.7), we used Damped least squares (DLS) or Levenberg-Marquardt stabilization (Wampler 1986). This approach was employed firstly for avoiding the singularity of inverse kinematics solution (Nakamura and Hanafusa 1986). Rather than merely obtaining the minimum vector $\Delta\theta_d$ that provides a best solution to Eq. (9.7), we determine the value of $\Delta\theta_d$ that minimizes the quantity:

$$\min_{\Delta\eta_d} ||J\Delta\theta_d - F_m||^2 + \lambda^2||\Delta\theta_d||^2 \tag{9.31}$$

where $0 < \lambda < 1$ is the damping factor and can be determined to be positive depending upon the accurate estimation specifications. Since the sum of Eq. (9.8) can be written as:

$$\left\| \begin{pmatrix} J \\ \lambda I \end{pmatrix} - \begin{pmatrix} F \\ 0 \end{pmatrix} \right\| \tag{9.32}$$

Consequently, the DLS solution is:

$$\Delta\theta_d = J^T(JJ^T + \lambda^2 I)^{-1}F \tag{9.33}$$

It is sufficient to substitute $\Delta\theta_d$ in Eq. (9.29) to obtain the estimated DMI.

9.4 Experimental and Comparative Study

9.4.1 Experiments Set-Up

Implementation was carried out on the ETS-MARSE system described below.The system consists of three processing units, the first is a PC where the top-level commands are sent to the robot using LabVIEW interface, i.e. the control scheme selection, joint or Cartesian space trajectory, gain adjustments, etc. This PC also receives the data after the robot task is executed to analyze its performance. The other two processing units are part of a National Instruments PXI platform. Firstly, a NI-PXI 8081 controller card with an Intel Core Duo processor; in this card, the main operating system of the robot and the top-level control scheme are executed. In our case, the proposed control as well as the estimation based on TDE approach, at a sampling time of 500 s. Finally, at input-output level, a NI PXI-7813R remote input-output card with an FPGA (Field programmable gate array) executes the low-level control; i.e. aPI current control loop (sampling time of 50 s) to maintain the current of the motors required by the main controller. Also, in this FPGA, the position feedback via Hall-sensors (joint position), and basic input-output tasks are executed.Force sensor feedback is important to accurately control the movement of the exoskeleton. A high linearity 6 axis force sensor [NANO17-R-1.8-M2-M1PCI,ATI industrial Automation] is so chosen to obtain accurate real-time force measurements.This sensor is mounted on the tip of the exoskeleton robot. The joint of the ETS-MARSE is powered by Brushless DC motors (Maxon EC-45, EC-90) combined with harmonic drives (gear ratio 120:1 for motor-1, motor-2, and motor-4 and gear ratio 100:1 for motor-3 and motors 5 and7) (Brahmi et al. 2018b; Rahman et al. 2013). The diagram of the architecture and overview of the ETS-MARSEsystem with a human subject is shown in Fig. 9.3. The control gains are chosen manually as follows: $k_1 = 18I_6, k_2 = 80I_6, k_4 = 5.7I_6, k_3 = 20I_7, L = 0.6$, $\beta = 1.5$ and $\lambda = 0.6$.

An experimental session was created to show the effectiveness of the designed control system. The physical therapy tasks are performed by three different healthy subjects (average age: $27 \pm 4{:}6$ years; average height: $170 \pm 8{:}75$ cm; average weight: 75 ± 18 kg). In the session, the subject was comfortably seated relax in a chair as we show in Fig. 9.3. Experimental session was separated into three scenarios. In the two scenarios, the subject-1 (age: 28 years; height: 177 cm; weight: 83 kg) performed the designed task as rectangular form (Initial position $\longrightarrow$ Target-

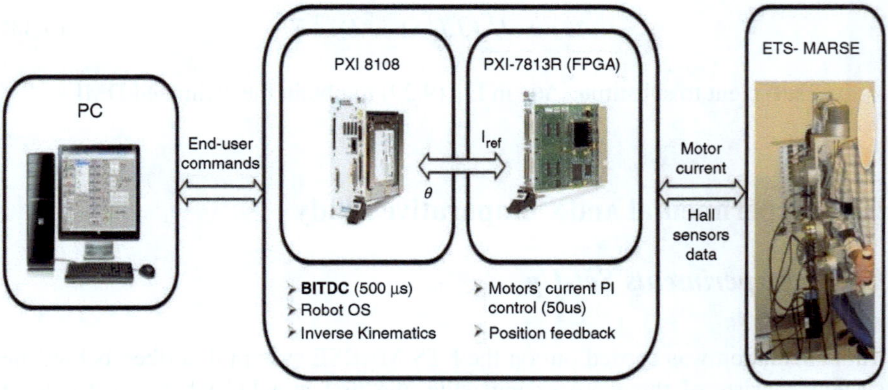

Fig. 9.3 The diagram of the experiment architecture and overview of the ETS-MARSEsystem with a human subject

A $\longrightarrow$ Target-B $\longrightarrow$ Initial position). The initial position of the exoskeleton robot is given where elbow joint position is at 90°. This part is followed directly by a comparison study through evaluating the proposed controller with another subject (age: 38 years; height: 173 cm; weight: 94 kg) which permits to influence the exoskeleton robot by different kinds of disturbances. Where, the external disturbances here are represented by different physiological conditions of subjects, such as non-linear biomechanical characteristics of the musculoskeletal system and the different payload of the upper-limb for each subject. In the last scenario, we evaluated the proposed control to achieve the active assistive motion performed by the subject-3 (age: 30 years; height: 178 cm; weight: 76 kg).

9.4.2 Results of Passive Rehabilitation Motion

The experimental results with the ETS-MARSE robot in Cartesian space conducted by Subject-1: (age: 28 years; height: 177 cm; weight: 83 kg) using the proposed control strategy are shown in Figs. 9.4, 9.5 and 9.6. From Fig. 9.4, we notice that the reference trajectory (red line) nearly interlocked with the measured trajectory (black line). In this case, we can say that these results are reasonably good. Figure 9.5 exhibits the Cartesian tracking errors as functions of time, where they are clearly converging and smaller along the reference trajectory. Figure 9.6 displays that the control input is bounded without any remarkable chattering.

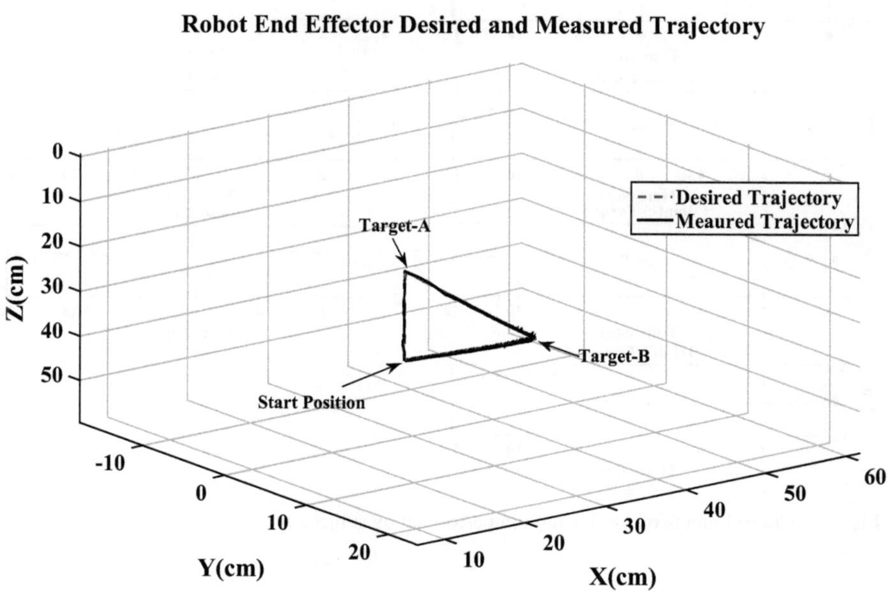

Fig. 9.4 Workspace performance of ETS-MARSE on 3D Cartesian space performed by subject-1: (age: 28 years; height: 177 cm; weight: 83 kg)

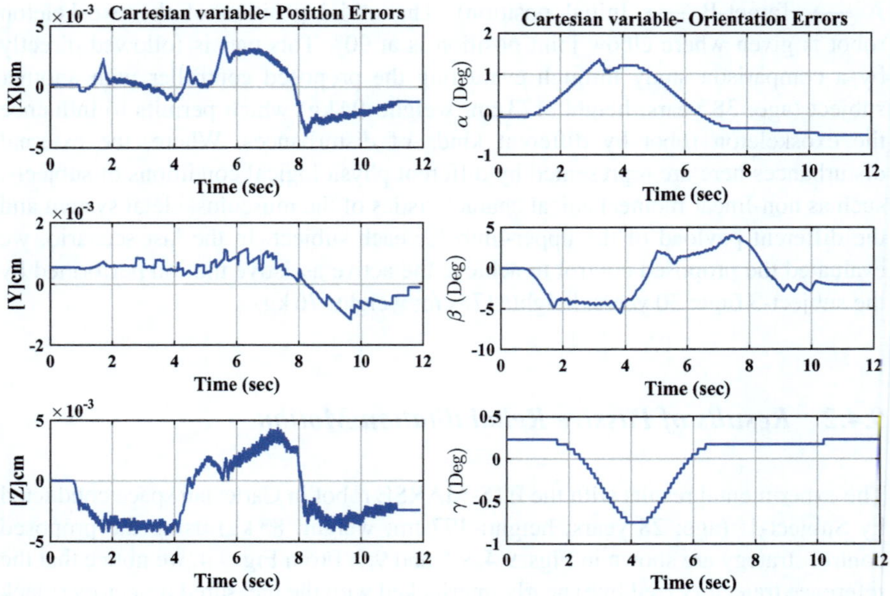

Fig. 9.5 Cartesian Error tracking of passive motion performed by subject-1

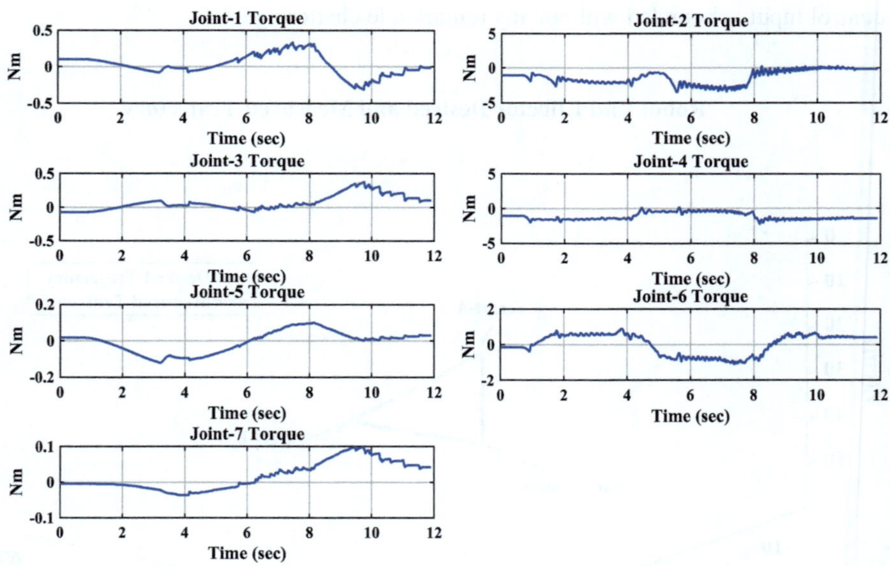

Fig. 9.6 Control inputs of passive motion performed by subject-1

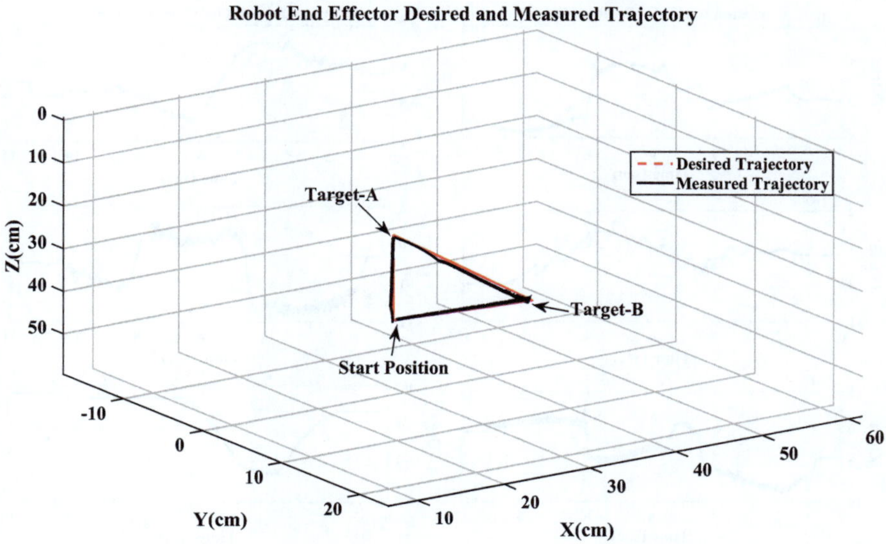

Fig. 9.7 Workspace performance of ETS-MARSE on 3D Cartesian space performed by subject-2: (age: 38 years; height: 173 cm; weight: 94 kg)

In order to show the efficiency and feasibility of the proposed scheme, we evaluated the proposed controller by using the exoskeleton with another subject (age: 38 years; height: 173 cm; weight: 94 kg), which means different payload of the upper limb of the human.

Figures 9.7, 9.8 and 9.9 present the Cartesian tracking (red is the reference trajectory, black is real trajectory) performed by Subject-2: (age: 28 years; height: 177 cm; weight: 83 kg) using the same gain's values that employed in the first trial. In fact, we remark from figures (Figs. 9.7, 9.8 and 9.9) that the proposed control maintains a good performance. It is clear from Fig. 9.8 that the Cartesian error is getting smaller with time. Figure 9.9 illustrates a smooth and bounded control input. From the comparison of the two experimental results, we can conclude that the proposed control algorithm strategy keeps providing a high level of precision and robustness to the nonlinear uncertain dynamics and unknown. So,these results confirm that the proposed control algorithm is proper to accomplish the desired passive rehabilitation performance even than the nonlinear dynamics of the exoskeleton robot are uncertain and the presence of external disturbances.

9.4.3 Results of Active Rehabilitation Motion

In this section, the subject exercises a free motion in 3D space. The objective of this section is to evaluate the active rehabilitation motion and to show the effectiveness of

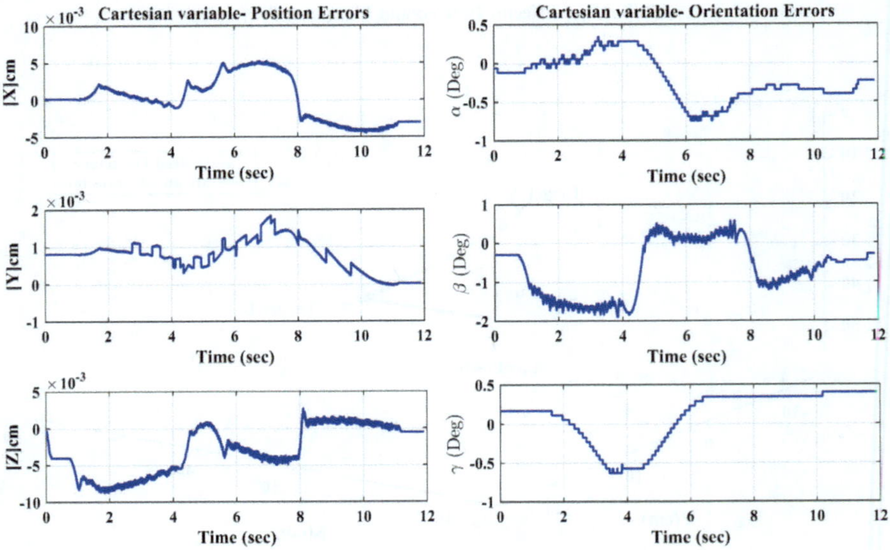

Fig. 9.8 Cartesian Error tracking of passive motion performed by subject-2

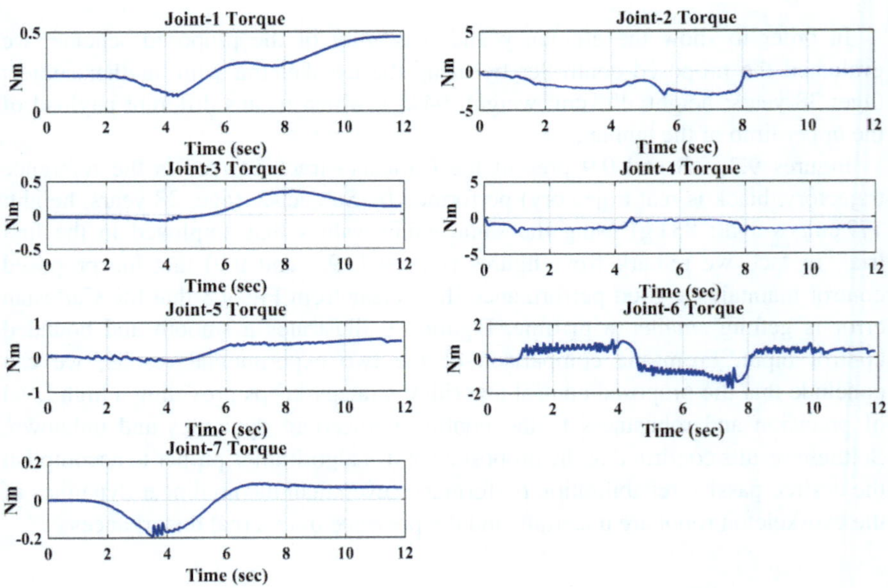

Fig. 9.9 Control inputs of passive motion performed by subject-2

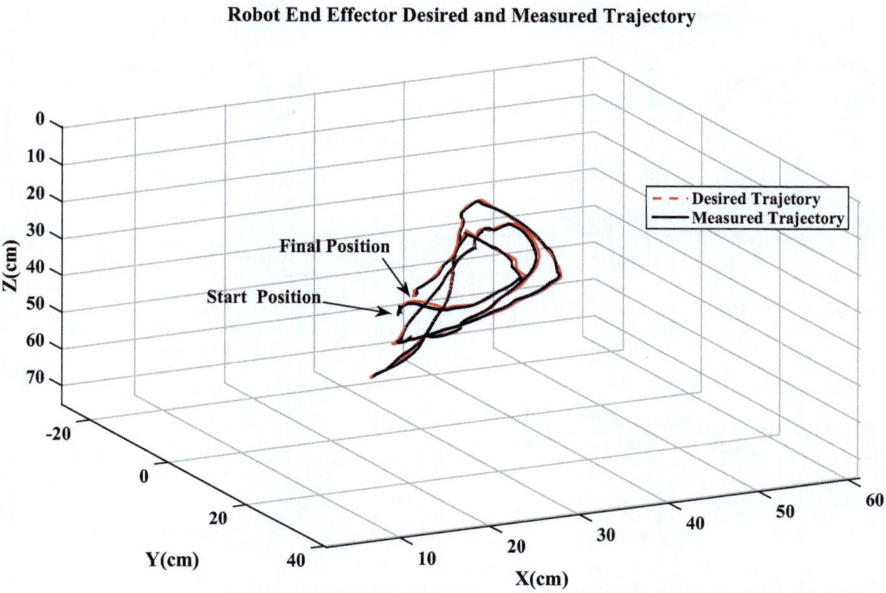

Fig. 9.10 Workspace performance of ETS-MARSE on 3D Cartesian space performed by subject-3(age: 30 years; height: 178 cm; weight: 76 kg)

the estimation DMI of the subject. In this case, the exoskeleton robot is completely passive and the subject is completely active. The Damped Least Square (DLS) algorithm is used in order to provide the estimation of the desired movement intention (DMI). Figures 9.10, 9.11 and 9.12 present the results of this experiment.

Figure 9.10 presents the performance of the subject-3 (age: 30 years; height: 178 cm; weight: 76 kg) with helping of ETS-MARSE exoskeleton robot (red line is desired and the blue line is the measured trajectory) using Damped Least Square (DLS) algorithm. It is clear from these plots (Figs. 9.10, 9.11 and 9.12) that the control strategy achieved the desired performance with small tracking errors and acceptable control input. From the good performance of the exoskleton-subject-3 (Fig. 9.11), we can infer that the Damped Least Square (DLS) algorithm was qualified to estimate accurately the desired movement intention Δx_d. It is easy to conclude from these plots (Figs. 9.10, 9.11 and 9.12) two points. The first one is that the proposed control approach achieved the desired active motion performance with high characteristics (small tracking errors and acceptable control input). The second one is the algorithm of DLS permits to estimate the DMI accurately.

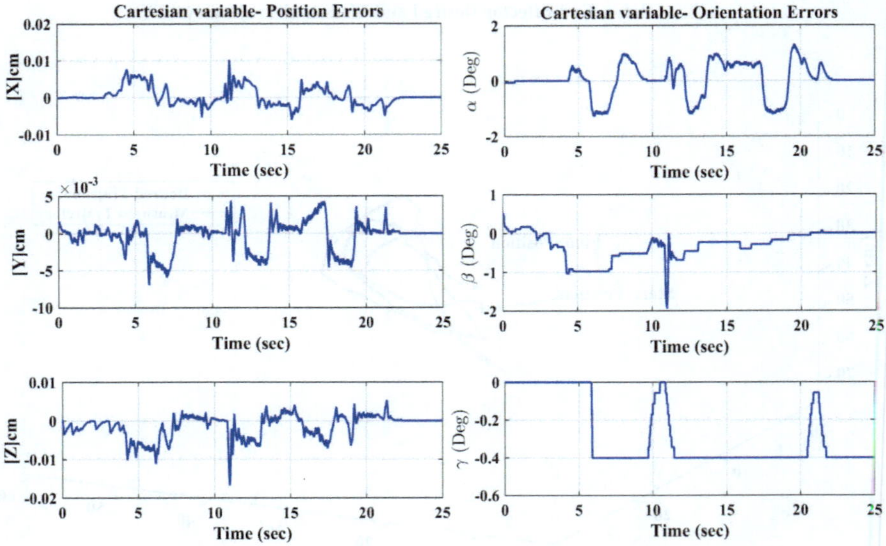

Fig. 9.11 Cartesian Error tracking of active motion performed by subject-3

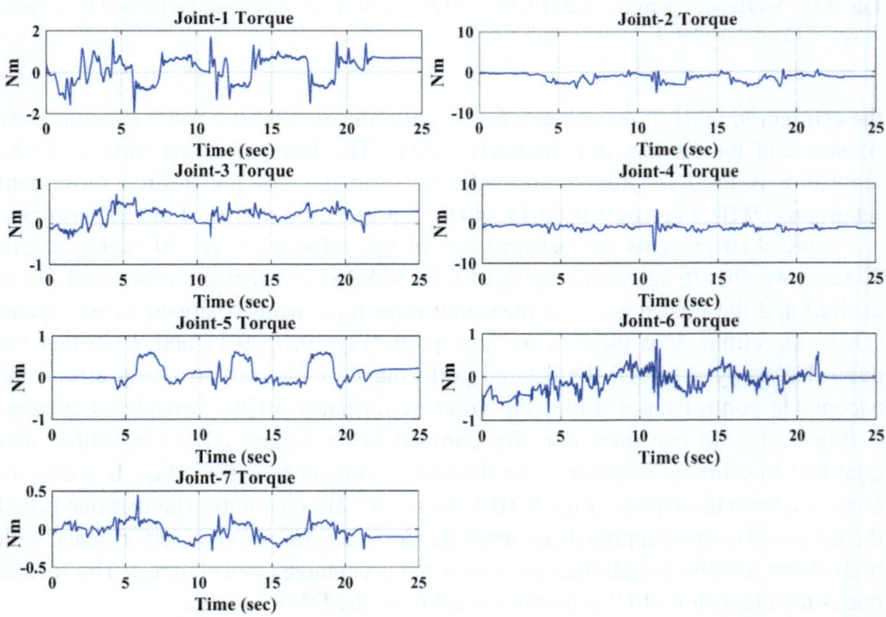

Fig. 9.12 Control inputs of active motion performed by subject-3

9.5 Conclusions

In this chapter, we have proposed a Cartesian adaptive control strategy based on time delay estimation and a robust proportional sliding mode controller which permits to a redundant exoskeleton robot dealing with nonlinear uncertain dynamics and to perform a passive and active rehabilitation treatment. The key advantage of the designed controller is that precise information about the dynamic of ETS-MARSE robot is not required. The robustness and the highest precision of the controller have been achieved along a desired therapeutic task. The stability has been ensured for the ETS-MARSE exoskeleton robot in closed loop, based on a Lyapunov theorem. The experimental results have confirmed the efficiency and feasibility of the proposed algorithm. In future work, we will seek to overcome the limitations of this approach. In particular, the value of the delayed acceleration for the controller; where the estimation of this variable may deteriorate its accuracy. We are also leaning to use this controller with a more complex scenario.

References

Baek, J., Jin, M., & Han, S. (2016). A new adaptive sliding-mode control scheme for application to robot manipulators. *IEEE Transactions on Industrial Electronics, 63*(6), 3628–3637.

Brahim, B., Maarouf, S., Luna, C. O., Abdelkrim, B., & Rahman, M. (2016a). Adaptive iterative observer based on integral backstepping control for upper extremity exoskelton robot. In *8th International Conference on Modelling, Identification and Control (ICMIC)* (pp. 886–891). Piscataway: IEEE.

Brahim, B., Rahman, M. H., Saad, M., & Luna, C. O. (2016b). Iterative estimator-based nonlinear backstepping control of a robotic exoskeleton. *World Academy of Science, Engineering and Technology, International Journal of Mechanical, Aerospace, Industrial, Mechatronic and Manufacturing Engineering, 10*(8), 1313–1319.

Brahmi, B., Saad, M., Lam, J. T. A. T., Luna, C. O., Archambault, P. S., & Rahman, M. H. (2018a). Adaptive control of a 7-DOF exoskeleton robot with uncertainties on kinematics and dynamics. *European Journal of Control, 42*, 77–87.

Brahmi, B., Saad, M., Luna, C. O., Rahman, M. H., & Brahmi, A. (2018b). Adaptive tracking control of an exoskeleton robot with uncertain dynamics based on estimated time delay control. *IEEE/ASME Transactions on Mechatronics, 23*, 575–585.

Chen, P., Chen, C.-W., & Chiang, W.-L. (2009). GA-based modified adaptive fuzzy sliding mode controller for nonlinear systems. *Expert Systems with Applications, 36*(3), 5872–5879.

Craig, J. J. (2005). *Introduction to robotics: Mechanics and control*. Upper Saddle River: Pearson/Prentice Hall.

De Morand, A. (2014). Pratique de la rééducation neurologique. Issy-les-Moulineaux: Elsevier Masson.

Fridman, L. (1999). The problem of chattering: An averaging approach. In *Variable structure systems, sliding mode and nonlinear control* (pp. 363–386). London: Springer.

Khalil, H. K., & Grizzle, J. (1996). *Nonlinear systems*. New Jersey: Prentice Hall.

Khan, A. M., Yun, D.-W., Ali, M. A., Zuhaib, K. M., Yuan, C., & Iqbal, J. (2016). Passivity based adaptive control for upper extremity assist exoskeleton. *International Journal of Control, Automation and Systems, 14*(1), 291–300.

Lewis, F. L., Dawson, D. M., & Abdallah, C. T. (2003). *Robot manipulator control: Theory and practice*. Boca Raton: CRC Press.
Nakamura, Y., & Hanafusa, H. (1986). Inverse kinematic solutions with singularity robustness for robot manipulator control. *Journal of Dynamic Systems, Measurement, and Control, 108*(3), 163–171.
Ochoa-Luna, C., Habibur Rahman, M., Saad, M., Archambault, P. S., & Bruce Ferrer, S. (2015) Admittance-based upper limb robotic active and active-assistive movements. *International Journal of Advanced Robotic Systems, 12*(9), 117.
Park, C.-W., & Cho, Y.-W. (2007). Robust fuzzy feedback linearisation controllers for Takagi-Sugeno fuzzy models with parametric uncertainties. *IET Control Theory & Applications, 1*(5), 1242–1254.
Rahman, M. H., Saad, M., Kenné, J.-P., & Archambault, P. S. (2013). Control of an exoskeleton robot arm with sliding mode exponential reaching law. *International Journal of Control, Automation and Systems, 11*(1), 92–104.
Rahman, M. H., Rahman, M. J., Cristobal, O., Saad, M., Kenné, J.-P., & Archambault, P. S. (2015). Development of a whole arm wearable robotic exoskeleton for rehabilitation and to assist upper limb movements. *Robotica, 33*(1), 19–39.
Reyes-Sierra, M., & Coello, C. A. C. (2005). A study of fitness inheritance and approximation techniques for multi-objective particle swarm optimization. In *IEEE Congress on Evolutionary Computation*, Edinburgh (pp. 65–72).
Shieh, H.-J., & Hsu, C.-H. (2008). An adaptive approximator-based backstepping control approach for piezoactuator-driven stages. *IEEE Transactions on Industrial Electronics, 55*(4), 1729–1738.
Slotine, J.-J. E., & Li, W. (1991). *Applied nonlinear control.* Englewood Cliffs: Prentice-Hall.
Wampler, C. W. (1986). Manipulator inverse kinematic solutions based on vector formulations and damped least-squares methods. *IEEE Transactions on Systems, Man, and Cybernetics, 16*(1), 93–101.
Wang, L., Chai, T., & Zhai, L. (2009). Neural-network-based terminal sliding-mode control of robotic manipulators including actuator dynamics. *IEEE Transactions on Industrial Electronics, 56*(9), 3296–3304.
Xie, S. (2016). *Advanced robotics for medical rehabilitation*. Cham: Springer.
Yim, J., & Park, J. H. (1999). Nonlinear H ∞ control of robotic manipulator. In *Proceedings of IEEE International Conference on Systems, Man, and Cybernetics, SMC'99. Proceedings 1999* (pp. 866–871). Piscataway: IEEE.
Youcef-Toumi, K., & Ito, O. (1990). A time delay controller for systems with unknown dynamics. *Journal of Dynamic Systems, Measurement, and Control, 112*(1), 133–142.
Young, K. D., Utkin, V. I., & Ozguner, U. (1999). A control engineer's guide to sliding mode control. *IEEE Transactions on Control Systems Technology, 7*(3), 328–342.

Chapter 10
Canonical Particle Swarm Optimization Algorithm Based a Hybrid Vehicle

Mohamed Elhedi Hmidi, Ines Ben Salem, and Lilia El Amraoui

Abstract This paper deals with the modeling and optimization of a PID 2 DOF controller design for a Hybrid vehicle. Such a Particle Swarm Optimization (PSO) based PID 2Dof controller is investigated in order to stabilize both the velocity of studied vehicle. The aim goal of this paper is to improve the effectiveness of synthesized control using the strategy of a canonical PSO optimization to tune its weighting matrices instead to configure it by a trials-errors method. This work reminds firstly to describe all aerodynamic forces and moments of the hybrid vehicle within an inertial frame and a dynamical model is obtained thanks to the Lagrange formalism. A 2Dof PID controller is then designed for the Velocity stabilization of the studied vehicle. Several PSO updating strategies are proposed to enhance the stability and the rapidity of our studied system through minimizing a definite cost function of' controller's weighting matrices. The obtained results are carried out in order to show the effectiveness and robustness of the different PSO updating strategies based the 2Dof PID Controller.

Keywords Hybrid vehicle · Modeling · PID two degree of freedom control · Driving cycle · Particle swarm optimization · PSO velocity stabilization

10.1 Introduction

In recent years, research and developments related to the vehicle have attracted much attention. Improvements in the autonomy, flexibility, and adaptability of these vehicles have been proposed during the thermal phase (Mensing and Trigui 2011). The hybrid vehicle is now one of the ways to reduce the energy consumption.

M. E. Hmidi (✉) · I. B. Salem · L. El Amraoui
Research Unit Signals and Mechatronics System, National Engineering School of Carthage (ENI-Carthage), University of Carthage, Carthage, Tunisia
e-mail: ines.benselem@enicarthage.rnu.tn; lilia.elamraoui@enicarthage.rnu.tn

N. Derbel et al. (eds.), *New Developments and Advances in Robot Control*, Studies in Systems, Decision and Control 175,
https://doi.org/10.1007/978-981-13-2212-9_10

This vehicle hybrid consists of Electric assembly and thermal with a tilting mechanism to toggle between these phase of driving.

This cycle of driving of the vehicle combines the advantages of electric and thermal vehicles allowed to experiment the best effects of the electric phase and to minimize the energy consumed in thermal phase.

This is very useful, compared to others vehicles classical, However, these structures design bring its own problems since the degradation in stability is usual-ly observed in the thermal phase of driving.

The history of hybrid vehicles (HV) dates back to the late nineteenth century (Trigui and Jeanneret 2003). However, because of their limited autonomy, the industry has further promoted the evolution and use of this new vehicle type, taking into ac-count the availability and price of oil. Following oil price inflation and taking into account environmental constraints, a great deal of research has been carried out to market a new generation of HVs equipped with cleaner sources of renewable energy. Different approaches have been studied to model the vehicles in order to increase their autonomy while minimizing the fuel consumption of the main source.

In recent years, these hybrid vehicles have seen a great evolution in terms of the optimization of energy consumption, the modeling and especially the driving control design (Romero et al. 2016). This explains the interest of researchers to study their driving dynamics and improve controller performances of this kind of hybrid vehicle thanks to different metaheuristic optimization algorithms. We find in the literature several works that focus to enhance controller's efficiencies thanks to tuning its decision parameters. Among these works (Amir Ghoreishi and Basiri 2011; Hamidi 2012) the authors presented an LQR controller designed that adjusted through its weighting matrices using both PSO and GA algorithms. An Intelligent Fuzzy Logic Control proposed by Zoric et al. (2014) based on PSO strategy to improve the robustness of parameters. The work of Prasad and Tyagi (2012) proposed a new algorithm (ABC) provided to adjust the weighting matrices of LQG controller for the antenna system.

Thus, a dynamic model of this type of hybrid vehicle is established thanks to the Lagrange formalism (Romero et al. 2016) We are interested in the dynamic modeling of the thermal phase based on the nonlinear model of this system studied obtained from a sub model to make the system easy to control. Control methods of these nonlinear systems have been received much attention and become one of the most important topics in this research fields, Many control schemes such as PID classical controller, Two-Degree-of-Freedom PID Controllers have been employed for the velocity control and stabilization of different structures of the driving vehicle.

For a conventional PID regulator structure, there are several versions, for example, the serial and parallel structure and other structures depending on the degree of freedom. The two main objectives of this controller are the follow-up of the Set-point and the rejection of the load disturbances. Using the 2DOF structure introduces other parameters that need to be defined precisely. The History of regulators is already long and it may be interesting to recall some important steps, the first centrifugal type regulators appeared around 1750 to regulate the speed of windmills, followed in 1788 by the famous speed controller of a steam engine Of James Watt.

The PID regulators with two degrees of freedom (2DOF) include a setpoint weighting on proportional terms and derivatives. A 2DOF PID controller is capable of quickly rejecting disturbances without increasing the overshoot to the set-point level. These controllers are also useful for attenuating the influence of variations of the reference signal on the control signal.

This control approach is a promising design technique that shows many applications in various engineering fields. Vilanova (2012) proposed a Model Reference Robust Tuning of PID Controllers. The PID 2DOF control strategy is also widely applied in the classical and modern vehicle field, due to its control methodology. It has been applied successfully to the regulation of disturbance rejections, it also remains at a given set-point and the follow-up of the controls whose controlled variable is good in monitoring the desired value (Araki 2003). So, in this document, such a PID 2DOF control strategy is used for controlling the velocity of the thermal phase of a hybrid vehicle, as well as the complete modeling of the traction chain also presented.

The remainder of this article is organized as follows. Section 10.2 presents the nonlinear mathematical model of the studied driving phase based on the Lagrange formalism. In Sect. 10.3, we present the formulation of the design problem of the 2DOF PID controller and provide the calculated control laws. Section 10.4 describes the demonstrative simulations that are performed to validate the proposed control approach. Finally, the conclusions are drawn from Sect. 10.5.

10.2 Dynamic System Model

10.2.1 Vehicle Model

The model of the thermal phase of the hybrid vehicle is a rigid body, in this model Let us consider an Internal Combustion Engine powered Vehicle (ICEV) and let us make the hypothesis is that it is possible to measure the speed $v(t)$ of the vehicle during a trip.

Using the fundamental principle of dynamics that is written:

$$M\vec{a} = \sum \vec{F} \tag{10.1}$$

where $\vec{a}$ is the vehicle acceleration, $\vec{F}$ the sum of streaks forces on the vehicle. We notice $\vec{F_x}$ the tractive force abandoned the wheels supplied by the motor and $\vec{F_{ext}}$ all external forces. So the main resistance forces are the aerodynamic force $\vec{F_A}$, rolling resistance $\vec{F_R}$ and the force of gravity $\vec{F_G}$. In Fig. 10.1, they are shown by their standards denoted respectively F_A, F_R and F_G and their directions shown by the

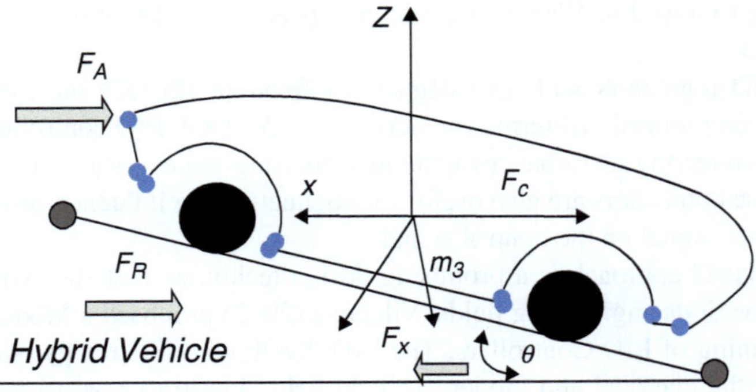

Fig. 10.1 Schematic of the system architecture

arrows. In the case of a rise with an angleθ, all those forces opposed to the direction of movement of the vehicle. The tractive force $\overrightarrow{F_x}$ is itself in the direction of path.

And each can be expressed as follows.

Aerodynamic Force is calculated as:

$$F_A = \frac{1}{2}\rho S C_x v_r^2 \tag{10.2}$$

Rolling Resistance force is formulates using:

$$F_R = M_g C_R \tag{10.3}$$

The gravity force using:

$$F_G = Mg\sin\theta \tag{10.4}$$

with S, the frontal surface C_x, the aerodynamic drag coefficient C_R, the rolling resistance coefficient g, the gravity acceleration and θ, the road slope angle.

10.2.2 Force of the Tires Provided by the Engine

The engine is directly connected to the transmission and to the wheels. It participates parallel to the movement of the vehicle, and transmits a mechanical power to the wheels.

The conventional vehicle is only equipped with a conventional combustion engine running on gasoline. The mechanical braking device is loading slow the vehicle during braking. The torque engine T_e at the output of the motor shaft is delivered to the input of the gearbox via the clutch. Considering I_e the engine rotational inertia, w_e the acceleration angular of engine, so the torque to the output of the clutch is:

$$T_c = T_e - I_e \underset{e}{\dot{w}} \tag{10.5}$$

At the output of the gearbox, the torque engine at the input is amplified by the gear ratio, then we noted:

$$\mathrm{T_d} = (\mathrm{T_c} - I_\mathrm{t} \underset{\mathrm{e}}{\dot{\omega}})\mathrm{N_t} \tag{10.6}$$

where N_t the value of the transmission ratio of the box and I_t the rotational inertia of the transmission, at the same principle, the torque T_a is needed to turn the wheels and providing the transmission of force:

$$\mathrm{T_a} = (\mathrm{T_d} - I_\mathrm{d} \underset{d}{\dot{\omega}})N_f \tag{10.7}$$

where I_d is the rotational inertia of the drive shaft, and N_f the differential gear ratio.
Finally combining all the above equations, the torque supplied by the engine through the transmission and given by:

$$\mathrm{T_a} = \big([\mathrm{T_e} - (I_\mathrm{e} + I_\mathrm{t}) \underset{\mathrm{e}}{\dot{\omega}}]N_t - I_d \omega_d\big)N_f \tag{10.8}$$

In the existence of the braking torque, the actual torque T_x created by this equation:

$$\mathrm{T_X} = rF_x + I_\omega \underset{\omega}{\dot{w}} = T_a - T_b \tag{10.9}$$

We conclude, the tractive force to the wheels delivered by the combustion engine can be written as:

$$F_x = \frac{1}{r}\left\{\eta T_e N - \{I_{et}\ N^2 + I_d\ N_f^2\right\} \underset{\omega}{\dot{w}} - T_b\Big\} \tag{10.10}$$

where: $I_{et} = I_e + I_t$, $N^2 = N_t^2 N_f^2$ and $N = N_t N_f$.

10.2.3 Equation of Movement

The tractive force is knowing F_x provided by the engine and the different forces opposing the motion of the vehicle, using the fundamental principle of dynamics, so we get the equation of motion of the vehicle approximated by the equation:

$$M\dot{v} = F_x - F_R - F_A - F_G \tag{10.11}$$

with the absence of longitudinal sliding, the angular velocity of the wheel, is given by the following equation:

$$\dot{v} = r\,\underset{\omega}{\dot{w}} \tag{10.12}$$

where r is the wheel radius, and W_w is the angular velocity of the wheel.

In conclude therefore the final equation of motion of the vehicle:

$$M\,\dot{v} = \frac{\eta T_e N}{r} - \left\{ I_{et} N^2 + I_d N_f^2 + I_\omega \right\} \frac{\dot{v}}{r^2} - H \tag{10.13}$$

So we note:

$$\left(M + \frac{I_{et} N^2 + I_d N_f^2 + I_\omega}{r^2} \right) \dot{v} = \frac{\eta T_e N - T_b}{r} - H \tag{10.14}$$

Finally we find:

$$\dot{v} = \frac{r}{Mr^2 + I_{et} N^2 + I_{d\omega}} \times (\eta T_e N - T_b - rH) \tag{10.15}$$

The longitudinal modeling of the vehicle and therefore described by the follow-ing system of equations:

$$\begin{cases} \dot{x} = v_x \\ \dot{v_x} = \dfrac{r}{Mr^2 + I_{et} N^2 + I_{d\omega}} \times (\eta T_e N - T_b - rH) \end{cases} \tag{10.16}$$

10.3 Control Design Strategies

10.3.1 PID Control

The field of hybrid vehicle control the main objective of all controllers is the reduction of error in the majority of cases. In this case, the PID controller is one of the solutions, because of its simplicity and its best performance in the majority of cases. Optimizing the PID controller is the main task for better performance. The PID control system is characterized by its three correction terms, proportional, integral and derivative (Haugen 2012; Taguchi 2000), as shown by the PID structure of Fig. 10.2.

Defining $u(t)$ as the controller output of the PID, is expressed by this following equation:

$$u(t) = Kp.e(t) + Ki \int e(t) + K_d \frac{d}{dt} e(t) \tag{10.17}$$

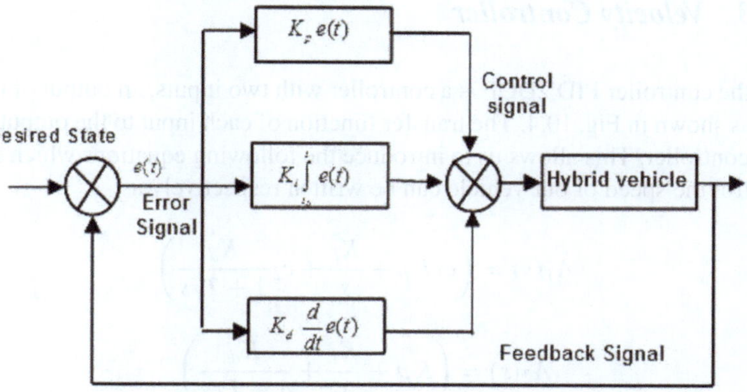

Fig. 10.2 PID controller structure of hybrid vehicle

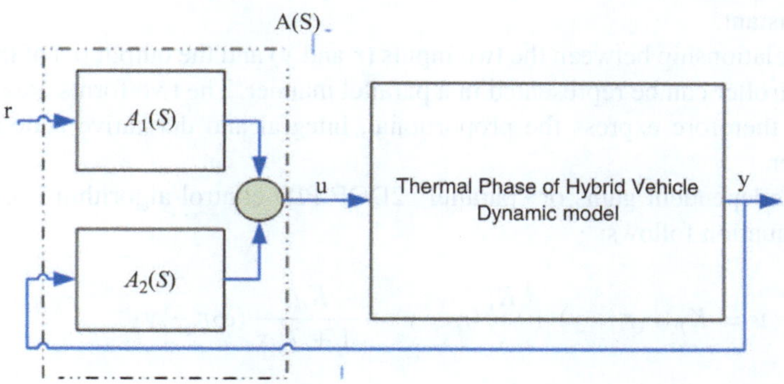

Fig. 10.3 2DOF PID controller structure of hybrid vehicle

where K_p, K_i,K_d are respectively the controller gain, the integral time constant and the derivative time

10.3.2 PID2-DOF Control Problem Formulation

Since the hybrid vehicle model thermal phase is strongly coupled and non-linear, the proposed PID 2DOF control system is a multi-objective problem. A general form of the 2DOF PID controller is shown in Fig. 10.3, where the controller consists of two compensators $A_1(s)$ and $A_2(s)$.

The PID 2DOF control laws design for the Thermal Phase model is achieved according to the procedure proposed by Kumar and Vandana (2015).

10.3.3 Velocity Controller

Since the controller PID 2DOF is a controller with two inputs, an output of the form $A(s)$, as shown in Fig. 10.4. The transfer function of each input to the output is itself a PID controller. This allows us to introduce the following equations which are used to control the speed of our vehicle can be written respectively as:

$$A_1(s) = \left(c_1 K_p + \frac{K_i}{s} + c_2 \frac{K_d}{1 + T_f s}\right) \tag{10.18}$$

$$A_2(s) = \left(Kp + \frac{K_i}{s} + \frac{K_d}{1 + T_f s}\right) \tag{10.19}$$

where: K_p, K_i,K_d are respectively the controller gain, the integral time constant and the derivative time, constant, c_1 and c_2 the set-point weights, and T_f the derivative filter constant.

The relationship between the two inputs (r and y) and the output (u) of the 2DOF PID controller can be represented in a parallel manner. The two forms are different, one can therefore express the proportional, integral and derivative actions of the controller.

The Independent gains or "parallel" 2DOF PID control algorithm is expressed by the equation follows:

$$u = K_p(c_1 r - y) + \frac{K_i}{s}(r - y) + \frac{K_d s}{1 + T_f s}(c_2 r - y) \tag{10.20}$$

where, u is the Manipulated variable, y is the controlled variable or output, and r is the set-point.

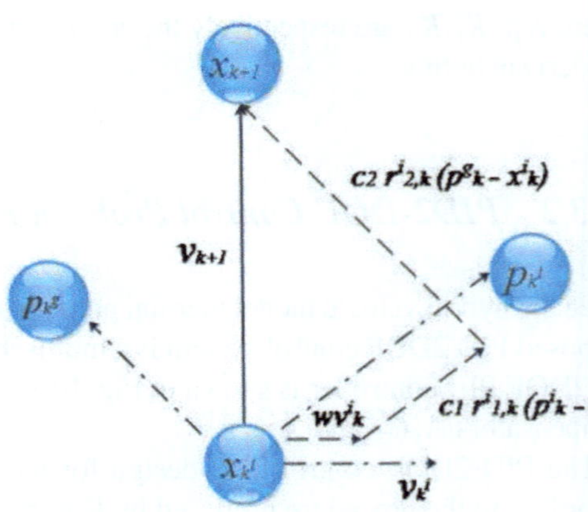

Fig. 10.4 Particle's position and velocity update

10.4 Proposed Particle Swarm Optimization Algorithms

10.4.1 Basic Concept

The Particle Swarm Optimization (PSO) metaheuristic is a swarm intelligence Algorithm based computation algorithm originally proposed by Eberhart and Shi (2012), This new stochastic and global optimization technique is inspired by the collaborative behaviour of biological populations and organisms such as the birds' flocks and fish's schools.

The information exchange between different population individuals allows solving various complex optimization problems in various engineering fields. These applications can be summarized around domains of robotics, image and signal processing, electronic circuits design, communication networks, but more especially the domain of plant control design.

10.4.2 Canonical PSO Algorithm

The PSO algorithm uses a swarm consisting of n_p particles, i.e., $x^1, x^2, \ldots, x^{n_p}$ randomly distributed in the considered initial search space, to find an optimal solution of a generic optimization problem (10.22). Each particle, that represents a potential solution, is characterised by a position and a ve-locity given by$x_k^i :=$ $\left(x_k^{i,1}, x_k^{i,2}, \ldots, x_k^{i,m}\right)^T$ and $v_k^i := \left(v_k^{i,1}, v_k^{i,2}, \ldots, v_k^{i,m}\right)^T$ where: $(i,k) \in [\![1, n_p]\!] \times [\![1, k_{\max}]\!]$.

At each algorithm iteration, the ith particle position, evolves based on the following update rules:

$$x_{k+1}^i = x_k^i + v_{k+1}^i \tag{10.21}$$

$$v_{k+1}^i = w v_k^i + c_1 r_{1,k}^i \left(p_k^i - x_k^i\right) + c_2 r_{2,k}^i \left(p_k^g - x_k^i\right) \tag{10.22}$$

where: w is the inertia factor, c_1 and c_2 are the cognitive and the social scaling factors respectively, $r_{1,k}^i$ and $r_{2,k}^i$ are random numbers uniformly distributed in the interval $[\![0, 1]\!]$,p_k^i is the best previously obtained position of the ith particle and p_k^g is the best obtained position in the entire swarm at the current iteration k. Hence, the principle of a particle displacement in the swarm is graphically shown in the Fig. 10.4, for a two dimensional design space.

In the PSO formalism, a particle can leave the search space initially defined during the motion of the swarm. Therefore, it should define constraints on the problem decision variables, given as follows:

$$x_{\min}^j \le x_k^{i,j} \le x_{\max}^j, \forall (i,j,k) \in [\![1, n_P]\!] \times [\![1, m]\!] \times [\![1, k_{\max}]\!] \tag{10.23}$$

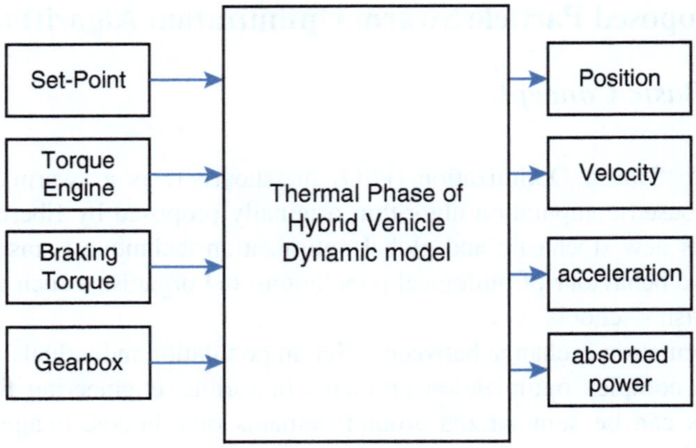

Fig. 10.5 Vehicle hybrid, thermal phase block

where i, j and k represent the indexes on the particle in the swarm, the components of the decision vector and the algorithm iterations, respectively (Fig. 10.5).

The following interval confinement mechanism is introduced on the decision variables when a particle leaves the range $\left[\!\left[x^j_{\min}, x^j_{\max}\right]\!\right]$ defined according to the optimization problem:

$$x_k^{i,j} = \min\left\{\max\left(x_k^{i,j} + v_k^{i,j}, x_{\min}^j\right), x_{\max}^j\right\} \tag{10.24}$$

In this case, Eq. (10.21) replaces the above PSO motion equation (10.24) and the new velocity of the corresponding particle is updated accordingly as follows:

$$\tilde{v}_k^{i,j} = -\mu v_k^{i,j}, \quad 0 \leq \mu < 1 \tag{10.25}$$

In order to improve the exploration and exploitation capacities of the proposed PSO algorithm, we choose for the inertia factor a linear evolution with respect to the algorithm iteration as given in Eberhart and Shi (2012):

$$w_{k+1} = w_{\max} - \left(\frac{w_{\max} - w_{\min}}{k_{\max}}\right) k \tag{10.26}$$

where $w_{\max} = 0.9$ and $w_{\min} = 0.4$ represent the maximum and minimum inertia factor values, respectively, $w_{\max}$ is the maximum iteration number Finally, a pseudo code of this swarm intelligence based algorithm is given for a minimization problem in Canonical PSO Algorithm:

- Initialize a population of n_{PART} particles having random positions and velocities on D dimensions of the search space
- At every iteration K and for each ith particle x_k^i evaluate the considered optimization fitness function on the D decision variables.

- Compare particle's fitness evaluation $f_k^i = f\left(x_k^i\right)$ with its $pbest_k^i = f\left(p_k^i\right)$. If $f_k^i \le pbest_k^i$ then $pbest_k^i = f_k^i$ and $p_k^i = x_k^i$
- Identify the particle in the neighborhood with the best success so far and as-sign its index to the variable g.
- Change the velocity and position of the particle according to the motion equations (10.24) and (10.25).

If the last criterion is met (satisfied fitness or a maximum number of iterations), the algorithm terminates with the solution, otherwise, go to Step 2:

$$x^* = \arg\min_{x_k^i} \left\{ f\left(x_k^i\right), \forall i, k \right\}$$

10.5 Obtained Results

The simulations of our dynamic model of the hybrid vehicle thermal phase are first carried out by a system involved in Simulink/Matlab. Our model of the hybrid vehicle thermal phase is detailed by the following block diagram (Fig. 10.6):

10.5.1 Simulation of the Model with the PID and the PID 2-DOF Controller

The control objectives of such a thermal phase of hybrid vehicle, whose physical parameters are given in Table 10.1, are the closed-loop stabilization of its velocity (Fig. 10.7)

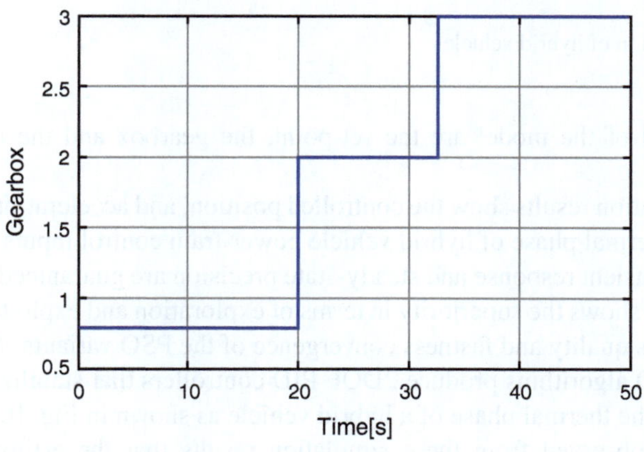

Fig. 10.6 The gearbox

Table 10.1 Control parameters of the PSO algorithms

Parameters	Values
Dimension of search space	$d = 15$
Population size	$n_p = 30$
Number of generations	$k_{max} = 100$
Cognitive coefficient	$c_1 = 0.5$
Social coefficient	$c_2 = 0.5$
Inertia weight	$w = 0.9$

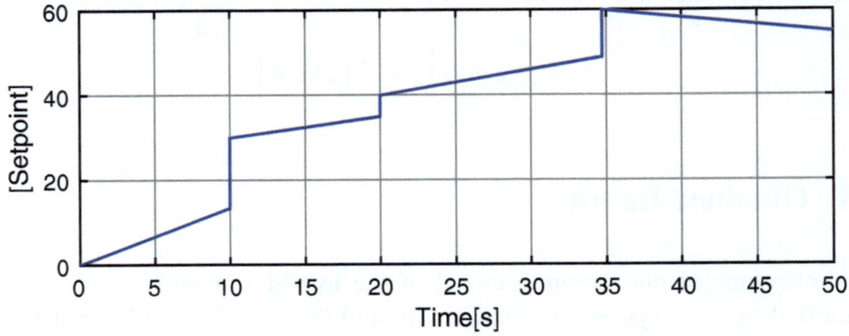

Fig. 10.7 The driving scenario

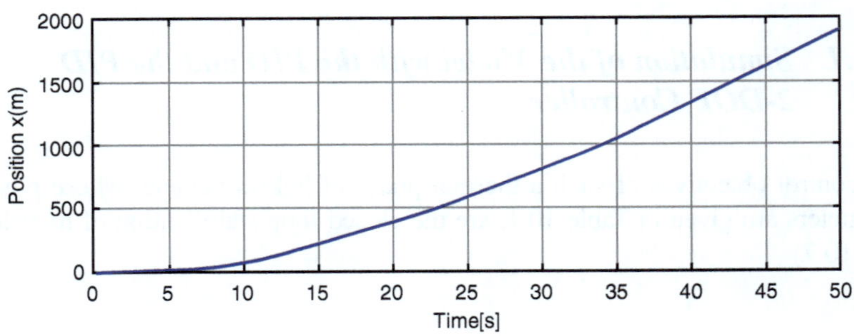

Fig. 10.8 Position of hybrid vehicle

The inputs of the model are the set-point, the gearbox and the torque engine (Fig. 10.8).

The simulation results show the controlled position, and acceleration dynamics as well as the thermal phase of hybrid vehicle power-train control inputs. High performances in transient response and steady-state precision are guaranteed (Table 10.2).

This result shows the superiority in terms of exploration and exploitation capabilities, solutions quality and fastness convergence of the PSO variants (Fig. 10.9). All proposed PSO algorithms produce 2DOF PID controllers that stabilize the velocity dynamics of the thermal phase of a hybrid vehicle as shown in Fig. 10.10.

It can be observed from these simulation results that the estimated and real system states are close and similar to the velocity dynamics. In addition, the system

Table 10.2 Model parameters of the thermal phase vehicle

Parameters	Description	Values
m	Mass of vehicle	1269 kg
r	Tire radius	0.269 m
I_{et}	Motor inertia	0.2630 m
I_d	Inertia drive	0.115 kg m^3
I_w	Inertia of wheel	2.8 kg m^3
SC_x	Aerodynamic	0.720
ρ	Air density	1.205
C_R	Rolling resistance	0.020 kg m^3
g	Acceleration of the gravity	9.81 m/s^2

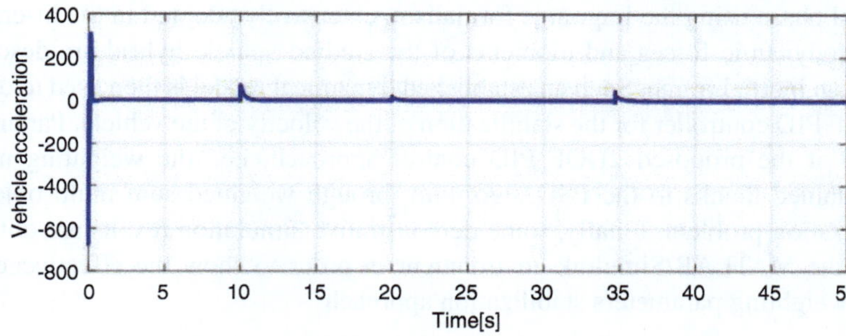

Fig. 10.9 Velocity of hybrid vehicle

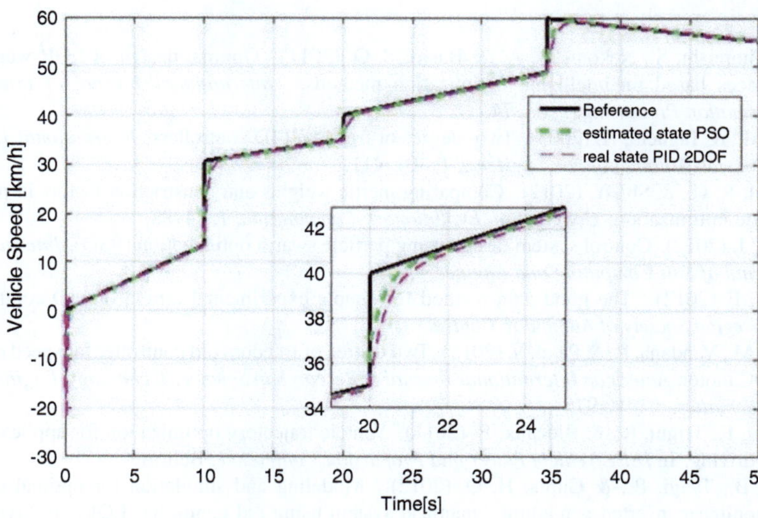

Fig. 10.10 System responses using PID 2DOF controller based PSO

became responds to its set points with a stable and rapid manner, due to the tuned weighting matrices through the different algorithms. On the other hand, the estimation errors are negligible and a good reconstitution of the system state is obtained. The stabilization objectives of the 2DOF PID controller are made with a satisfied tracking performance and the effectiveness of the proposed 2DOF PID control approach is guaranteed.

10.6 Conclusion

In this paper, we established a nonlinear dynamical model of a Hybrid vehicle in the thermal phase using the Lagrange formalism, extensively adopted in the lit-erature. All aerodynamic forces and moments of the studied vehicle hybrid are described within an inertial frame. Such an established dynamical model is then used to design a 2DOF PID controller for the stabilization of the velocity of the vehicle. Parameters design of the proposed 2DOF PID control approach, i.e., the weighting matrix are obtained thanks to the PSO algorithm through weighted sum multi-objective optimization problem. Finally, some demonstrative simulation results are obtained under the MATLAB/Simulink environment in order to show the effectiveness of tuned weighting parameters stabilization approach

References

Amir Ghoreishi, S., Nekoui, M. A., & Basiri, S. O. (2011). Optimal design of LQR weighting matrices based on intelligent optimization methods. *International Journal of Intelligent Information Processing, 2*, 63–74.

Araki, M., & Taguchi, H. (2003). Two- degree of freedom PID controllers. *International Journal of Control, Automation, and Systems, 1*, 401–411.

Eberhart, R. C., & Shi, Y. (2012). Comparing inertia weights and constriction factors in particle swarm optimization. *Congress on Evolutionary Computation, 1*, 84–88.

Hamidi, J. (2012). Control system design using particle swarm optimization (PSO). *International Journal of Soft Computing and Engineering, 1*.

Haugen, F. (2012). The good gain method for simple experimental tuning of PID controllers. *Norwegian Society of Automatic Control*.

Kumar, M., Vandana, P., & Patel, V. (2015). Two degree of freedom pid controller for speed control of DC motor. *American International Journal of Research in Science, Technology, Engineering, Mathematics, 39*, 94–97.

Mensing, F., Trigui, R., & Bideaux, E. (2011). Vehicle trajectory optimization for application in eco-driving. In *IEEE Vehicle Power and Propulsion Conference*, Belfort.

Prasad, B., Tyagi, B., & Gupta, H. O. (2012). Modeling and simulation for optimal control of nonlinear inverted pendulum dynamical system using pid controller, LQR. In *Sixth Asia Modeling Symposium*, Kuala Lumpur.

Romero, J. A., Lozano-Guzman, A. A., Betanzo-Quezada, E., & Arroyo Contreras, G. M. (2016). Cargo securement methods and vehicle braking performance. *International Journal of Vehicle Performance, 4*, 353–373.

Taguchi, H., & Araki, M. (2000). Two-degree-of-freedom pid controllers their functions and optimal tuning. *IFAC Digital Control, 33*, 5–7.

Trigui, R., Jeanneret, B., & Badin, F. (2003). Systemic modelling of hybrid vehicles in order to predict dynamic performance and energy consumption. *VEHLIB Library of Models*.

Vilanova, V. A. R. (2012). Conversion formulae and performance capabilities of two-degreeof-freedom PID control algorithms. In: *17th IEEE International Conference on Emerging Technlogies and Factory Automation (ETFA)*, Kraków (pp. 17–21).

Zoric, N. D., Simonovic, A. M., Mitrovic, Z. S., Stupar, S. N., Obradovic, A. M., & Lukic, N. S. (2014). Free vibration control of smart composite beams using particle swarm optimized self-tuning fuzzy logic controller. *Journal of Sound and Vibration, 333*, 5244–5268.

Chapter 11
Digital Stabilizing and Control for Two-Wheeled Robot

Bachir Nail, Abdellah Kouzou, and Ahmed Hafaifa

Abstract In this chapter a design control is proposed for a two-wheeled robot system based on a combined between the matrix fraction description (MFD) theory and digital PID controller, the model of the system is presented in nonlinear differential equations and by using the Taylor series expansion method, the dynamics model is linearized to be linear multivariable with two inputs (voltages) and two outputs (the rotational speed and tilt angle) model. after the check of the block controllability applied on the model, the transformation of the model into block controllable form have been done. In order to stabilized the behavior dynamics of robot model, a set of block poles are chosen diagonally were consists from stable and fast eigenvalues, are assigned by matrix gain K. To make the outputs of the robot system track its set references and to ensure a good controlling, two PID controllers are proposed, one for the rotational speed output and the second for the tilt angle, the parameters of the PID controllers (K_p, K_i and K_d), are tuning and selected based on Ziegler-Nichols method, through the simulations results performed illustrated the effectiveness of the proposed design controller and it is more suitable and robustness against the disturbances which were injected and shown that the robot can be confirmed as mobile platform for transporting human and goods.

Keywords Two-wheeled robot · Multivariable system · Matrix fraction description (MFD) · Block poles · PID controller · Ziegler-Nichols

Nomenclature

τ_m	Motor torque(Nm)
T_{mL},T_{mR}	Applied torque from motors to wheels

B. Nail (✉) · A. Kouzou · A. Hafaifa
Applied Automation and Industrial Diagnostics Laboratory, Faculty of Sciences and Technology, University of Djelfa, Djelfa, Algeria
e-mail: b.nail@univ-djelfa.dz; kouzouabdellah@ieee.org; a.hafaifa@univ-djelfa.dz

N. Derbel et al. (eds.), *New Developments and Advances in Robot Control*, Studies in Systems, Decision and Control 175,
https://doi.org/10.1007/978-981-13-2212-9_11

τ_a	Applied torque(Nm)
H_{fL},H_{fR}	Friction forces
R	Nominal resistance(Ω)
r	Radius of wheels
L	Rotor inductance(H)
θ	Angle of body offset from vertical
k_f	frictional constant$\left(\frac{Nms}{rad}\right)$
$\dot{\theta}$	Angular velocity
k_m	Torque constant$\left(\frac{Nm}{A}\right)$
$\ddot{\theta}$	Angular acceleration
k_e	Back emf coefficient$\left(\frac{Vs}{rad}\right)$
θ_w	Rotation angle of the wheels
V_a	Applied voltage (V)
w	Angular velocity of shaft
V_e	Back emf voltage (V)
α,$\dot{w}$	Angular acceleration of shaft
i	Armature current (A)
$\dot{\theta}_w$	Angular velocity of wheels
I_R	Rotor inertia (Kgm^2)
$\dot{x}$	Linear speed
M_w	Wheel mass
$\ddot{x}$	Linear acceleration
M_p	Robot's chassis mass
T_m	Torque of the driving wheels
I_w	Moment of inertia of the wheels
T_r	Torque of the reaction wheels
I_p	moment of inertia of the body
V_{Dw}	Input voltage of the driving wheel
H_L,H_R,P_L,P_L	Reaction forces
V_{Rw}	Input voltage of the reaction wheel

11.1 Introduction

There are various definitions for designating a mobile robot, but in general it can be presented as a machine capable of autonomously carrying out a displacement from one point to another in an environment not perfectly known a priori. To do this, the robot must have the functionality of perception, decision on the movement to be carried out, and realization of this movement.

Nowadays, mobile robotics is experiencing a real expansion with more and more attractive applications, of which we can cite as examples: planetary exploration,

exploration or intervention in environments hostile to humans (radioactive zone, polar, submarine depths, presence of fire…), or simply difficult to access, in the industry, applications mainly concern transport or distribution, whether in factories, mines, hospitals or workshops. An example is conveying in a often congested environment, which consists of storing or shipping spare parts, medicines, waste, or automatic soil cleaning. Another example is the automatic conveyance of vehicles used in road transport.

The design of automated vehicles, or mobile robots, especially with wheels is a rapidly expanding field of research. These robots are used in industry as a means of transport, inspection or operation and are particularly suited to interventions in hostile environments confined areas, high construction, underwater depths, etc.

Mobile robots with wheels are characterized by their simplicity of design. However, these systems have raised several difficult problems that have been the subject of several research studies. These are problems of control, planning, stability and trajectory tracking.

Many important papers and chapters published in recent years talk about the modeling, stability and tracking in the two wheeled robot among them we find: A transformable wheel-legged mobile robot: Design, analysis and experiment (TaoSun et al. 2017), Mechanical Design and Dynamic Modeling of a Two-Wheeled Inverted Pendulum Mobile Robot (Jingtao et al. 2007), Posture self-stabilizer of a biped robot based on training platform and reinforcement learning (Weiguo and Liyang 2017), modelling and control of two-wheeled robots (Ronald et al. 2013), A robust adaptive fuzzy variable structure tracking control for the wheeled mobile robot: Simulation and experimental results (Mauricio et al. 2017), MIMO Fuzzy Control for Autonomous Mobile Robot (Mac et al. 2016), Experimental Validation of an Under-actuated Two-Wheeled Mobile Robot (Oryschuk et al. 2009), Intelligent block spectral factors relocation in a quadrotor unmanned aerial vehicle (Bekhiti et al. 2017), Development of a Self-Balancing Human Transportation Vehicle for the Teaching of Feedback Control (Lin and Tsai 2009), Adaptive Neural Network Control of a Self-Balancing Two Wheeled Scooter (Tsai et al. 2010), Modeling and LQR Control of a Multi-DOF Two-wheeled Robot (Shigong et al. 2014), Wireless Controlled Two Wheel Balancing Robot (Yong and Foong 2011), Autonomous Dual Wheel Self Balancing Robot Based on Microcontroller (Adeel et al. 2013), Two-wheeled self balanced pendulum workspace improvement via under-actuated robust nonlinear control (Guilherme et al. 2015), Adaptive neural network control of a self-balancing two-wheeled scooter (Tsai et al. 2010), Motion control for a two-wheeled vehicle using a self-tuning PID controller (Ren et al. 2008), Maneuverability and path following control of wheeled mobile robot in the presence of wheel skidding and slipping (Low and Wang 2010), Equiangular navigation and guidance of a wheeled mobile robot based on range-only measurements (Teimoori and Savkin 2010), Bio-inspired algorithm for wheeled robot's navigation (Kherici and Ali 2011), Design and parametric control of co-axes driven two-wheeled balancing robot (Memarbashi and Chang 2011) and others researches, which are indicate the importance of this robot.

The current concern of robotics is to realize robots capable of reacting fast without impairing desired performance, stability and precision. One of the first solutions to the problem of control, in the history of robotics, consisted in using linear PID-type fixed-gain drives.

The advantage of such a method is its simplicity of implementation and low installation cost. It has given good performance because they were slow movements and actuators with a high reduction ratio. The effects of the load (robot arm with the mass of the load) on the motors are practically negligible, in comparison with the inertia of the engines, since they are divided by a high reduction ratio.

In this chapter a combination between the stability and the tracking are suggested to control the two-wheeled robot system, and the main contributions in this chapter is organized as the following steps:

The Sect. 11.1 an introduction, the Sect. 11.2 the problem statement the difficulties in the model behavior…etc., in the Sect. 11.3 presented the nonlinear dynamics model of the two-wheeled robot, and the linearized model based on their equilibrium point using the Taylor series expansion (Jingtao et al. 2007; Shigong et al. 2014), and the discretizing (digitalizing) the obtained continuous linearized model (Ogata 1994), the Sect. 11.4 checking the block controllability of the dynamic model (Chen 1984; Kailath and Li 1980; Nail et al. 2016b), and transforming the system to block controllable form (Chen 1984; Kailath and Li 1980; Nail et al. 2016b), in the Sect. 11.5 the stabilization by assigning block poles choosing diagonally, where calculating the feedback matrix gain K based on MFD (Bekhiti et al. 2017) using the Vandermonde matrix (Bekhiti et al. 2017), in Sect. 11.6. The tuning of the two digital PID controllers via Ziegler-Nichols method rules (Ogata 1994, 2010) in order to obtained best set-points tracking, and finally in Sect. 11.7 a conclusion summarized the advantages and the limitations of proposed controller.

11.2 Problem Statement

The problem arise in this work is to achieve and ensure best control to two-wheeled robot with minimal cost and with simple mathematical tools, by using the matrix fraction description theory which is used generally in the linear systems, on the other hand the robot model is a non-linear behavior, also the choice of the best block poles in which forms gives the best performances, also the interactions between the inputs/outputs of the multivariable model which are influenced negatively in the selection of the two PID controllers parameters.

11.3 Model of Two-Wheeled Self-Balancing Robot System

In this chapter we have used the final model of the two-wheeled inverted pendulum and the linear model for a DC motor, where the model behavior is multivariable unstable dynamics, with a very high non-linearity and coupling, for more detail see Jingtao et al. (2007) and Shigong et al. (2014).

The nonlinear dynamics model for a two wheeled inverted pendulum and the body is:

$$\begin{cases} \ddot{\theta} = \dfrac{\alpha\beta g \sin\theta - \left(\beta + \dfrac{\alpha}{r}\cos\theta\right) T_m - \beta T_r - \alpha^2\dot{\theta}^2 \sin\theta\cos\theta}{\beta\gamma - \alpha^2\cos^2\theta} \\ \ddot{x} = \dfrac{\dot{\theta}^2\alpha\gamma\sin\theta - \alpha^2 g\sin\theta\cos\theta + T_m\left(\alpha\cos\theta + \dfrac{\gamma}{r}\right) + T_r\alpha\cos\theta}{\beta\gamma - \alpha^2\cos^2\theta} \end{cases} \tag{11.1}$$

where: $\alpha = M_p\, l,\ \ \beta = 2M_w + \dfrac{2I_w}{r^2} + M_p,\ \ \gamma = I_p + M_p\, l^2$

11.3.1 Linearization of the Plant

The system plant has to be linearized about the equilibrium point, which in the case of inverted pendulum is the upright position $\theta = 0°$. Hence the system is only valid in the region of operation about this equilibrium point. This linearization is done by assuming θ as a small angle in the Taylor series expansion.

Therefore: $\sin\theta \approx \theta, \quad \cos\theta \approx 1$ and $\left(\dfrac{d\theta}{dt}\right)^2 \approx 0$

the resulting equations are given below:

$$\begin{cases} \ddot{\theta} = \dfrac{\alpha\beta g\theta - \left(\beta + \dfrac{\alpha}{r}\right) T_m - \beta T_r}{\beta\gamma - \alpha^2} \\ \ddot{x} = \dfrac{-\alpha^2 g\theta + \left(\alpha + \dfrac{\gamma}{r}\right) T_m + \alpha T_r}{\beta\gamma - \alpha^2} \end{cases} \tag{11.2}$$

$$T = -\frac{k_m k_e}{R} w + \frac{k_m}{R} V_a \tag{11.3}$$

the relationship between the input voltage to the motor and the output torque is given by Eq. (11.3) and knowing that the motors have the same specifications leads to:

$$\begin{cases} \ddot{\theta} = a_1\theta + a_2 w + b_1 V_{Dw} + b_2 V_{Rw} \\ \ddot{x} = a_3\theta + a_4 w + b_3 V_{Dw} + b_4 V_{Rw} \end{cases} \tag{11.4}$$

Where:

$$a_1 = \frac{\alpha\beta g}{\beta\gamma - \alpha^2}, \quad a_2 = \frac{k_m k_e\left(2\beta + \frac{\alpha}{r}\right)}{R(\beta\gamma - \alpha^2)}$$

$$a_3 = -\frac{\alpha^2 g}{\beta\gamma - \alpha^2}, \quad a_4 = -\frac{k_m k_e\left(2\alpha + \frac{\gamma}{r}\right)}{R(\beta\gamma - \alpha^2)}$$

$$b_1 = -\frac{k_m\left(\beta + \frac{\alpha}{r}\right)}{R(\beta\gamma - \alpha^2)}, \quad b_2 = -\frac{\beta k_m}{R(\beta\gamma - \alpha^2)}$$

$$b_3 = \frac{k_m\left(\alpha + \frac{\gamma}{r}\right)}{R(\beta\gamma - \alpha^2)}, \quad b_4 = \frac{\alpha k_m}{R(\beta\gamma - \alpha^2)}$$

For the reaction wheel: $\sum M = T_r = I_R \dot{w}$. Therefore: $\dot{w} = -\frac{k_m k_e}{I_R R} w + \frac{k_m}{I_R R} V_{Rw}$.

Thus from Eq. (11.4) the state space model is represented by a dynamic and control matrices A and B respectively, where its formulas are given as follows:

$$\begin{pmatrix} 0 & -\frac{\alpha^2 g}{\beta\gamma - \alpha^2} & 0 & -\frac{k_m k_e\left(2\alpha + \frac{\gamma}{r}\right)}{R(\beta\gamma - \alpha^2)} \\ 0 & 0 & 1 & 0 \\ 0 & \frac{\alpha\beta g}{\beta\gamma - \alpha^2} & 0 & \frac{k_m k_e\left(2\beta + \frac{\alpha}{r}\right)}{R(\beta\gamma - \alpha^2)} \\ 0 & 0 & 0 & -\frac{k_m k_e}{I_R R} \end{pmatrix}, \begin{pmatrix} \frac{k_m\left(\alpha + \frac{\gamma}{r}\right)}{R(\beta\gamma - \alpha^2)} & \frac{\alpha k_m}{R(\beta\gamma - \alpha^2)} \\ 0 & 0 \\ -\frac{k_m\left(\beta + \frac{\alpha}{r}\right)}{R(\beta\gamma - \alpha^2)} & -\frac{\beta k_m}{R(\beta\gamma - \alpha^2)} \\ 0 & \frac{k_m}{I_R R} \end{pmatrix}$$

where: $x_1 = \dot{x}$ is the linear speed, $x_2 = y_1 = \theta$ is the tilt angle of the body, $x_3 = \dot{\theta}$ is the angular velocity of the body and $x_4 = y_2 = w$ is the rotational speed of the reaction wheel.

The robot state space equations should be transformed to the discrete time for the digital implementation control:

$$\begin{cases} X(k+1) = A_d X(k) + B_d u(k) \\ Y(k) \quad\;\, = C_d X(k) + D_d u(k) \end{cases} \tag{11.5}$$

where

$X \in R^n, Y \in R^p, u \in R^m, A_d \in R^{n\times n}, B_d \in R^{n\times m}, C_d \in R^{p\times n}$, and $D_d \in R^{p\times m}$.

The conversion will be done by the following formulas based on Ogata (1994):

$$A_d = e^{A.t_s} \approx I + A.t_s, \quad B_d = \int_0^{t_s} e^{A.\delta} B d\delta, \quad C_d = C. \tag{11.6}$$

We assume that our sampling time t_s is very short compared with mechanical dynamics of the system.

11.4 Transformation to Block Canonical Forms

The system presented in Eq. (11.5) is block controllable of index l if:

- $\Omega_c = [B_d, A_d B_d, {A_d}^2 B_d, \ldots, {A_d}^{l-1} B_d]$ has full rank
- $l = \dfrac{n}{m}$ is an integer

If both conditions are satisfied, then the change of coordinates $X_c(t) = T_c X(t)$ transforms the system into the following Block controller form:

$$\begin{cases} \dot{X}_c(t) = A_c X_c(t) + B_c u(t) \\ Y(t) = C_c X_c(t) + D_c u(t) \end{cases} \tag{11.7}$$

where

$$T_c = \begin{pmatrix} T_{c1} \\ T_{c1} A \\ \vdots \\ T_{c1} A^{l-1} \end{pmatrix}, T_{c1} = {B_c}^T {\Omega_c}^{-1}, \begin{cases} A_c = T_c A {T_c}^{-1} \\ \quad = \begin{pmatrix} O_m & I_m & \cdots & O_m \\ O_m & O_m & \cdots & O_m \\ \vdots & \vdots & \ldots & O_m \\ O_m & O_m & \ldots & I_m \\ -A_l & -A_{l-1} & \cdots & -A_1 \end{pmatrix} \\ B_c = T_c B = \begin{pmatrix} O_m & O_m & \cdots & I_m \end{pmatrix}^T \\ C_c = C {T_c}^{-1} = \begin{pmatrix} C_l & C_{l-1} & \cdots & C_1 \end{pmatrix} \end{cases}$$

The system described by general state space Eq. (11.5) with $n = 4$: state number, $m = 2$: input number, and $p = 2$: output number satisfy the condition $n/m = l = 2 =$ integer and controllable system then can be transformed to block controller form using the following similarity transformation $\mathbf{x_c} = T_c \mathbf{x}$ Where:

$$T_c = \begin{pmatrix} T_{c1} \\ \\ T_{c1} A_d \end{pmatrix}, \quad T_{c1} = \begin{bmatrix} O_m & I_m \end{bmatrix} \begin{bmatrix} B_d & A_d B_d \end{bmatrix}^{-1} \tag{11.8}$$

For more detail we orient the reader to see Nail et al. (2018a). Referring to Nail et al. (2018b), we will end up with the following model

$$\begin{cases} A_c = T_c A_d T_c^{-1}, \quad B_c = T_c B_d, \quad C_c = C_d T_c^{-1} \\ A_c = \begin{pmatrix} 0 & 0 & 1 & 0 \\ 0 & 0 & 0 & 1 \\ 0.1289 & 0.5944 & 0.6803 & -1.0812 \\ 0.5865 & 0.0576 & -0.5911 & 0.9884 \end{pmatrix}, \; O_m = \begin{pmatrix} 0 & 0 \\ 0 & 0 \end{pmatrix}, \; I_m = \begin{pmatrix} 1 & 0 \\ 0 & 1 \end{pmatrix} \\ B_c = \begin{pmatrix} O_m & I_m \end{pmatrix}^T \\ C_c = \begin{pmatrix} 0.2637 & -0.0013 & 0.0025 & 0.5102 \\ -0.1070 & 0.0524 & 0.0802 & -0.1209 \end{pmatrix} \end{cases}$$

The right matrix fraction description (RMFD) form is:

$$y(q^{-1}) = N(q^{-1})D^{-1}(q^{-1})u(q^{-1}) \tag{11.9}$$

The left matrix fraction description (LMFD) form is:

$$y(q^{-1}) = A^{-1}(q^{-1})B(q^{-1})u(q^{-1}) \tag{11.10}$$

Where

$$D(q^{-1}) = I + D_1 q^{-1} + D_2 q^{-2}$$

$$N(q^{-1}) = N_0 + N_1 q^{-1} + N_2 q^{-2}$$

$$A(q^{-1}) = I + A_1 q^{-1} + A_2 q^{-2}$$

$$B(q^{-1}) = B_0 + B_1 q^{-1} + B_2 q^{-2}$$

The coefficient of $D(q^{-1}), N(q^{-1})$ are well known and directly obtained from A_c, C_c see Nail et al. (2016a,b, 2017):

$$D_1 = \begin{pmatrix} -0.6803 & 1.0812 \\ 0.5911 & -0.9884 \end{pmatrix}, \quad D_2 = \begin{pmatrix} -0.1289 & -0.5944 \\ -0.5865 & -0.0576 \end{pmatrix}$$

$$N_0 = O_{2\times2}, \quad N_1 = \begin{pmatrix} 0.0025 & 0.5102 \\ 0.0802 & -0.1209 \end{pmatrix}, \quad N_2 = \begin{pmatrix} 0.2637 & -0.0013 \\ -0.1070 & 0.0524 \end{pmatrix}$$

From Eqs. (11.9) and (11.10) we can write:

$$A(q^{-1})N(q^{-1}) = B(q^{-1})D(q^{-1}) \tag{11.11}$$

Expanding the last equation we can write in more compact form:

$$\begin{pmatrix} B_1^T \\ B_2^T \\ A_1^T \\ A_2^T \end{pmatrix} = \begin{pmatrix} I_2 & O_2 & O_2 & O_2 \\ D_1^T & I_2 & -N_1^T & O_2 \\ D_2^T & D_1^T & -N_2^T & -N_1^T \\ O_2 & D_2^T & O_2 & -N_2^T \end{pmatrix}^{-1} \begin{pmatrix} N_1^T \\ N_2^T \\ O_2 \\ O_2 \end{pmatrix} \tag{11.12}$$

Solving this last equation yield to:

$$A_1 = \begin{pmatrix} -0.9939 & 1.1919 \\ 0.7845 & -0.6749 \end{pmatrix}, \quad A_2 = \begin{pmatrix} 0.1146 & -0.4768 \\ -0.7148 & -0.0035 \end{pmatrix}$$

$$B_0 = O_{2\times 2}, \quad B_1 = \begin{pmatrix} 0.0025 & 0.5102 \\ 0.0802 & -0.1209 \end{pmatrix}, \quad B_2 = \begin{pmatrix} 0.0569 & -0.1510 \\ -0.0330 & 0.3280 \end{pmatrix}$$

11.5 Dynamics Stabilization Using Block Poles Placement

This algorithm is an extension of classical pole placement using the theory of matrix fraction description, in this application we reinstruct to give the necessary steps and for more detail see our contribution (Nail et al. 2016a, 2017).

✓ Right block poles

Right block poles chosen diagonally based on selected eigenvalues:

$$R_1 = \begin{pmatrix} -.5 & 0 \\ 0 & -.2 \end{pmatrix}, \quad R_2 = \begin{pmatrix} -.35 & 0 \\ 0 & -.7 \end{pmatrix}$$

✓ Right block Vandermonde matrix

$$V_R = \begin{pmatrix} I & I \\ R_1{}^1 & R_2{}^1 \end{pmatrix} = \begin{pmatrix} 1 & 0 & 1 & 0 \\ 0 & 1 & 0 & 1 \\ -0.5 & 0 & -0.35 & 0 \\ 0 & -0.2 & 0 & -.7 \end{pmatrix}$$

✓ Constructing from a complete set of right block poles

Consider a complete set of right poles $\{R_1, R_2\}$ for the desired matrix polynomial $D_s(\lambda)$, If R_i is a right poles of $D_s(\lambda)$ so:

$$R_i{}^2 + Ds_1 R_i{}^1 + D_2 = O_2 \tag{11.13}$$

$$Ds_1 R_i{}^1 + Ds_2 = -R_i{}^2 \tag{11.14}$$

Replacing i from 1 to 2, we get the following:

$$[Ds_{d2}, Ds_{d1}] = -\left[R_1^2, R_2^2\right] V_R{}^{-1}$$

Where V_R is the right block Vandermonde matrix.

✓ State Feedback matrix gain

Consider the linear discret-time invariant dynamic system described by Eq. (11.21). Applying the state feedback $u = -Kx(t)$ to this system, where K is a 2×4 gain matrix, after using the block controller form transformation for the system, we get $u = -K_c x_c(t)$.

$$K = K_c T_c = \left[K_{c(2)}, K_{c(1)}\right] \tag{11.15}$$

T_c and $K_{ci} \in R^{m \times m}$ For $i = 1, 2$, the closed loop system is shown below:

$$\begin{cases} \dot{x}_c = (A_c - B_c K_c)\, x_c \\ y_c = C_c x_c \end{cases} \tag{11.16}$$

Where

$$(A_c - B_c K_c) = \begin{bmatrix} O_m & I_m \\ -(A_2 + K_{c2}) & -(A_1 + K_{c1}) \end{bmatrix} \tag{11.17}$$

The characteristic matrix polynomial of this closed loop system is:

$$D(\lambda) = I_m \lambda^2 + (A_1 + K_{c1})\, \lambda^1 + (A_2 + K_{c2}) \tag{11.18}$$

From a set of desired eigenvalues, we construct the set of Block poles, the desired characteristic matrix polynomial of the block poles in the form, by putting

$$Ds_d(\lambda) = D(\lambda) \tag{11.19}$$

We get the coefficients $K_{ci} = Ds_{di} - A_i$ as follows:

$$K_{ci} = Ds_{di} - A_i \text{ for } i = 1, 2 \tag{11.20}$$

The gain matrix given by the following formula:

$$K = K_c T_c \tag{11.21}$$

Finally and after numerical application the feedback static matrix gain is given as follows:

$$K = \begin{pmatrix} 0.3039 & 0.5944 & 1.5303 & -1.0812 \\ 0.5865 & 0.1976 & -0.5911 & 1.8884 \end{pmatrix}$$

11.6 Design Digital Controller for the Tracking

How to tuning the two PID Control of Plant. If a mathematical model of the plant can be derived, then it is possible to apply various design techniques for determining parameters of the controller that will meet the transient and steady-state specifications of the closed-loop system. However, if the plant is so complicated that its mathematical model cannot be easily obtained, then an analytical or computational approach to the design of a PID controller is not possible. Then we must resort to experimental approaches to the tuning of PID controllers. The process of selecting the controller parameters to meet given performance specifications is known as controller tuning. Ziegler and Nichols suggested rules for tuning PID controllers (meaning to set values k_p, k_i and k_d) based on experimental step responses or based on the value of k_p that results in marginal stability when only proportional control action is used. Ziegler-Nichols rules, which are briefly presented in the following, are useful when mathematical models of plants are not known. Such rules suggest a set of values of k_p, k_i and k_d that will give a stable operation of the system. However, the resulting system may exhibit a large maximum overshoot in the step response, which is unacceptable. In such a case we need series of fine tunings until an acceptable result is obtained. In fact, the Ziegler-Nichols tuning rules give an educated guess for the parameter values and provide a starting point for fine tuning, rather than giving the final settings for k_p, k_i and k_d in a single shot (Ogata 1994, 2010).

Ziegler-Nichols method Rules is used for Tuning the two PID Controllers shown in the schematic diagram in Fig. 11.1 we determining the values of the proportional gain k_p integral time k_i and derivative time k_d of the two PID based on the transient response characteristics of a given linearized plant. Such determination of the parameters of PID controllers or tuning of PID controllers can be made by engineers on-site by experiments on the plant (Ogata 1994, 2010). Figure 11.1 shown the schematic bloc diagram of the two digital PID controllers connected with the stabilized linearized dynamics plant by the feedback matrix gain. Table 11.1 shown the parameters of the two PID controllers.

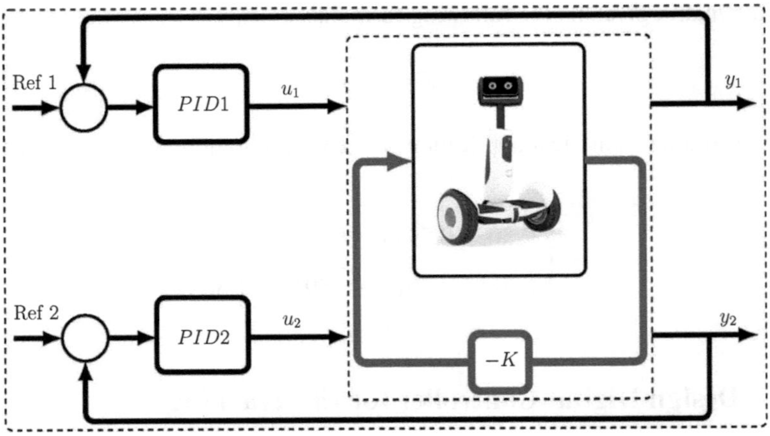

Fig. 11.1 Schematic bloc diagram of the proposed controller

Table 11.1 Digital PID controllers parameters

	K_p	K_i	K_d
PID controller 1	0.95	200.00	0.001
PID controller 2	−1.38	−276.61	0.100

11.6.1 Simulation Results and Interpretations

From the obtained results as shown in the next figures, we see that the angular speed of the motor tracks its reference trajectory in Fig. 11.2 with very small error in Fig. 11.3, no overshooting, no static error is obtained at both transient and steady state regimes, the convergence is ensured based on the smooth error and tracking. In Fig. 11.5, the real and the desired tilt angle signals are coincided in all time with negligible static error in Fig. 11.6. From the Figs. 11.4 and 11.7, the control signals voltages not exceed the norm voltage of the two DC motors, which are drive the two-wheeled robot.

11.7 Conclusion

In this chapter digital controller is proposed and implemented in two-wheeled robot system in order to control two inputs: angular speed and tilt angle, based on MFD theory, which its effect has been clear in the stabilization of the behavior and the dynamics of the two-wheeled robots, and also by the feedback matrix gain K which is designed by the assigning of two robust stable block poles in diagonal form, and for the tracking, two digital PID controllers are tuning using Ziegler-Nichols method to acquired the best parameters, the obtained results are well satisfactory

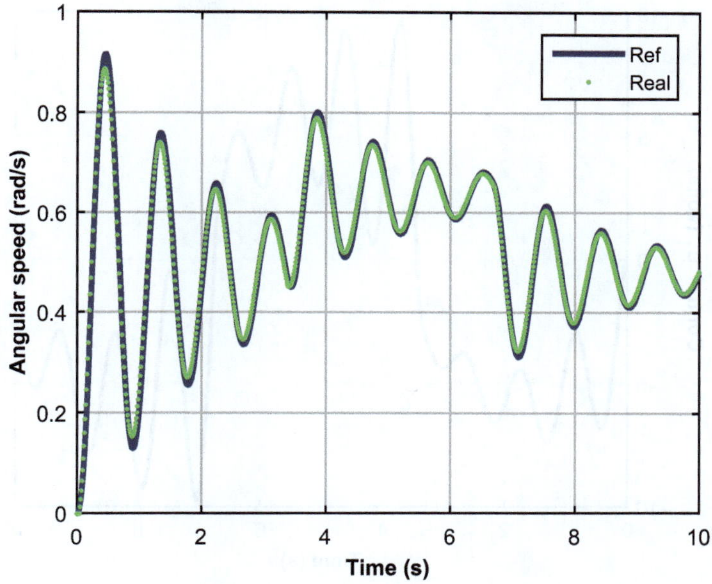

Fig. 11.2 The signals of the real and the desired rotational speed

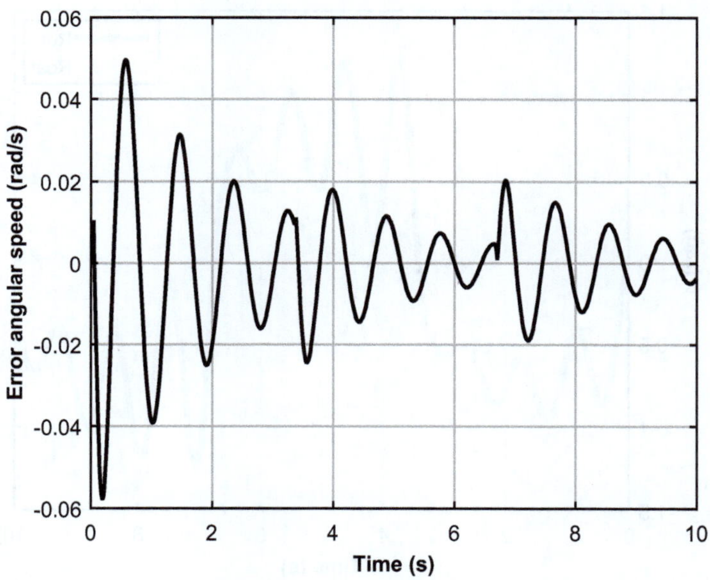

Fig. 11.3 The error between the real and the desired rotational speed

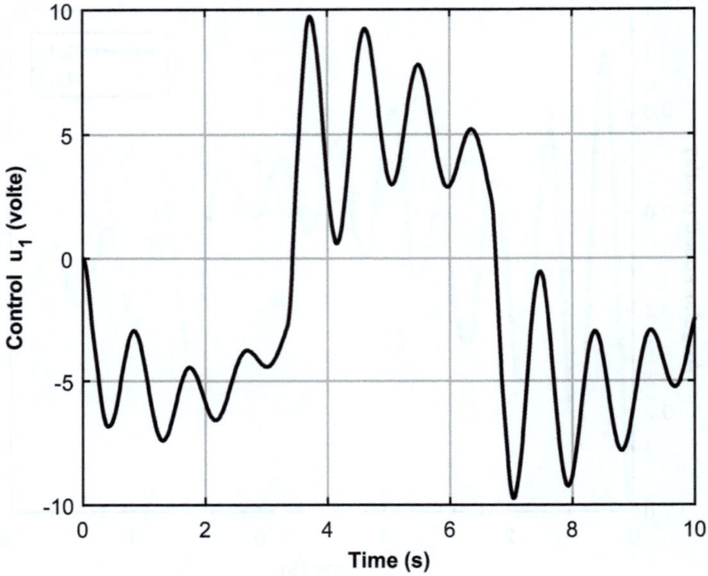

Fig. 11.4 The input signal control u_1

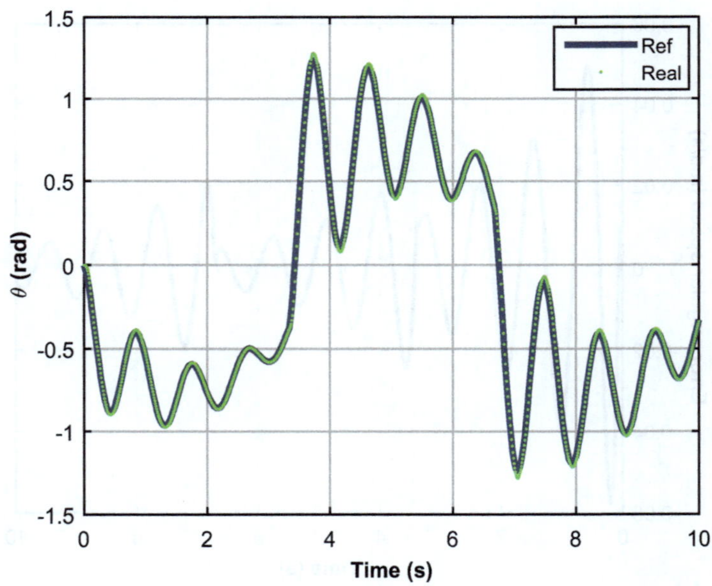

Fig. 11.5 The signals of the real and the desired tilt angle

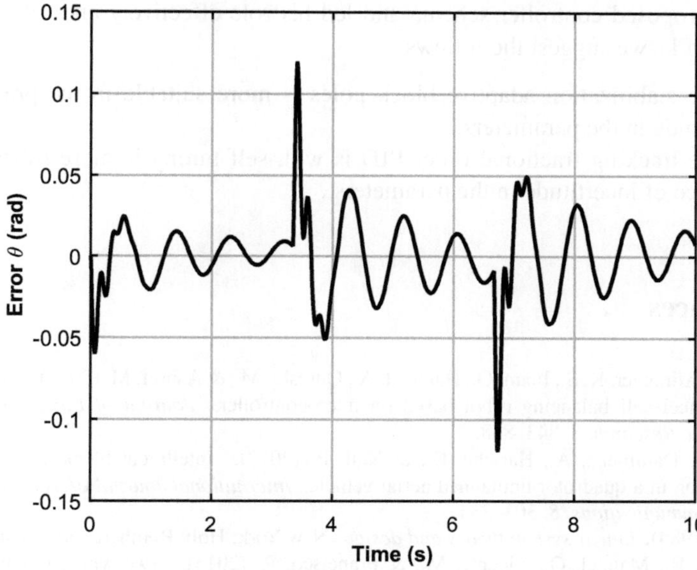

Fig. 11.6 The error between the real and the desired tilt angle

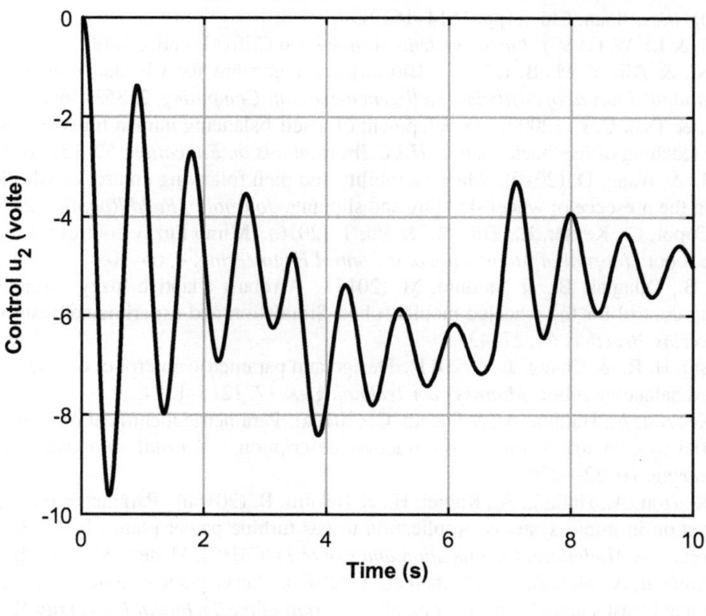

Fig. 11.7 The input signal control u_2

and the proposed controller schema and led his role effectively and reliability, as further works we suggest the follows:

- For the stabilization adaptive block poles is more suitable in the presence of incertitude in the parameters.
- For the tracking fractional order PID is with self tuning is more robust in the presence of incertitude in the parameters..

References

Adeel, U., Alimgeer, K. S., Inam, O., Hameed, A., Qureshi, M., & Ashraf, M. (2013). Autonomous dual wheel self balancing robot based on microcontroller. *Journal of Basic and Applied Scientific Research, 3*, 843–848.

Bekhiti, B., Dahimene, A., Hariche, K., & Nail, B. (2017). Intelligent block spectral factors relocation in a quadrotor unmanned aerial vehicle. *International Journal of Systems, Control and Communications, 8*, 303–334.

Chen, C. (1984). *Linear system theory and design*. New York: Holt, Reinhart and Winston.

Guilherme, R., Manuel, O., Vicente, M., & Francisco, R. (2015). Two-wheeled self-balanced pendulum workspace improvement via underactuated robust nonlinear control. *Control Engineering Practice, 44*, 231–242.

Jingtao, L., Xueshan, G., & Qiang, H. (2007). Mechanical design and dynamic modeling of a two-wheeled inverted pendulum mobile robot. In *IEEE International Conference on Automation and Logistics*, Jinan, China (pp. 1614–1619).

Kailath, T., & Li, W. (1980). *Linear systems*. Englewood Cliffs: Prentice Hall.

Kherici, N., & Ali, Y. M. B. (2011). Bio-inspired algorithm for wheeled robot's navigation. *International Journal of Artificial Intelligence and Soft Computing, 2*, 353–366.

Lin, S.-C., & Tsai, C.-C. (2009). Development of a self-balancing human transportation vehicle for the teaching of feedback control. *IEEE Transactions on Education, 52*, 157–168.

Low, C. B., & Wang, D. (2010). Maneuverability and path following control of wheeled mobile robot in the presence of wheel skidding and slipping. *Journal of Field Robotics, 27*, 127–144.

Mac, T., Copot, C., Keyser, R., Tran, T., & Vu, T. (2016). Mimo Fuzzy control for autonomous mobile robot. *Journal of Automation and Control Engineering, 4*, 65–70.

Mauricio, B., Douglas, B., & Nardnio, M. (2017). A robust adaptive fuzzy variable structure tracking control for the wheeled mobile robot: Simulation and experimental results. *Control Engineering Practice, 64*, 27–43.

Memarbashi, H. R., & Chang, J.-Y. (2011). Design and parametric control of co-axes driven two-wheeled balancing robot. *Microsystem Technologies, 17*, 1215–1224.

Nail, B., Kouzou, A., Hafaifa, A., & Djeddi, C. (2016a). Parametric identification of multivariable industrial system using left matrix fraction description. *Journal Automation & Systems Engineering, 10*, 221–230.

Nail, B., Kouzou, A., Hafaifa, A., Kamel, H., & Bekhiti, B. (2016b). Parametric output feedback stabilization in mimo systems: Application to gas turbine power plant. In *8th International Conference on Modelling, Identification and Control (ICMIC)*, Medea, Algeria (pp. 971–976).

Nail, B., Kouzou, A., Hafaifa, A., & Chaibet, A. (2017). Power plant system identification based on extended least square kronecker operator. *Journal of the Technical University-Sofia Plovdiv Branch, Bulgaria Fundamental Sciences and Applications, 23*, 15–20.

Nail, B., Kouzou, A., & Hafaifa, A. (2018a). Robust block roots assignment in linear discrete-time sliding mode control for a class of multivariable system: *Gas turbine power plant application. Transactions of the Institute of Measurement and Control*. https://doi.org/10.1177/0142331218774615.

Nail, B., Kouzou, A., Hafaifa, A., & Chaibet, A. (2018b). Parametric identification and stabilization of turbo-compressor plant based on matrix fraction description using experimental data. *Journal of Engineering Science & Technology, 13*(6), 1850–1868.

Ogata, K. (1994). *Discrete-time control systems*. Englewood Cliffs/Upper Saddle River: Prentice-Hall.

Ogata, K. (2010). *Modern control engineering* (Prentice-Hall Electrical Engineering Series). Upper Saddle River: Instrumentation and Controls Series.

Oryschuk, P., Salerno, A., Al-Husseini, A. M., & Angeles, J. (2009). Experimental validation of an underactuated two-wheeled mobile robot. *IEEE/ASME Transactions on Mechatronics, 14*, 252–257.

Ren, T.-J., Chen, T.-C., & Chen, C.-J. (2008). Motion control for a two-wheeled vehicle using a self-tuning pid controller. *Control Engineering Practice, 16*, 365–375.

Ronald, C., Karl, S., & Roger, H. (2013). Review of modelling and control of two-wheeled robots. *Annual Reviews in Control, 37*, 89–103.

Shigong, J., Fuquan, D., Ling, L., & Xueshan, G. (2014). *IEEE International Conference on Robotics and Biomimetics Robio Modeling and LQR Control of a Multi-dof Two-Wheeled Robot* (pp. 1964–1969).

Sun, T., Xiang, X., Su, W., Wu, H., & Song, Y. (2017). A transformable wheel-legged mobile robot: Design, analysis and experiment. *Robotics and Autonomous Systems, 98*, 30–41.

Teimoori, H., & Savkin, A. V. (2010). Equiangular navigation and guidance of a wheeled mobile robot based on range-only measurements. *Robotics and Autonomous Systems, 58*, 203–215.

Tsai, C.-C., Huang, H.-C., & Lin, S.-C. (2010). Adaptive neural network control of a self-balancing two-wheeled scooter. *IEEE Transctions on Industrial Electronics, 57*, 1420–1428.

Weiguo, W., & Liyang, G. (2017). Posture self-stabilizer of a biped robot based on training platform and reinforcement learning. *Robotics and Autonomous Systems, 98*, 42–55.

Yong, C., & Foong, C. (2011). Wireless controlled two wheel balancing robot. *International Journal Network and Mobile Technologies, 2*, 88–98.

Chapter 12
Mobile Robot Path Planning Based on Optimized Fuzzy Logic Controllers

L. Cherroun, M. Boumehraz, and A. Kouzou

Abstract The path planning task is defined as the process to compute the motion sequence allowing the robot to move from the start position to the final destination autonomously without human actions. The path planning is one of the popular tasks encountered by imprecision and uncertainties and it has been studied using fuzzy logic systems (FLS). The construction of a well performing fuzzy controller is not always easy. The problem of finding appropriate membership functions and fuzzy rules is a difficult task. However, the design of fuzzy rules is often reliant on heuristic experience and it lacks systematic methodology, therefore these rules might not be correct and consistent. The design can prove to be long and delicate due to the important number of parameters to determine, and can lead then to a solution with poor performance. To cope with this difficulty, many researchers have been working to find and apply learning algorithms for fuzzy controller design. These automatic methods enable to extract information when the knowledge is not available. The most popular approach to optimize fuzzy logic controllers may be a kind of supervised learning where the training data is available. However, in real applications, extraction of training data is not always easy and become impossible when the cost to obtain training data is expensive. For these problems, reinforcement learning (RL) is more suitable than supervised learning. A control strategy with a learning capacity can be carried out by using Q-learning for tuning fuzzy logic controllers; which the robot receives only a scalar signal likes a feedback. This information makes to adjust the robot behavior in order to improve their performances. The basic idea in Q-learning algorithm of RL is to maximize the received rewards after each interaction with the environment. In this chapter,

L. Cherroun (✉) · A. Kouzou
Laboratory of Applied Automatic and Industrial Diagnostics (LAADI), University of Djelfa, Djelfa, Algeria
e-mail: cherroun_lakh@yahoo.fr; kouzouabdellah@ieee.org

M. Boumehraz
LMSE Laboratory, University of Biskra, Biskra, Algeria
e-mail: medboumehraz@netcourrier.com

N. Derbel et al. (eds.), *New Developments and Advances in Robot Control*, Studies in Systems, Decision and Control 175,
https://doi.org/10.1007/978-981-13-2212-9_12

Q-learning algorithm is used to optimize Takagi-Sugeno fuzzy logic controllers for autonomous path planning of a mobile robot. These optimized fuzzy controllers are used for the different robot tasks: goal seeking, obstacle avoidance and wall-following. The obtained results of this optimization method present significant improvements of the robot behaviors.

Keywords Mobile robot · Behavior · Fuzzy controller · Optimization · Q-learning · Fuzzy Q-learning

12.1 Introduction

The existence of robots in various types became very significant in the industrial and service sectors. Due to the growing interest of the service robots, they can achieve their tasks in environments contains multiple obstacles. Autonomous mobile robot navigation and path planning are very important issues in the field of unmanned robot control. The ability to autonomous navigation is one of the standing challenging aspects in mobile robotic field. It is a difficult task, which requiring a complete modeling of the environment and intelligent controllers (Cuesta and Ollero 2005). Autonomous path planning is considered as the task of determining a collision-free path that enables the robot to move through an obstacle course, starting from an initial position to reach a goal position in unknown spaces. Path planning must respect the constraints kinematics of the robot without human intervention. It can be classified into two categories: global path planning and local path planning (Khatib 1986; Latombe 1991). Global path planning methods are usually conducted off-line in a completely known environments (Cuesta and Ollero 2005; Latombe 1991), where an exact environment model is required for planning the path. Although these approaches have an exact solution, their time complexity grows with the geometry complexity and grows exponentially with the number of degrees of freedom in the vehicle's motion (Latombe 1991). This fact has led to the emergence of numerous heuristic or approximating approaches, which rely on either calculating the potential Fields (Borestein and Koren 1989) or performing a search through a state space model (Cuesta and Ollero 2005). As a pre-specified environment is required for these methods to plan the path, they fail when the environment is not fully known.

On the other hand, the local path planning techniques, also known as the obstacle avoidance methods, are more efficient in autonomous mobile robot navigation in unknown or partially known environment (Algabri et al. 2015; Khatib 1986; Seraji and Howard 2002; Wang and Liu 2008). These reactive methods use the on-line sensory information to tackle the uncertainty.

Fuzzy logic systems (FLSs) are the robust approaches used for autonomous mobile robot navigation and path planning (Antonelli and Fusco 1986; Seraji and Howard 2002; Wang and Liu 2008; Yang et al. 2005). These systems have been applied in many engineering problems, especially in control systems and robotic

(Abadi and Khooban 2015; Al Yahmedi and Fatmi 2011; Cherroun and Boumehraz 2013a; Vadakkepa et al. 2004). The modeling process is based on the behavior of the operator and their knowledge to handle uncertainties. However the performances of FLSs depend on the formulation of the fuzzy rules and the numerical specification of all linguistic terms. The conception of a powerful FLS is not always easy. Each design process can prove to be long and difficult due to the number of parameters of membership functions of inputs-outputs, the fuzzy rules, consequent parameters (Cherroun and Boumehraz 2013b; Jallouli et al. 2010; Passino and Yurkovich 1998; Timothy 2004).

Tuning and adjusting parameters of FLC are interesting issues that are studied by many researchers. The most applied are based on learning algorithms and optimization method as the use of Neural Networks (NN) in (Algabri et al. 2014a; Bakdi et al. 2016; Pothal and Parhi 2015). Using Genetic Algorithms (GA) as in Begum et al. (2008) for simultaneous localization and mapping of mobile robots, or for robot navigation as the paper (Zhao et al. 2015). FLC Optimization is too done by using particle swarm optimization algorithms (PSO) as in Algabri et al. (2014a), and by applying Ant Colony optimization (ACO) in Hsu et al. (2013). The optimization task can be done using hybrid approaches between the previous methods. Other soft computing techniques for mobile robot navigation are presented in Algabri et al. (2015).

Reinforcement learning (RL) (Kaelbling et al. 1996; Sutton and Barto 1998) is an automatic tuning method used by some authors to optimize FLC parameters: premisse and conclusion parts (Berenji and Khedkar 1992; Jouffle 1998; Nemra and Rezine 2008). This hybrid system is attractive in the field of robotic as an optimized fuzzy navigators for the different robot behaviors as in Joo and Deng (2005); Juang and Hsu (2009), and Cherroun and Boumehraz (2013b). Or by combination with neural networks for robot obstacle avoidance as presented in the paper (Mihai and Gheorghe 2016).

These automatic methods enable to extract information when the experts' prior knowledge is not available.

A control strategy with a learning capacity can be applied using the reinforcement learning and fuzzy logic approaches; which the robot receives only a scalar signal likes a feedback. This reinforcement makes it possible the navigator to adjust its strategy in order to improve their performances (Glorennec and Jouffle 1997; Kaelbling et al. 1996; Sutton and Barto 1998). RL is considered as an automatic modification of the robot behavior in its working environment (Cherroun and Boumehraz 2013b; Nemra and Rezine 2008). The reinforcement learning is a method of optimal control, when the agent starts from an ineffective solution which gradually improves according to the experience gained to solve a sequential decision problem. The basic idea in reinforcement learning is that, the agent is placed in its environment and it can observe the results of their actions. The objective of the agent is to maximize the received rewards (Sutton and Barto 1998). To use reinforcement learning, several approaches are possible (Jouffle 1998). The first consists in discrediting the problem using Q-tables, and the second method is based on working at continuous spaces by the estimation of the quality functions.

In this chapter, reinforcement learning is used to optimize fuzzy logic controllers (FLCs) for mobile robot path planning. The obtained fuzzy navigators are able to perform a successful autonomous path planning.

The present chapter is organized as follows: Sect. 12.2 gives the necessary background of Fuzzy Logic Systems (FLS). Reinforcement learning algorithms are presented in Sect. 12.3 by explaining Q-learning and the optimization algorithm of fuzzy systems. Section 12.4 discusses the application of optimized fuzzy controllers for the different tasks of a mobile robot with the obtained results. Section 12.5 presents conclusions and future research directions of the work presented in this chapter.

12.2 Fuzzy Logic System

Fuzzy logic system is defined as a mathematical function used to imitate the capacity of reasoning and decision making from the human operator, using uncertainties and/or unclear information (Passino and Yurkovich 1998). In fuzzy logic applications, the model of a system is ensured by linguistic rules (fuzzy rules) between the input and output variables. A fuzzy logic system consists basically of four blocks: fuzzification, rule-base, inference and defuzzification (Passino and Yurkovich 1998; Timothy 2004). In Fig. 12.1, we present the basic structure of a FLS. We can define each block as follows.

12.2.1 Fuzzification

This part converts the real inputs (X) into information easily used to activate and apply fuzzy rules. The fuzzy sets of all inputs must be defined. This interface maps inputs of the associated fuzzy sets in order to calculate the membership degree of each input variable $\mu_A(x)$.

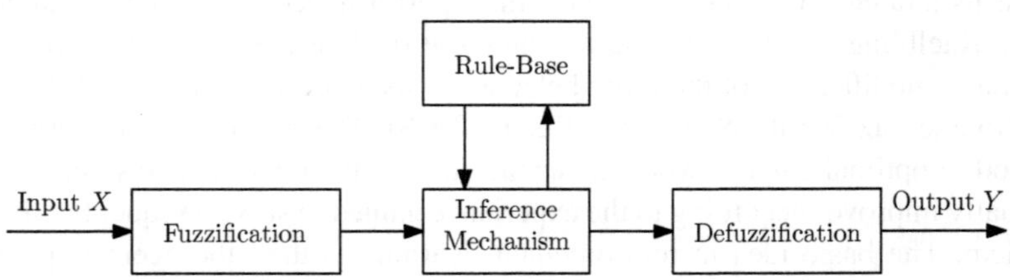

Fig. 12.1 Structure of fuzzy logic system

12.2.2 Rule-Base

The linguistic description of the expert to quantify the input-output relationship is done by a set of "If-Then" rules. For a fuzzy logic system with p inputs ($x_1 \in X_1, \ldots\ x_p \in X_p$) and one output ($y \in Y$). Using M fuzzy rules, the rth rule for $r = 1 \ldots M$ has the following forms for a FLS:

$$\text{If } (x_1 \text{ is } F_1^r) \text{ and } (x_2 is F_2^r) \ldots \text{ and } (x_p \text{ is } F_p^r) \text{ Then } (y \text{ is } G^r) \tag{12.1}$$

The fuzzy rule-base of the robot controller can be described as follows:

R_1:	If	(goal is Far)	Then	(speed is Big)	and	(steering is zeros)
R_2:	If	(goal is right)	Then	(speed is Medium)	and	(steering is right)
:						
R_n:	If	(goal is near)	Then	(speed is zeros)	and	(steering is zeros)

12.2.3 Inference Mechanism

The role of this block is to interpret and apply the operator knowledge to control the system based on If-Then fuzzy rules. The inference mechanism combines the input fuzzy sets and output fuzzy sets by compositions of the relationships.

For a FLS with ($x_1 \in X_1, \ldots\ x_p \in X_p$) inputs and one output ($y \in Y$) using M rules as mentioned in Eq. (12.2). The firing strength of the ith rule is calculated by:

$$F^i(x) = \mu_{F_1^i(x_1)} \star \mu_{F_2^i(x_2)} \star \ldots \star \mu_{F_p^i(x_p)} \tag{12.2}$$

12.2.4 Defuzzification

converts the results of the inference mechanism into real-applied control values (Y). The output is calculated by Eq. (12.3) using the centroid method:

$$y = \frac{\sum_{i=1}^{M} F^i(x) y^i}{\sum_{i=1}^{M} F^i(x)} \tag{12.3}$$

In fuzzy logic applications, the two most popular fuzzy logic systems used in control fields are the Mamdani and Takagi-Sugeno (TS) models (Passino and Yurkovich

1998; Timothy 2004). The difference between TS fuzzy logic controller and a Mamdani fuzzy logic consist essentially of the definition of outputs (consequent part of rules).

In this chapter, we used optimized Takagi-Sugeno Fuzzy Logic Controllers (TSFLC) for the design of mobile robot behaviors. All inputs and outputs variables (angle, distances, speed and steering…) have degree of membership functions. Each functionality is designed using an optimized zero order TSFLC.

12.3 Reinforcement Learning (RL)

Reinforcement learning (RL) is a machine learning method. It makes possible the solution of the problem in a finite time based on its experimental learned knowledge. In reinforcement learning approach, an agent learns to optimize an interaction with a dynamic environment through trial and error process. The RL structure is shown in Fig. 12.2. The agent receives a scalar value or reward with every action it executes. The objective of the agent (in our case, it is a mobile robot) is to learn a strategy for selecting actions such that the expected sum of discounted rewards is maximized (Kaelbling et al. 1996; Sutton and Barto 1998). In the standard reinforcement learning model, the robot is connected to its environment via perception and action. At any given time step t, the robot perceives the state s_t of the environment and selects an action a_t. The environment returns by a scalar reinforcement signal r_t and changing into state s_{t+1} (Sutton and Barto 1998). The robot should choose actions that tend to increase the long run sum of values of the reinforcement signal. It can learn to do this overtime by systematic trial and error process.

The agent task is to find an optimal policy, $\pi : \{S, A\} \longrightarrow [0, 1]$ which maps states to actions in order to maximize the received reinforcements. In general, the robot's actions determine not only its immediate rewards, but also the next state of the environment. As a result, when applying the generated action, the robot must take the next state into account. Generally, the function value is defined at the form of PDM by:

$$V_\pi(s) = E_\pi(R_t/s_t = s) = E_\pi\left(\sum_{k=1}^{\infty} \gamma^k r_{t+k}/s_t = s\right) \tag{12.4}$$

where $\gamma \in]0, 1[$ is a factor to regulate the importance of future returns.

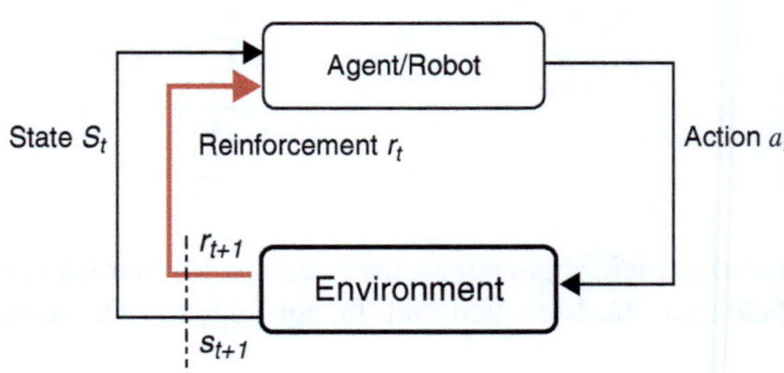

Fig. 12.2 Reinforcement learning structure

The most algorithms of reinforcement learning use a quality function noted Q-function, representing the value of each pair state-action to obtain an optimal behavior (Sutton and Barto 1998). It gives for each state, the future return if the agent pursues this policy π:

$$Q^{\pi}(s, a) = E_{\pi}(R_t / s_t = s, a_t = a) \tag{12.5}$$

The optimal quality is calculated as:

$$Q^{\star}(s, a) = \max_{\pi} Q^{\pi}(s, a) \tag{12.6}$$

We obtain then:

$$Q^{\star}(s, a) = E(r_{t+1} + \gamma V^{\star}(s_{t+1}) / s_t = s, a_t = a) \tag{12.7}$$

12.3.1 Q-Learning Algorithm (QLA)

Q-Learning is an artificial intelligent technique that has been successfully used for solving complex Markovian Decision Process (MDP) that model realistic system. It is a technique to solve specific problems by using the reinforcement learning approach. Q-Learning is an incremental version of dynamic programming for solving multistage decision problems. It is the popular of temporal difference algorithms proposed by Watkins and Dayan (1992). It is based on three main functions: an evaluation function, a reinforcement function and an update function (Cherroun and Boumehraz 2013b). The idea of Q-learning is to learn a Q-function that maps the current state s_t and action a_t to $Q^{\pi}(s_t, a_t)$, that predicts the total future discounted reward that will be received from the current action a_t. In that, it learns the optimal policy function incrementally after each transition (s_t, a_t, r_t, s_{t+1}).

The update of $Q^{\star}(s, a)$ is done by observation of the instantaneous transitions and their rewards associated by the following equation (Kaelbling et al. 1996; Sutton and Barto 1998):

$$Q(s_t, a_t) =\longleftarrow Q^{\pi}(s_t, a_t) + \alpha[r_t + \gamma \max_{a \in A(s)} Q(s_{t+1}, a_t) - Q(s_t, a_t)] \tag{12.8}$$

where $\alpha \in [0, 1]$ is a learning rate.

The quality functions are stored in the form of table: the qualities of all actions for a given state. Firstly, when the table does not contain sufficient data, a random component is added in order to restrict the eligible actions with the small number of the actions already tested. As the table fills, this random component is reduced in order to allow the exploitation of received information and for obtaining a good performance.

In the next section, we will present a method has as advantages two properties: extending the Q-learning algorithm to continuous state-action spaces, and considered as an optimization method of FLC parameters (conclusion parts).

12.3.2 *Optimization of FLS Using Q-Learning Algorithm (FQLA)*

Fuzzy inference systems (FIS) are promising solutions for representing the quality functions with continuous spaces of states and actions (Glorennec and Jouffle 1997; Nemra and Rezine 2008). The idea of this optimization is to propose several conclusions for each rule, and to associate each conclusion by a quality function which will be evaluated during the time. The training process permit to obtaining the best rules that maximizing the future reinforcements (Jouffle 1998). This fuzzy version of the Q-learning algorithm is named fuzzy Q-learning algorithm (FQLA) summarized in Table 12.1. The initial proposed rule-base using a zero order Takagi-Sugeno model is composed therefore of rules and N conclusions such as Eq. (12.9) from Glorennec and Jouffle (1997), Nemra and Rezine (2008), and Cherroun and Boumehraz (2013b):

$$\begin{aligned} \text{If } s \text{ is } S_i \text{ Then } & \; y = a[i,1] \text{ with } q[i,1] = 0 \\ \text{or } & \; y = a[i,2] \text{ with } q[i,2] = 0 \\ & \; \vdots \\ \text{or } & \; y = a[i,N] \text{ with } q[i,N] = 0 \end{aligned} \tag{12.9}$$

where $q(i, j)$ with $i = 1 \ldots m$ and $j = 1 \ldots N$, are potential solutions whose values are initialized to 0. During the learning process, the conclusion part of each

Table 12.1 Optimization of FLSs using Q-learning

1.	Choose the FIS structure
2.	Initialize randomly $q[i, j]$, $i = 1, \ldots m$ (m: rule number)
	$j = 1, \ldots N$ (N: Number of proposed conclusions)
3.	$t = 0$, observe the state s_t
4.	For each rule i, compute $w_i(s_t)$
5.	For each rule i, choose a conclusion with the EEP
6.	Compute the action $A(s_t)$ and correspondence quality $Q[s_t, A(s_t)]$
7.	Apply the action $A(s_t)$. Observe the new state s_t'
8.	Receive the reinforcement r_t
9.	For each rule i, compute $w_i(s_t')$
10.	Compute a new evaluation of the state value
11.	Update parameters $q[i, j]$ using this evaluation
12.	$t \leftarrow t + 1$, Go to 5

rule is selected by means of an exploration-exploitation policy noted (EEP), where $EEP(i) \in \{1, \ldots, N\}$ such as the ε greedy:

$$A_{\varepsilon-greedy} = \left| \begin{array}{l} \arg\max_{a \in A(s_t)} Q(s_t, a) \quad \text{with} \quad \varepsilon \quad \text{probability} \\ \text{Random action } A(s_t) \text{ with } 1 - \varepsilon \text{ probability} \end{array} \right. \tag{12.10}$$

In this case, the inferred action is given by:

$$A(s) = \sum_{i=1}^{m} w_i(s)\, a[i, EEP(i)] \tag{12.11}$$

and the quality of this action will be:

$$\widehat{Q}[s, A(s)] = \sum_{i=1}^{m} w_i(s)\, q[i, EEP(i)] \tag{12.12}$$

12.4 Application of Optimized Fuzzy Controllers and Results

Some typical cases are simulated in this section to illustrate the effectiveness of the designed fuzzy controllers. In which, the robot task is to move from a current position (x_0, y_0) to a desired goal (x_g, y_g) in various unknown environments (free or with encounter obstacles).

12.4.1 Goal Seeking Task

The role of the goal seeking behavior is to align the mobile robot with the direction of the desired goal to reach their coordinates (x_g, y_g) by considering all types of environments (containing or without obstacles).

The application of optimized FLC is summarized in the implementation of a Takagi-Sugeno fuzzy controller for mobile robot navigation. The initialized rule-base is adjusted on-line using a reinforcement signal (reward or punishment). In this optimization algorithm, the robot has an initial rule-base which defines the possible situations with the proposed actions for the desired robot task (optimization of conclusion parts).

12.4.1.1 Steering Controller

For each time step, the robot must define the state in which it is by calculating the angle between its orientation and that of the goal noted (θ_{rg}), and the distance between the robot and the target noted (d_{rg}) defining the position error. Using these

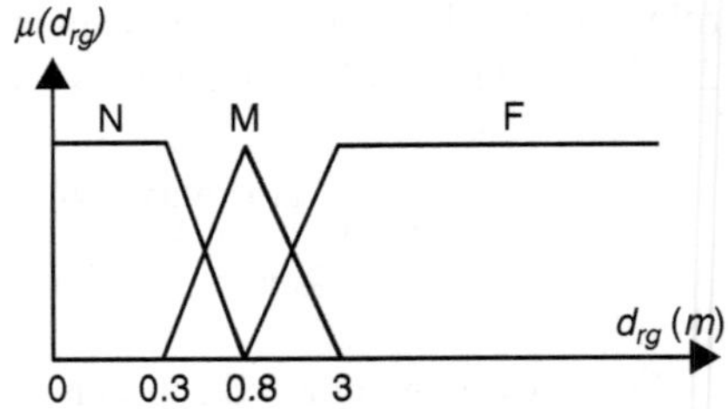

Fig. 12.3 The MFs of d_{rg}

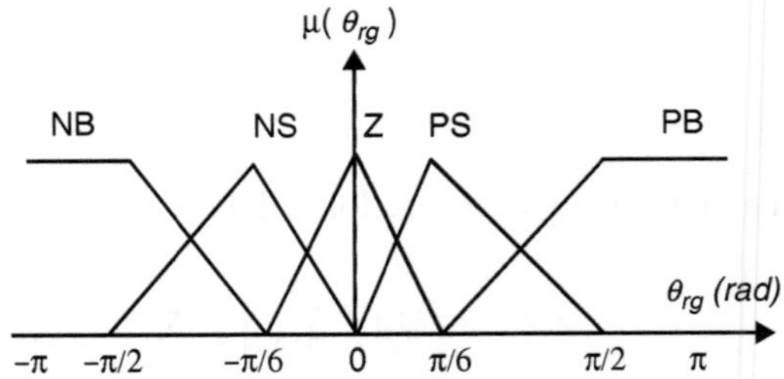

Fig. 12.4 The MFs of θ_{rg}

two variables as inputs, the optimized FLC must generate the appropriate control actions allowing the robot to move toward the desired goal. In this first application, we assume the robot motion with a fixed velocity.

The used membership functions to define the states of this behavior are shown in Figs. 12.3 and 12.4. This subdivision defines 15 fuzzy rules with the following fuzzy labels: (N: Near, M: Medium, F: Far, Z: Zero, PS: Positive Small, PB: Positive Big, NB: Negative Big, NS: Negative Small).

The fuzzy rule-base of the robot goal seeking behavior is as follows:

$$\begin{array}{llll} \text{if } s \text{ is } S_i & \text{Then } \alpha = \alpha_{i1} & \text{with the quality} & q[i,1] \\ & \text{or } \quad \alpha = \alpha_{i2} & \text{with the quality} & q[i,2] \\ & \text{or } \quad \alpha = \alpha_{i3} & \text{with the quality} & q[i,3] \end{array} \tag{12.13}$$

for: $i = 1 \ldots 15$.

For each rule, three conclusions are proposed: $\alpha_1 = -\frac{\pi}{5}$, $\alpha_2 = 0$, and $\alpha_1 = \frac{\pi}{5}$ (corresponding to the steering angle α of the front robot wheel). After a training time, the robot should choose for each fuzzy-rule the best conclusion corresponding to the best Q-function $q[i, j]_{j=1}^{N}$.

During the optimization process, the robot is guided by the following reinforcement signals:

$$r = \left| \begin{array}{rl} 4 & \text{if the robot reaches the target} \\ 3 & \text{if } d_{rg} \text{ decreases and } \theta_{rg} = 0 \\ 2 & \text{if } d_{rg} \text{ and } \theta_{rg} \text{ decrease} \\ -1 & \text{if } d_{rg} \text{ decreases and } \theta_{rg} \text{ increases} \\ -2 & \text{if } d_{rg} \text{ increases} \\ -3 & \text{if a collision is occurred with the environment} \end{array} \right. \tag{12.14}$$

Starting from the actual perceived state, the robot must take a decision on the action to be carried out (the steering angle). According to the result obtained during the execution of the selected action, it is either punished to decrease the probability of execution of the same action in the future, or rewarded, to support this behavior in the similar situations. These applied actions are chosen by the exploration-exploitation policy (EEP) shown in Eq. (12.10) in order to explore the robot state spaces.

12.4.1.2 Optimized FLC for a Fixed Position

Firstly, we tested the optimized FLC with fixed positions of the robot and the goal. Figure 12.5a, b present the robot paths during the learning process (episodes 1 and 40). As seems, the robot can't reach its goal, but after a sufficient time of optimization (number of episodes), the robot moves toward the desired goal successfully (it can reach their final coordinates) by executing continuous actions (steering angle). This result is depicted in Fig. 12.6. So, the conclusion part of the fuzzy rule-base is optimized by the selection of the best conclusion for each state condition. The optimized fuzzy controller is effective and well performed to accomplish this task.

The average rewards are maximized to a fixed value indicating the best optimization of parameters as shown in Fig. 12.7.

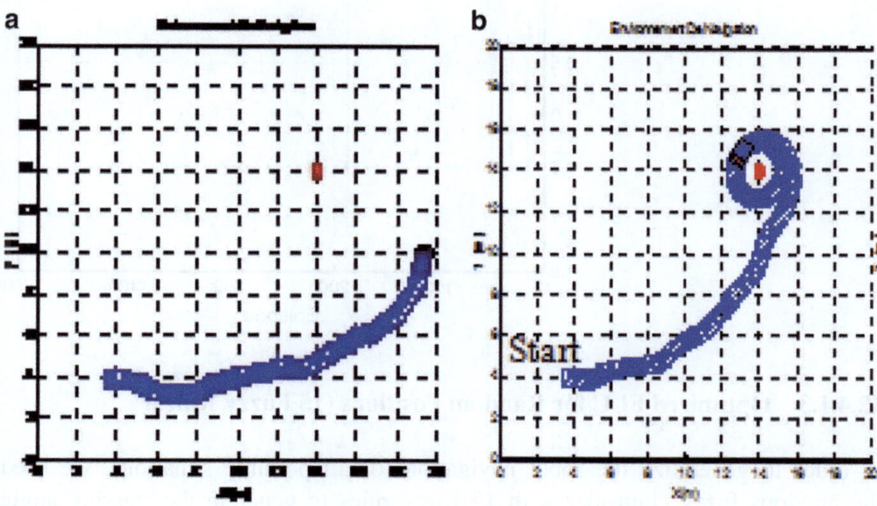

Fig. 12.5 (**a**) Episode 1. (**b**) Episode 40

Fig. 12.6 Mobile robot path

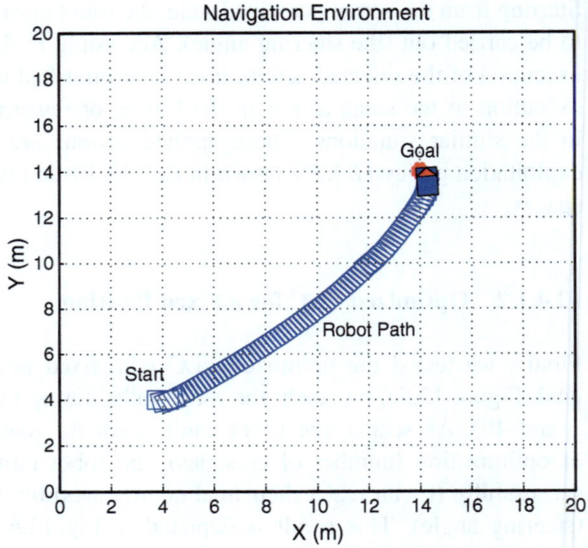

Fig. 12.7 Average rewards

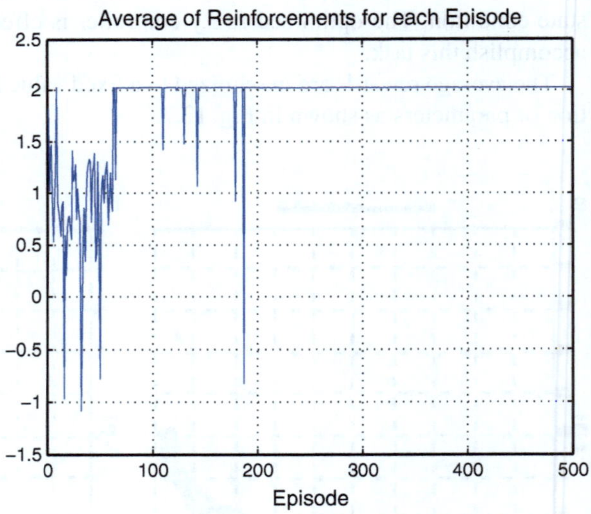

12.4.1.3 Optimized FLC for Random Positions (15 Fuzzy Rules)

In order to generalize the robot navigation for all possible situations, we used the previous fuzzy controller with 15 fuzzy rules to generate the steering angle α. The same parameters are used (number of membership functions: 3×5 and 3 conclusions, reinforcement values,...). The optimization process is made with a random initial positions of the robot and the goal for each episode. Each episode starts with a random position of the robot and finished when it is reached the target or strikes the limits of its environment. After an optimization task, the robot behavior

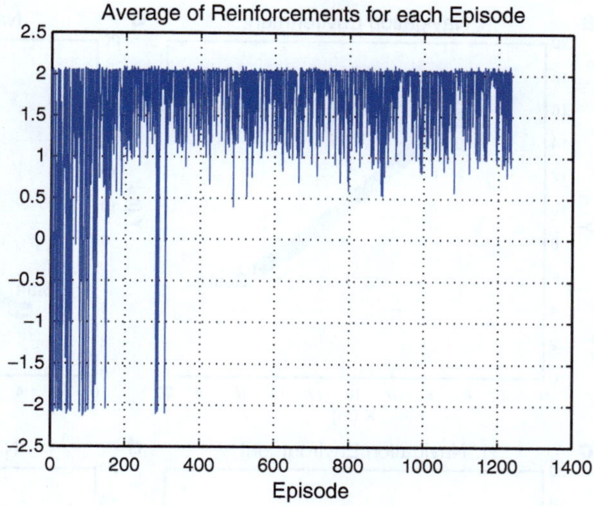

Fig. 12.8 Maximization of the robot rewards

is improved during the time. The average rewards are maximized indicating the best selected conclusions for each episode (Fig. 12.8). Figure 12.9a–d shows the obtained robot paths. As depicted, in all cases, the robot moves toward the goal effectively by executing continuous actions.

12.4.1.4 Optimized FLC (15 Fuzzy Rules-3 Conclusions)

In our contribution, we have applied this optimization method in two steps: the first to optimize the FLC to generate the appropriate robot steering angle (α), and the second to learn the optimal motion velocity (v_r).

Velocity Controller

When the robot is able to reach his final goal with a fixed speed (as done previously by the steering controller with the same parameters), we have applied the second algorithm (FQLA) to optimize the FLC of velocity. The premisse parts of the fuzzy rules are the same used for the calculation of the steering angle, but the difference is in the conclusion part. The proposed fuzzy rule-base is in the form given by the following equations:

$$\begin{aligned} \text{if } s \text{ is } S_i \text{ Then } & v_r = v_{i1} \text{ with the quality } q[i,1] \\ \text{or } & v_r = v_{i2} \text{ with the quality } q[i,2] \\ \text{or } & v_r = v_{i3} \text{ with the quality } q[i,1] \end{aligned} \tag{12.15}$$

for: $i = 1 \ldots 15$.

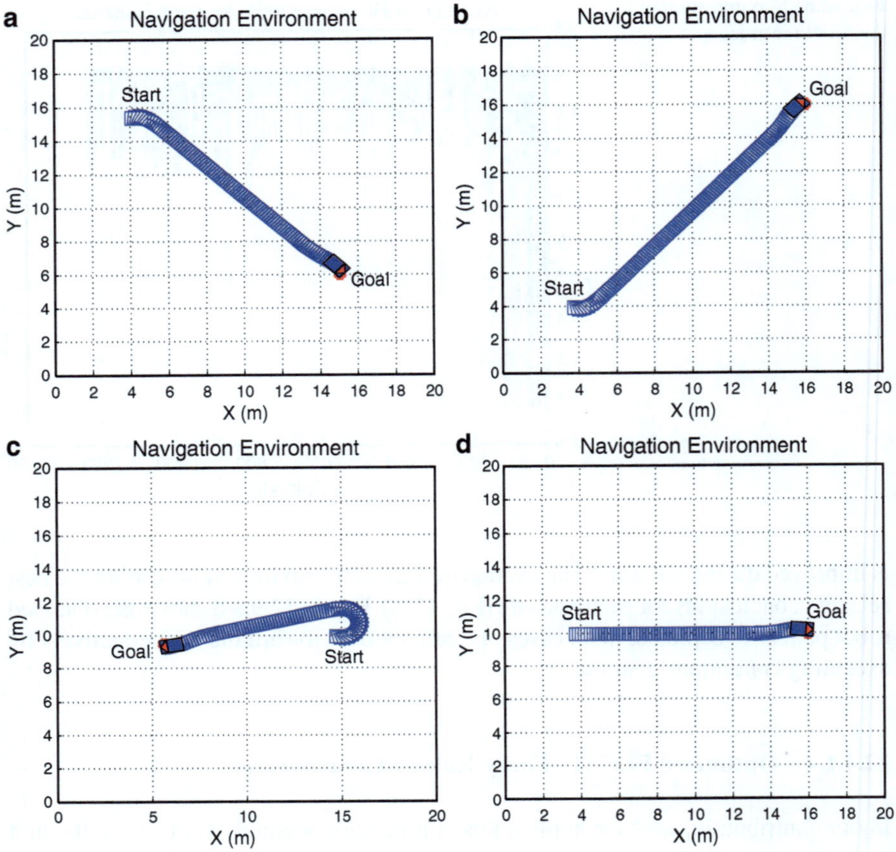

Fig. 12.9 Goal seeking using optimized FLC (15 fuzzy rules-3 conclusions)

To learn the robot velocity action, the mobile robot receives the following reinforcement values as punishments and rewards:

$$r = \left| \begin{array}{l} -1 \text{ if } d_{rg} \text{ and } \theta_{rg} \text{ decrease} \\ -1 \text{ if } \theta_{rg} \text{ decreases and the velocity and } d_{rg} \text{ increase} \\ \;\;\, 1 \text{ if the robot reaches the target with low velocity} \\ \;\;\, 0 \text{ elsewhere} \end{array} \right. \tag{12.16}$$

In Fig. 12.10a, b, we illustrate the average values of the obtained reinforcements in each episode to learn the two control values (steering angle and the linear velocity). As shown on these two curves, the behavior of the robot improves over time. The reinforcements are maximized illustrating the convergence of the optimized FLCs to accomplish this behavior. The trajectories of the mobile robot for this functionality with different start and end situations after learning are shown in Fig. 12.11a–d.

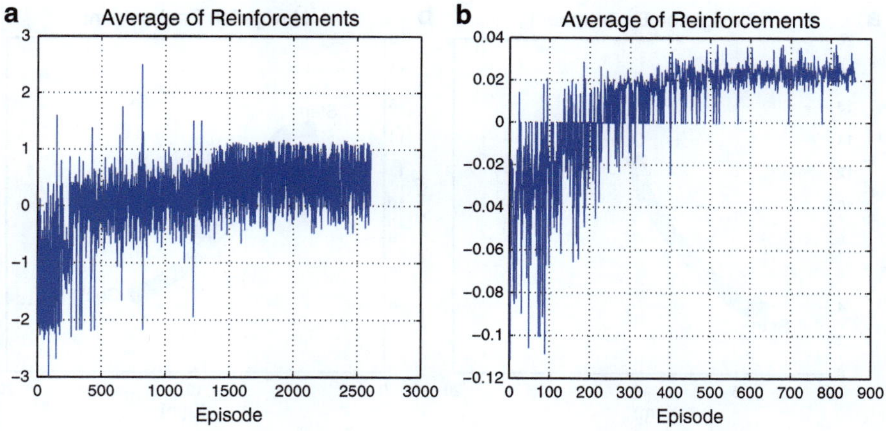

Fig. 12.10 Average values of the received reinforcements: steering and velocity

We observe that the robot is always able to move toward the desired goal from their initial positions autonomously. The robot stops when it reaches the goal by executing continuous actions generated by the optimized FLC.

12.4.1.5 Optimized FLC with 35 Fuzzy Rules

In the second test, we will optimize a Takagi-Sugeno FLCs for the goal seeking behavior by adjusting the conclusion parts for these two control actions (Steering controller and velocity controller). The two fuzzy systems are defined by 35 fuzzy rules with 5 proposed conclusions for each rule.

Steering Controller

The premisse part of the FLC is defined by the membership functions depicted in Figs. 12.12 and 12.13 for the distance (d_{rg}) and the angle (θ_{rg}) respectively. The used fuzzy labels are: Z(Zeros), S(Small), M(Medium), B(Big), VB(Very Big), NB(Negative Big), NM(Negative Medium), NS(Negative Small), PS(Positive Small), PM(Positive Medium) and PB(Positive Big). For the proposed steering angle values, it is suggested to uniformly distributed in the interval $\left[-\frac{\pi}{5}, \frac{\pi}{5}\right]$. The fuzzy rule-base used to define this FLC is as follows:

$$\begin{aligned} \text{if } s \text{ is } S_i \text{ Then } \alpha &= \alpha_{i1} \text{ with the quality } q[i,1] \\ \text{or } \alpha &= \alpha_{i2} \text{ with the quality } q[i,2] \\ \text{or } \alpha &= \alpha_{i3} \text{ with the quality } q[i,3] \\ \text{or } \alpha &= \alpha_{i4} \text{ with the quality } q[i,4] \\ \text{or } \alpha &= \alpha_{i5} \text{ with the quality } q[i,5] \end{aligned} \tag{12.17}$$

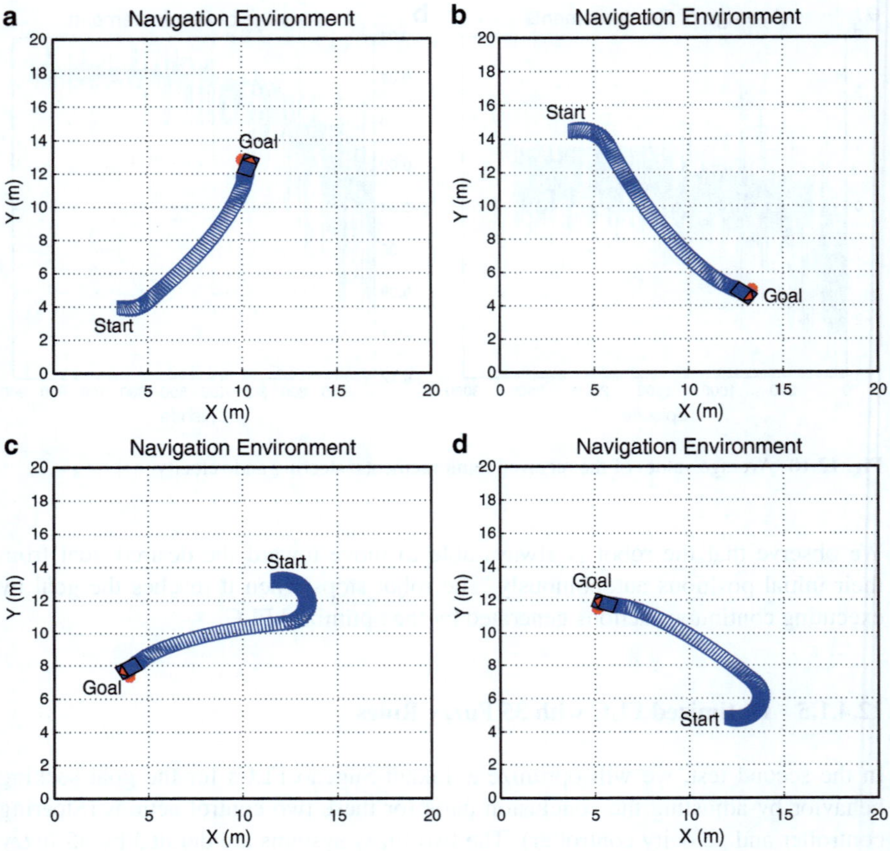

Fig. 12.11 Goal seeking using optimized FLC (15 fuzzy rules-3 control actions)

Fig. 12.12 Membership functions of d_{rg}

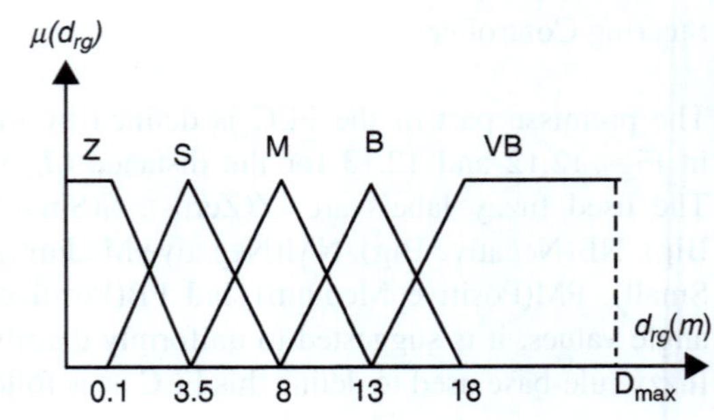

where: $\alpha_1 = -\frac{\pi}{5}, \alpha_2 = -\frac{\pi}{10}, \alpha_3 = 0, \alpha_4 = \frac{\pi}{10}$ and $\alpha_5 = \frac{\pi}{5}$.

As the same application done previously, this FLC is tuned on-line firstly for the basic control action (steering angle). After that, when the robot is able to reach the goal, this controller is optimized too to learn the required motion velocity. In both

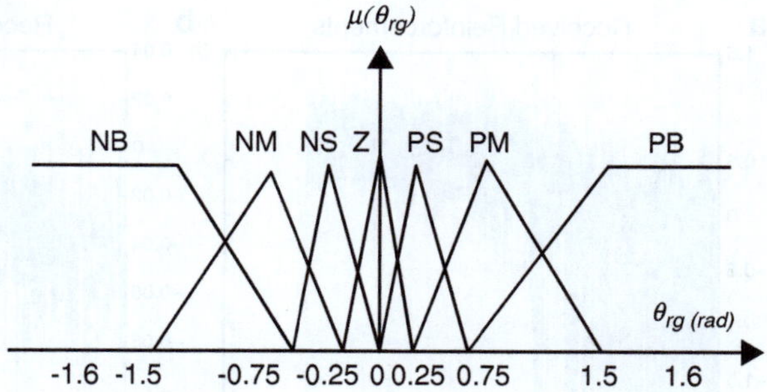

Fig. 12.13 Membership functions of θ_{rg}

optimization phases, the robot chooses for each fuzzy rule the optimal conclusion with the best quality function $q[i, j]_{j=1}^{N}$.

Velocity Controller

The fuzzy rule-base used to define this second FLC to learn the velocity is shown in Eq. (12.18). This controller contains 35 fuzzy rules and 5 proposed velocities distributed from 0 as a stop action to vmax. This last corresponds the maximum motion velocity.

$$\begin{array}{llll} \text{if } s \text{ is } S_i & \text{Then } v_r = v_{i1} & \text{with the quality} & q[i,1] \\ & \text{or } \quad v_r = v_{i2} & \text{with the quality} & q[i,2] \\ & \text{or } \quad v_r = v_{i3} & \text{with the quality} & q[i,3] \\ & \text{or } \quad v_r = v_{i4} & \text{with the quality} & q[i,4] \\ & \text{or } \quad v_r = v_{i5} & \text{with the quality} & q[i,5] \end{array} \tag{12.18}$$

The inferred control actions for robot movement are calculated by:

$$\alpha(s) = \sum_{i=1}^{35} w_i(s)\, \alpha[i,k] \tag{12.19}$$

$$v_r(s) = \sum_{i=1}^{35} w_i(s)\, v_r[i,k] \tag{12.20}$$

where: $k = 1 \ldots 5$, $w_i(s)$ is he firing strength of the ith rule in the state s.

After the two optimization passes, the FLCs are able to generate the correct control actions (steering and velocity) to move toward the desired destination. The indicators of optimization are the mean of the received reward maximized in each episode as depicted in Fig. 12.14a, b for the two actions respectively. The robot behavior is improved during the time to obtain the optimal parameters of the steering and the velocity Controllers. Figure 12.15a–d present the robot paths for different

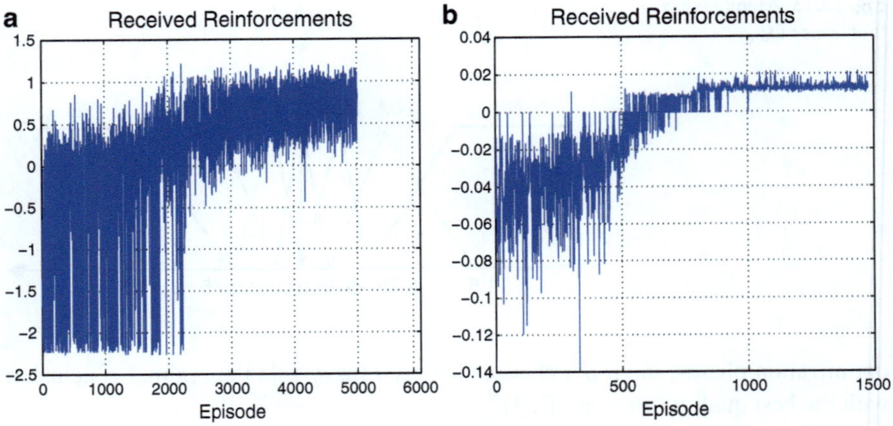

Fig. 12.14 Average values of the received reinforcements: steering and velocity

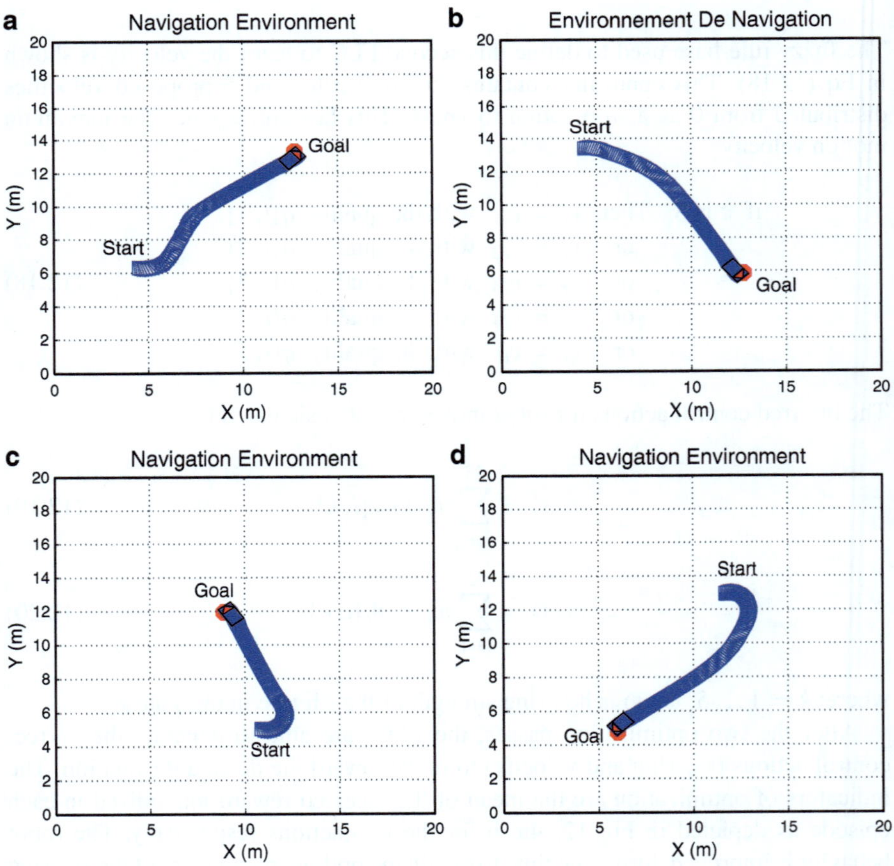

Fig. 12.15 Goal seeking using optimized FLC (35 fuzzy rules-5 control actions)

initial positions of the robot and the goal. The robot is able to reach the target successfully by executing the optimal actions. The obtained behavior is acceptable for this task.

As a comparison with the previous controller, the convergence of this algorithm (35 rules-5conclusions) to optimize the conclusion parts needs more time than the first (15 fuzzy rules-3 conclusions). But the robot behavior is more suitable.

12.4.2 Obstacle Avoidance Task

Obstacle avoidance (OA) is one of the basic missions of a mobile robot. It is a significant task that must have all the robots. This behavior permits these vehicles to move in unknown environments without collisions. The obstacle avoidance tends to steer the robot in such a way as to avoid collision with the nearest obstacles and objects that happens to be in the vicinity of the robot. It consists to keep a safe distance that the robot can move effectively.

To realize this task, it is assumed that the robot is able to determine the distance to an obstacle noted (D) and the angle between the obstacle and the orientation of the robot noted (θ_{r0}). These two variables define an area around the nearest obstacle. This area is considered critical. If the robot is located in this space, it is penalized and considered that a collision is occurred. The distance and the angle error to an obstacle are fuzzified using the membership functions presented in Figs. 12.16 and 12.17 respectively. Where: N: Near, M: Medium, and F: Far, Z: Zeros, NB: Negative Big, NM: Negative Medium, NS: Negative Small, PS: Positive Small, PM: Positive Medium and PB: Positive Big.

This fuzzy controller is based on 21 fuzzy rules. The reinforcement signals are used to select one conclusion from the set of the proposed conclusions: $\alpha_1 = -\frac{\pi}{5}$, $\alpha_2 = 0$ and $\alpha_3 = \frac{\pi}{5}$. In the tuning process, the robot receives these values as rewards and punishments:

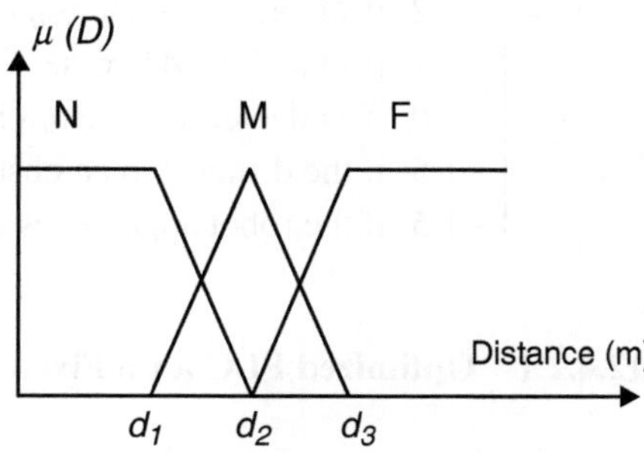

Fig. 12.16 MFs of the distance to obstacles (D)

Fig. 12.17 MFs of θ_{r0}

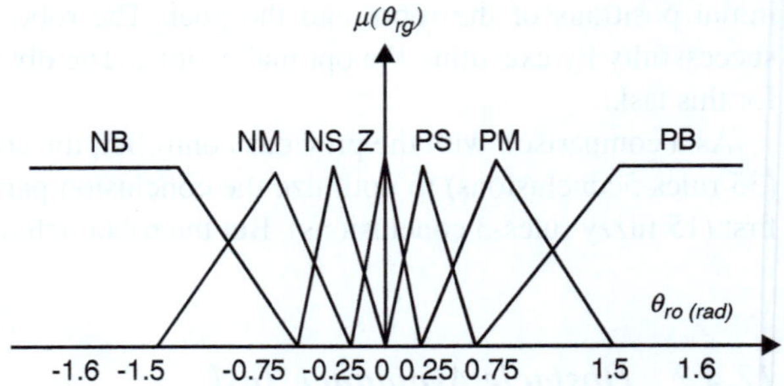

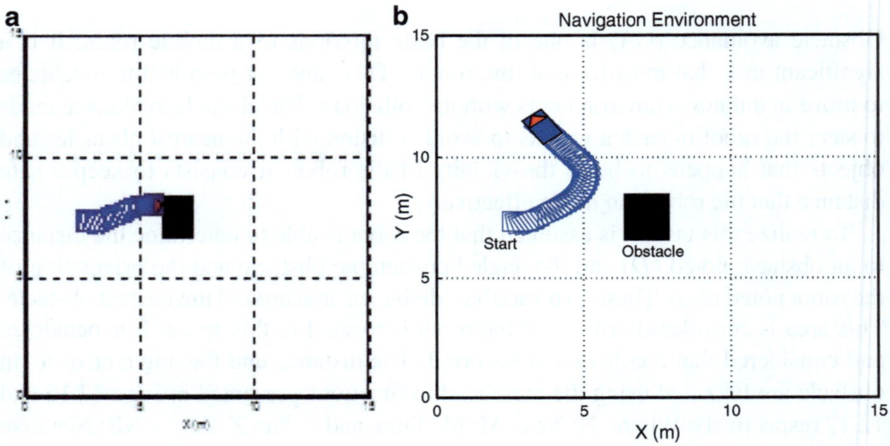

Fig. 12.18 (**a**) Episode 1. (**b**) Path after optimization

$$
r = \begin{cases} -3 & \text{if the robot sticks an obstacle} \\ -2 & \text{if an obstacle is near area} \\ 2 & \text{if an obstacle is detected on right or on left,} \\ & \quad \text{or it exists in a free area} \\ -2 & \text{if } D \text{ decreases and } \theta_{r0} = 0 \\ -1 & \text{if } D \text{ and } \theta_{r0} \text{ decrease} \\ 0 & \text{if } D \text{ decreases and } \theta_{r0} \text{ increases} \\ 1.5 & \text{if the distance to an obstacle increases} \\ -1.5 & \text{if the robot approaches the limits of the environment} \end{cases} \tag{12.21}
$$

12.4.2.1 Optimized FLC for a Fixed Position

For a fixed situation of the robot and the obstacle to avoid in their environment. In the first episode, the robot can't avoid the collision (so the FLC is not yet adjusted) as depicted in Fig. 12.18a. After a sufficient time of optimization (learning), the robot is able to avoid collision with the detected obstacle effectively and autonomously as shown in Fig. 12.18b. The mean values of the received reinforcement are maximized

Fig. 12.19 Reinforcements

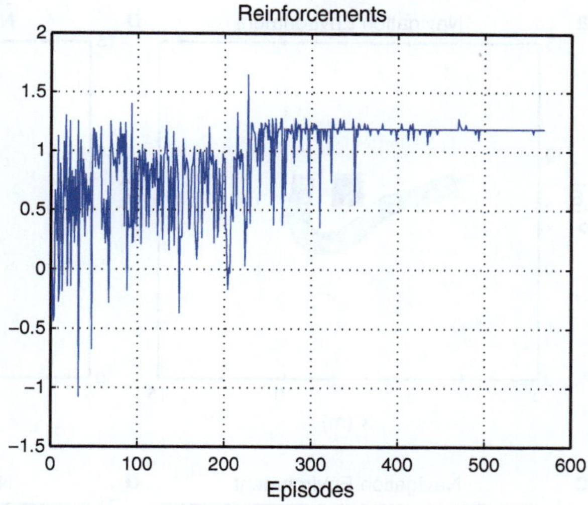

to a constant value (Fig. 12.19) indicating the optimal FLC parameters and the improvement of the robot behavior.

12.4.2.2 Optimized FLC for Random Positions

With random initial positions of the robot in its environment in each episode. Figure 12.20a–d presents the paths of the mobile robot to learn the obstacle avoidance behavior using an optimized FLC with 21 fuzzy rules and 3 proposed conclusions. After the optimization process, the fuzzy controller can guide the robot to avoid collisions with the obstacle effectively by generating correct and optimal control actions (steering with a fixed velocity).

The optimization task is verified by the maximization of the average reinforcement at each episode as depicted in Fig. 12.21.

12.4.2.3 Obstacle Avoidance Using Optimized FLC

Steering Controller

In this application, the optimized fuzzy controller uses as inputs the distances to obstacles in the three directions of mobile robot (on front d_F, on the right d_R and on the left d_L). Figure 12.22 represents the fuzzy sets used to fuzzify the readings of sensors around the robot with two linguistic variables, Near (N) and Far (F). To achieve this behavior, the rule-base consists of 8 basic situations. The optimization phase permits to tuning the FLC parameters in order to accomplish the obstacle avoidance task. For each fuzzy rule, we associate a quality function $q[i, j]_{j=1}^{J}$ for

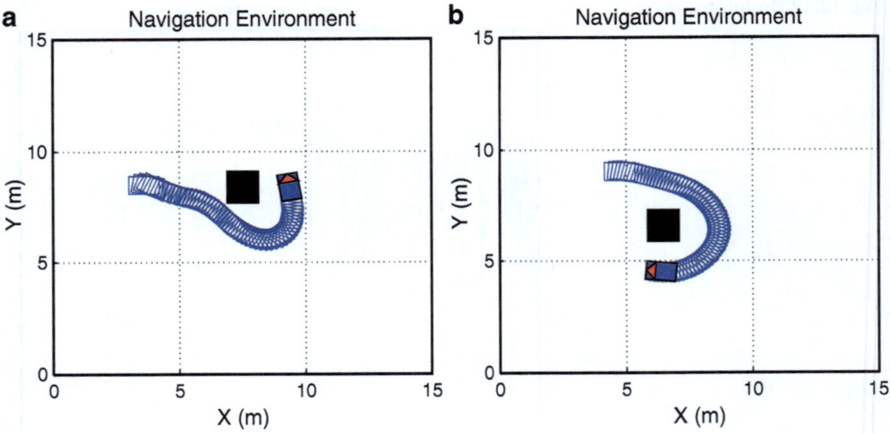

Fig. 12.20 Obstacle avoidance (21 rules-3 conclusions)

Fig. 12.21 Reinforcement values (random positions) (21 rules-3 conclusions)

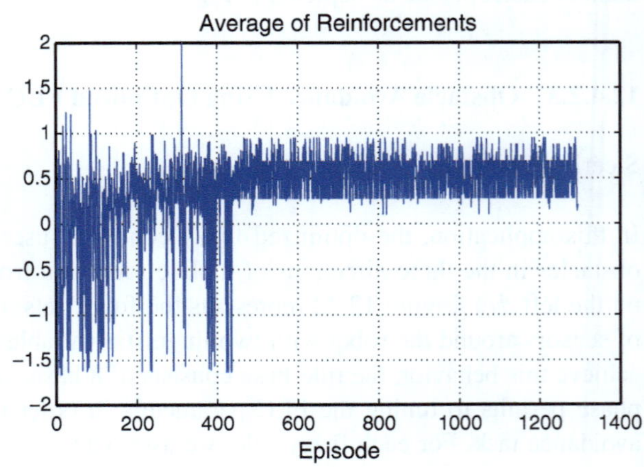

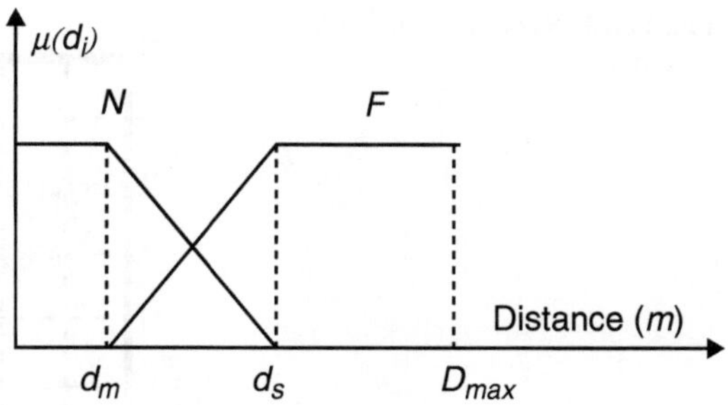

Fig. 12.22 MFs of the distances: d_F, d_R and d_L

the interpretation j of the rule i, where: $i = 1 \ldots 8$ and $j = 1 \ldots 3$. The used fuzzy rule-base to tune the steering angle is illustrated as:

$$\begin{array}{lll} \text{if } s \text{ is } S_i \text{ Then } & \alpha = \alpha_{i1} & \text{with the quality } q[i,1] \\ \text{or} & \alpha = \alpha_{i2} & \text{with the quality } q[i,2] \\ \text{or} & \alpha = \alpha_{i3} & \text{with the quality } q[i,3] \end{array} \tag{12.22}$$

where: $i = 1 \ldots 8$, $j = 1 \ldots 3$, $\alpha_1 = -\frac{\pi}{5}$, $\alpha_2 = 0$ and $\alpha_3 = \frac{\pi}{5}$.

The global quality of the inferred output is given by:

$$Q(s, \alpha) = \sum_{i=1}^{8} w_i(s)\, q[, j^{\star}(i)] \tag{12.23}$$

The reinforcement's values presented in the following equation are used to the optimization of the steering angle.

$$r = \left| \begin{array}{l} -4 \text{ if a collision is occured} \\ -1 \text{ if } d_i \text{ decreases and } d_i < \frac{1}{2} D_{max} \\ \;\;0 \text{ elsewhere} \end{array} \right. \tag{12.24}$$

This signal punished the activated conclusion permits the robot to enter in collision with the obstacles. In this case, the rewards values are used to define the best conclusion part from the three proposed conclusions: α_1, α_2 and α_3 as steering angles of the front wheel, where d_i, for $i = 1 \ldots 3$ are the distances in three directions, d_m is the minimum distance, d_s is distance of security and D_{max} is the maximum reading of the sensors.

Velocity Controller

After adjusting the steering values, we apply the same algorithm (FQLA) to generate the appropriate motion velocity, we used the following reinforcement values:

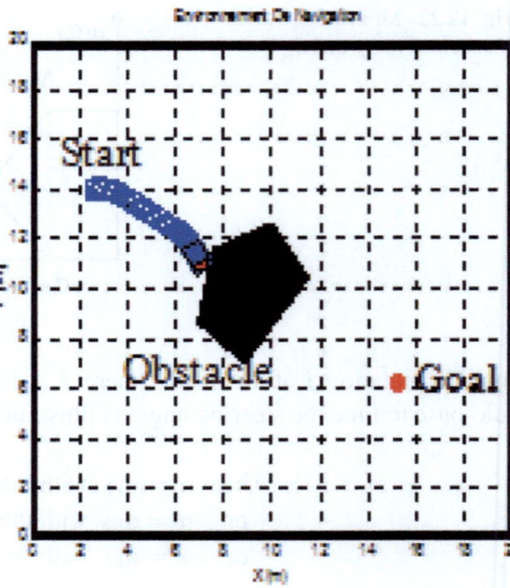

Fig. 12.23 Robot path in episode 1

$$r = \begin{vmatrix} -4 & \text{if the robot sticks an obstacle} \\ -1 & \text{if } d_i > D_{max} \text{ and } v_r \text{ is low, with } i = 1 \ldots 3 \\ -1 & \text{if } d_i < D_{max} \text{ and } v_r \text{ increases, with } i = 1 \ldots 3 \\ 0 & \text{elsewhere} \end{vmatrix} \quad (12.25)$$

The rule-base employed to learn the motion velocity is:

$$\begin{array}{lll} \text{if } s \text{ is } S_i \text{ Then} & v_r = v_{i1} & \text{with the quality } q[i, 1] \\ \text{or} & v_r = v_{i2} & \text{with the quality } q[i, 2] \\ \text{or} & v_r = v_{i3} & \text{with the quality } q[i, 1] \end{array} \quad (12.26)$$

In the first steps (during learning), the robot can't avoid collisions with obstacles as depicted in Fig. 12.23. After an optimization phase, the robot obtains the best behavior to reach the goal (by selecting the conclusion that maximize the total rewards). The robot can avoid the obstacles and moves in the direction of the target. If there is a near obstacle (detected by their sensors), it chooses the turn right action as depicted in Fig. 12.24a, b. The obtained behavior is more acceptable and the robot path is considered as optimal. For this situation, the average values of the received reinforcements are maximized for the two actions: Steering and velocity as seem in Fig. 12.25a, b. This maximization indicates the best optimization process.

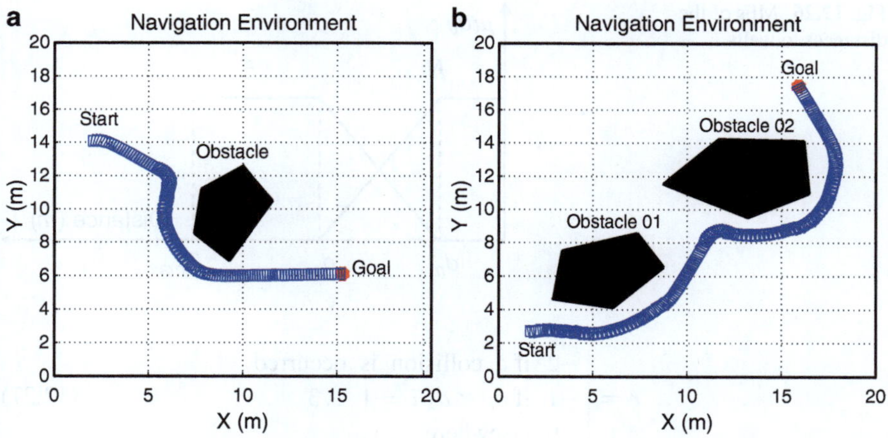

Fig. 12.24 Robot paths after optimization

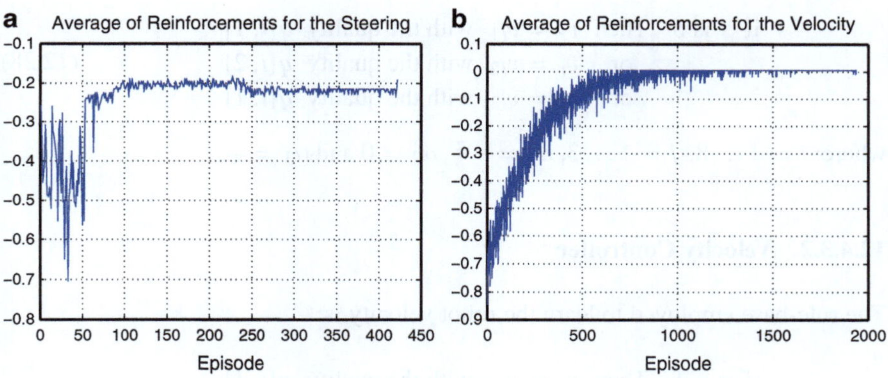

Fig. 12.25 Received reinforcements: steering and velocity (8 rules-3 conclusions)

12.4.3 Wall-Following Task

The objective of the wall following behavior is to keep the robot at a safe close distance to the nearest right or left wall. This task is accomplished using sensor readings about its environment presenting distances to obstacles and walls in various directions (front, right or left). The fuzzy sets used to fuzzify the three distances are illustrated in Fig. 12.26.

12.4.3.1 Steering Controller

In this application, we use an on-line optimization of this fuzzy rule-base with an imprecise knowledge using the following reinforcement signals:

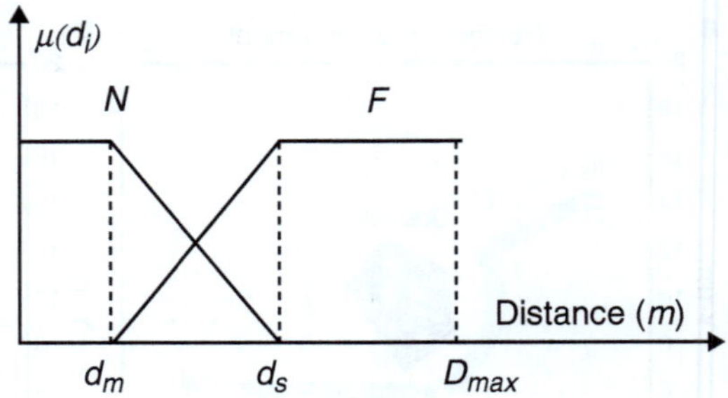

Fig. 12.26 MFs of the distances to walls

$$r = \begin{vmatrix} -2 & \text{if a collision is occurred} \\ -1 & \text{if } d_i < d_1 \; i = 1 \ldots 3 \\ 0 & \text{elsewhere} \end{vmatrix} \tag{12.27}$$

The fuzzy rule-base used to tune the steering angle fuzzy controller is as follows:

$$\begin{array}{lll} \text{if } s \text{ is } S_i \text{ Then } & v_r = v_{i1} & \text{with the quality } q[i,1] \\ \text{or} & v_r = v_{i2} & \text{with the quality } q[i,2] \\ \text{or} & v_r = v_{i3} & \text{with the quality } q[i,1] \end{array} \tag{12.28}$$

where: $i = 1 \ldots 8$, $j = 1 \ldots 3$, $\alpha_1 = -\frac{\pi}{5}$, $\alpha_2 = 0$ and $\alpha_3 = \frac{\pi}{5}$.

12.4.3.2 Velocity Controller

The rule-base employed to learn the robot velocity is:

$$\begin{array}{lll} \text{if } s \text{ is } S_i \text{ Then } & v_r = v_{i1} & \text{with the quality } q[i,1] \\ \text{or} & v_r = v_{i2} & \text{with the quality } q[i,2] \\ \text{or} & v_r = v_{i3} & \text{with the quality } q[i,1] \end{array} \tag{12.29}$$

After a learning process, the optimization results are shown in Fig. 12.27a, b presenting the mobile robot paths in unknown environment. The robot behavior is improved and the robot is able to navigate in its environment without collision with walls. It can keep a safe distance to the nearest left wall autonomously. The robot wall-following behavior is satisfactory in all cases.

The mean values of the reinforcements are maximized as depicted in Fig. 12.28a, b indicating the optimal FLC parameters for the two control actions: steering and velocity.

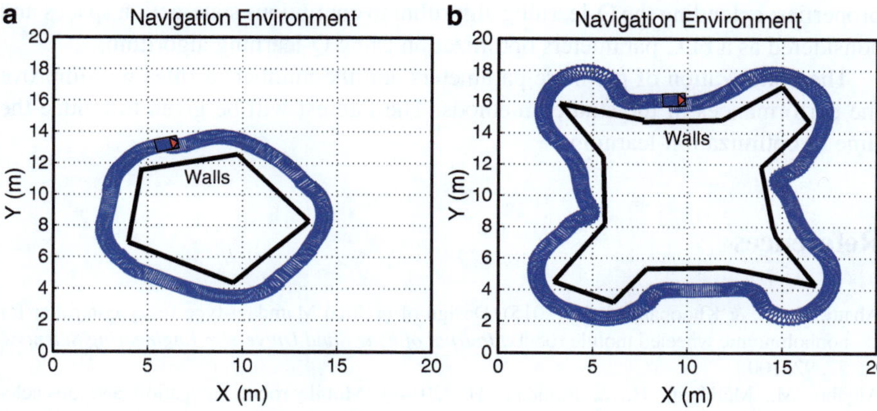

Fig. 12.27 Wall-following with different forms

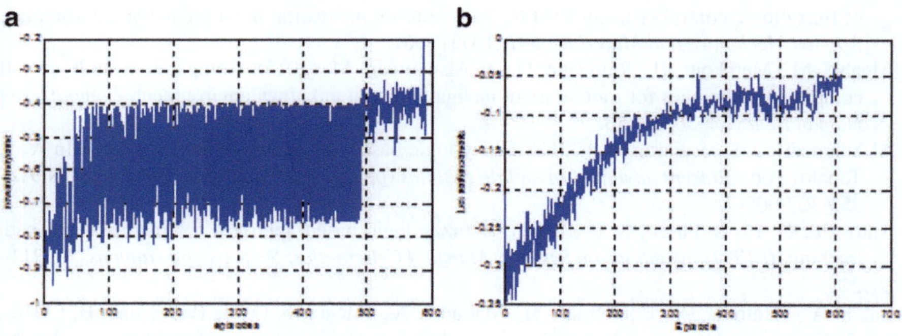

Fig. 12.28 Average values of the received reinforcements: steering and velocity

12.5 Conclusions

In this chapter, we presented a study of navigation methods for an autonomous mobile robot using optimized fuzzy logic controllers. The Q-learning algorithm is a powerful tool to obtain an optimal fuzzy behavior which requires only one scalar signal as a feedback indicating the quality of the applied action. Fuzzy controller promises an efficient tool for mobile robot path planning. It is tolerant to noise and error in the information coming from the sensory system and uncertainties, and most importantly; it is a factual reflection of the behavior of human expertise based on If-Then inference rules. Fuzzy Controller can be effectively tuned via reinforcement learning using FQLA. The idea of fuzzy Q-Learning algorithm consists at fuzzy inference systems optimization using a reinforcement signal. This signal makes the navigator able to adjust his strategy in order to improve their performances. This technique is based on the optimization of conclusion parts of FLC in order to maximize the return function. The consequent value is selected from predefined values set which is kept unchanged during learning. This used method has two properties: extending the Q-learning algorithm to continuous state-action spaces and considered as a FLC parameters optimization using Q-learning algorithm.

The optimization of premisse parameters and the number of rules will improve the performances of the studied methods. The interest will be given to reduce the time of optimization-learning.

References

Abadi, N. M., & Khooban, M. H. (2015). Design of optimal Mamdani-type fuzzy controller for nonholonomic wheeled mobile robots. *Journal of King Saud University-Engineering Sciences, 27*, 92–100.

Algabri, M., Mathkour, H., & Ramdane H. (2014a). Mobile robot navigation and obstacle-avoidance using ANFIS in unknown environment. *International Journal of Computer Applications, 91*, 36–41.

Algabri, M., Ramdane, H., Mathkour, H., Al-Mutib, K., & Alsulaiman, M. (2014b). Optimization of fuzzy logic controller using PSO for mobile robot navigation in an unknown environment. *Applied Mechanics and Materials, 541*, 1053–1061.

Algabri, M., Mathkour, H., Ramdane, H., & Alsulaiman, M. (2015). Comparative study of soft computing techniques for mobile robot navigation in an unknown environment. *Computers in Human Behavior, 50*, 42–56.

Al Yahmedi, A. S., & Fatmi, A. (2011). Fuzzy logic based navigation of mobile robots. In A. V. Topalov (Ed.), *Recent advances in mobile robotics* (pp. 287–310). Croatia: InTech. ISBN:978-953-307-909-7.

Antonelli, G. C., & Fusco, S. G. (2007). A fuzzy logic based approach for mobile robot path tracking. *IEEE Transactions on Systems, Man and Cybernetics: Systems and Humans, 15*, 211–221.

Bakdi, A., Hentouta, A., Boutamiab, H., Maoudja, A., Hachourb, O., & Bouzouiaa, B. (2016). Optimal path planning and execution for mobile robots using genetic algorithm and adaptive fuzzy-logic control. *Robotics and Autonomous Systems, 73*, 95–109.

Begum, M., Mann, G. K., & Gosine, R. G. (2008). Integrated fuzzy logic and genetic algorithmic approach for simultaneous localization and mapping of mobile robots. *Applied Soft Computing, 8*, 150–165.

Berenji, H. R., & Khedkar, P. (1992). Learning and tuning fuzzy logic controllers through reinforcements. *IEEE Transactions on Neural Network, 3*, 724–740.

Borestein, J., & Koren, Y. (1989). Real-time obstacle avoidance for fast mobile robot. *IEEE Transactions on Systems, Man, and Cybernetics, 19*, 1179–1186.

Cherroun, L., & Boumehraz, M. (2013a). Fuzzy behavior based navigation approach for mobile robot in unknown environment. *Journal of Electrical Engineering, 13*, 284–291.

Cherroun, L., & Boumehraz, M. (2013b). Fuzzy Logic and reinforcement learning based approaches for mobile robot navigation in unknown environment. *Mediterranean Journal of Measurement and Control, 9*, 109–117.

Cuesta, F., & Ollero, A. (2005). *Intelligent mobile robot navigation*. Heidelberg: Springer.

Glorennec. P. Y., & Jouffle. L. (1997). Fuzzy Q-learning. In *6th IEEE International Conference on Fuzzy Systems*, Barcelona (pp. 659–662).

Hsu, C. C., Hou, R. Y., & Wang, W. Y. (2013). Path planning for mobile robots based on improved Ant Colony optimization. In *IEEE International Conference on Systems, Man, and Cybernetics (SMC)*, Manchester (pp. 2777–2782).

Jallouli, M., Rekik. C., Chtourou. M., & Derbel. N. (2010). Optimised fuzzy Logic controller for a mobile robot navigation. *International Journal of Modelling, Identification and Control, 9*, 400–408.

Joo, M., & Deng, C. (2005). Obstacle avoidance of a mobile robot using hybrid learning approach. *IEEE Transactions on Industrial Electronics, 52*, 898–905.

Jouffle, L. (1998). Fuzzy inference system learning by reinforcement methods. *IEEE Transactions on Systems, Man, and Cybernetics-Part C, 28*, 338–355.

Juang, C. F., & Hsu, C. H. (2009). Reinforcement ant optimized fuzzy controller for mobile-robot wall-following control. *IEEE Transactions on Industrial Electronics, 56*, 3931–3940.

Kaelbling, L. P., Littman, M. L., & Moore. A. W. (1996). Reinforcement learning: A survey. *Journal of Artificial Intelligence Research, 4*, 237–285.

Khatib, O. (1986). Real-time obstacle avoidance for manipulators and mobile robots. *International Journal of Robotics Research, 5*, 90–98.

Latombe, J. C. (1991). *Robot motion planning*. Norwell: Kluwer.

Mihai, D., & Gheorghe, M. (2016). Neural networks based reinforcement learning for mobile robots obstacle avoidance. *Expert Systems with Applications, 62*, 104–115.

Nemra, A., & Rezine, H. (2008). Genetic reinforcement learning algorithms for on-line fuzzy inference system tuning, application to mobile robotic. In P. Pecherková, M. Flídr, & J. Duník (Eds.), *Robotics, automation and control* (pp. 228–256). Vienna: InTech. ISBN:978-953-7619-18-3.

Passino, K. M., & Yurkovich, S. (1998). *Fuzzy control*. Menlo Park: Addison Wesley.

Pothal, J. K., & Parhi, D. R. (2015). Navigation of multiple mobile robots in a highly clutter terrains using adaptive neuro-fuzzy inference system. *Robotics and Autonomous Systems, 72*, 48–58.

Seraji, H., & Howard, A. (2002). Behavior-based robot navigation on challenging terrain: A fuzzy Logic approach. *IEEE Transactions on Robotics and Automation, 18*, 308–321.

Sutton, R. S., & Barto, A. G. (1998). *Reinforcement learning: An introduction*. Cambridge: The MIT Press.

Timothy, J. R. (2004). *Fuzzy logic with engineering applications* (2nd ed.). Hoboken: Wiley.

Vadakkepa, P., Miin, O. C., Peng, X., & Lee, T. H. (2004). Fuzzy behavior-based control of mobile robots. *IEEE Transactions on Fuzzy Systems, 12*, 559–564.

Wang, M., & Liu, J. N. K. (2008). Fuzzy logic based real-time robot navigation in unknown environment with dead ends. *Robotics and Autonomous Systems, 56*, 625–643.

Watkins, C., & Dayan, P. (1992). Q-Learning. *Machine Learning, 8*, 279–292.

Yang, S. X., Moallem, M., & Patel, R. V. (2005). A layered goal-oriented fuzzy motion planning strategy for mobile robot navigation. *IEEE Transactions on Systems, Man, and Cybernetics-Part B: Cybernetics, 35*, 1214–1224.

Zhao, R., Lee, D. H., & Hong, K. L. (2015). Mobile robot navigation using optimized fuzzy controller by genetic algorithm. *International Journal of Fuzzy Logic and Intelligent Systems, 15*, 12–19.

Chapter 13
Design and Implementation of a Reactive Navigation System for a Smart Robot Using Udoo Quad

Mohamed Njah and Ridha El-Hamdi

Abstract This chapter describes a process of design and implementation of a reactive navigation system for a smart mobile robot. Equipped with a webcam and distance sensors, the autonomous robot will explore an arena to locate a number of sites in a limited time all while avoiding the arena boundary and any obstacles it might encounter. A fuzzy behavior-based control scheme with adaptive membership functions has been taken as a proposed reactive navigation system. The tests of the proposed method were performed in a real robot using a UDOO Quad board, which is a single-board computer equipped with two CPU, and experiments demonstrated that this embedded system was able to successfully complete the autonomous navigation task in a real arena.

Keywords Reactive navigation system · Behavior-based control · Mobile robot · Fuzzy logic · Adaptive membership function · UDOO Quad · OpenCV

13.1 Introduction

Considering the tremendous progress in technology, autonomous robots are attracting more and more attention worldwide, and there are a lot of ongoing research and development activities in both industry and academia. There are important benefits in using autonomous mobile robot system to perform different task in AI area (Arena et al. 2004; Assis et al. 2016; Hashikawa and Morioka 2015; Siegwart et al. 2011). For example, robots may be used in prospecting missions in deep underwater sites, hostile environments, or outer space, where human operators could be at risk (Lin and Tzeng 2014).

M. Njah (✉) · R. El-Hamdi
Control and Energy Management laboratory (CEM), Digital Research Center of Sfax, Technopole of Sfax, Sakiet Ezzit, Tunisia
e-mail: mohamed.njah@enis.tn; ridha.elhamdi@enig.rnu.tn

N. Derbel et al. (eds.), *New Developments and Advances in Robot Control*, Studies in Systems, Decision and Control 175,
https://doi.org/10.1007/978-981-13-2212-9_13

Navigating a mobile robot in a known environment requires little or no interaction with the environment. However, a robot may have to operate in partially known or completely unknown environments, like planetary exploration (Mathers et al. 2012). In such circumstances the robot should be able to react appropriately to unforeseen changes in the environment taking into account noise and the inaccuracy of sensor information: it results in a great advantage for the implementation of the reactive navigation system to adopt the best behavior-based architecture to take each time the right decisions on the basis of its perception of the external environment and to avoid possible damage of the robot.

The work reported in this chapter is motivated by the need to develop an efficient reactive navigation system for an automated robot able to explore an arena to locate a number of sites in a limited time all while avoiding the arena boundary and any obstacles it might encounter.

The rest of this chapter is organized as follows: Sect. 13.2 gives a summary of work in the field of reactive navigation. In Sect. 13.3, the proposed reactive navigation system is presented. It is based on Fuzzy Inference Systems (FIS) with an adaptive membership function scheme so as to avoid revisiting the same areas twice. For solving the technical challenge of integrating such system elements and the webcam processing, we use UDOO Quad to implement the software behavior-based control system. Details of the design and its implementation, including examples of experimental results, are presented in Sect. 13.4. The last section contains conclusions and directions for future work.

13.2 Review of Related Studies

The reaction of autonomous mobile robot to unforeseen changes in the environment is one of the most difficult issues in the control of intelligent autonomous robot movements. Reactive navigation systems typically have low computational complexity, often on the order of $O(n)$ (Murphy 2000). Several approaches have been proposed to address the problem of the optimal design of reactive navigation systems. The Potential Field algorithm has been a popular reactive approach for low speed, highly manoeuvrable mobile robots (Siegwart and Nourbakhsh 2004). The attractiveness of this approach is largely due to the simplicity in summing attractive and repulsive forces to generate a desired steering direction and speed. However, in its simple form, Potential Fields is sensitive to large, or closely spaced, obstacles. In addition, if multiple obstacles are in close proximity, Potential Fields can exhibit oscillatory navigation. Even worse, the robot can stop at local minima, where all the attractive/repulsive forces sum to zero (Koren and Borenstein 1991).

In Xu et al. (2015) proposed a deployment algorithm based on the electrostatic field theory for mobile wireless sensor networks. According to this algorithm, the obstacles and nodes in the deployment area are taken as the charged particles and the particles will move due to the Coulomb's force from other particles or obstacles. At the beginning of the deployment, there is a relatively large Coulomb's force and acceleration between nodes due to their higher densities. The velocity of the moving nodes is increasing quickly, making the nodes spread over the deployment region. When nodes hit the boundary, the repulsive force from the boundary will exert on them and slow down the velocity until they will reach equilibrium.

A new potential field technique is proposed by Zhang (2013) to eliminate largely the needless obstacle avoidance. The disadvantage of this algorithm is the collision when the velocity of the hindrance increases since the repulsive force is too minor to escape the moving obstacle. This shortcoming may be largely overcome by using dynamically expanding perception zones to help track objects of immediate interest. The Dynamic Expanding Zones (DEZ) reactive navigation algorithm (Putney 2006) is believed to represent a fundamental contribution to the body of knowledge in the area of high-speed reactive navigation. This method was implemented on the Virginia Tech DARPA Grand Challenge vehicles (Putney 2006).

Kunwar and Benhabib (2008) have proposed a time-optimal interception method for the autonomous robotic interception of moving targets in the presence of dynamic obstacles. The interception manoeuvre is computed using a five-state Extended Kalman Filter (EKF). The proposed algorithm, then, uses the novel Advanced Predictive Guidance Law (APGL) to obtain the required acceleration commands for rendezvous with the target. In the presence of obstacles, the algorithm uses the novel Obstacle-Avoidance Navigation Law (OANL), which first predicts the likelihood of a collision and, then defines a collision cone for each potentially colliding obstacle (within close proximity of) the pursuer. Based on this information the algorithm directs the relative velocity between the pursuer and the obstacle outside the collision cone. Tang et al. (2013) suggested a new reactive obstacle avoidance method that minimizes the tasks complexity and improves the reactivity, but it cannot gain the shortest route. This technique is based on the "situated-activity paradigm" and a "divide and conquer" approach, which navigate the robot to travel among unidentified obstacles and towards a goal without hitting.

13.3 Fuzzy Behavior-Based Control Scheme

As we have mentioned in the introduction, a fuzzy behavior-based control scheme with adaptive membership functions has been taken as a reactive navigation system. Figure 13.1 illustrates the block diagram of the proposed system.

The robot is equipped with:

- A USB Webcam, which provides the localization of the sites identified.
- Two independent continuous rotation servo for steering and driving.
- Four ultrasonic distance sensors, which provide information on the distance between the robot and obstacles or the side of the arena.

The robot receives a continuous flow of information, from the webcam and distance sensors, about occurring events and generates new commands in response to the incoming events, while previously planned motions are being executed.

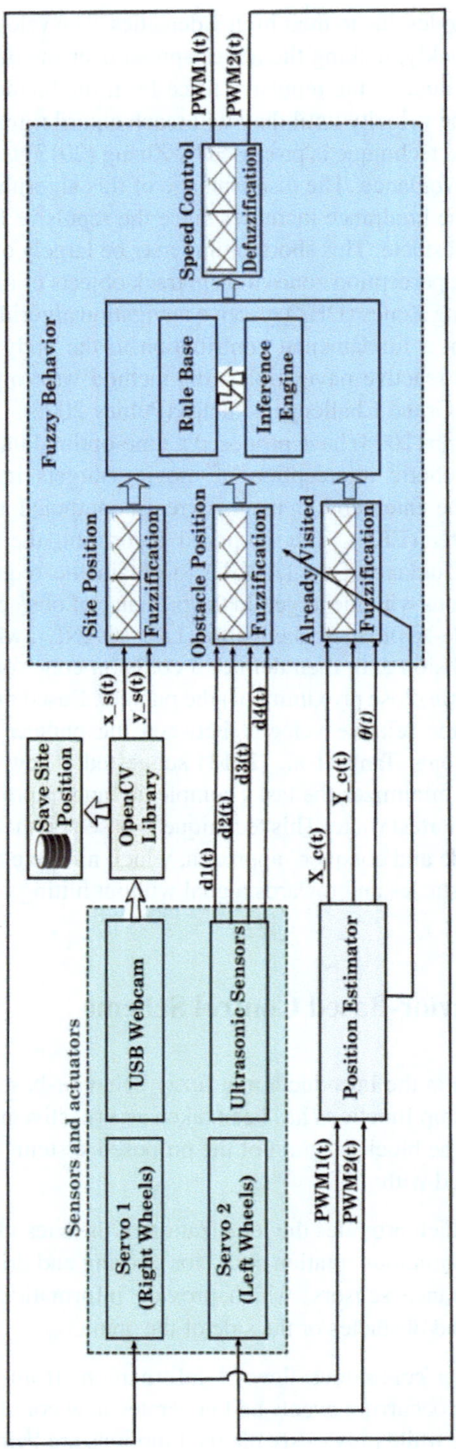

Fig. 13.1 Block diagram of the proposed reactive navigation system

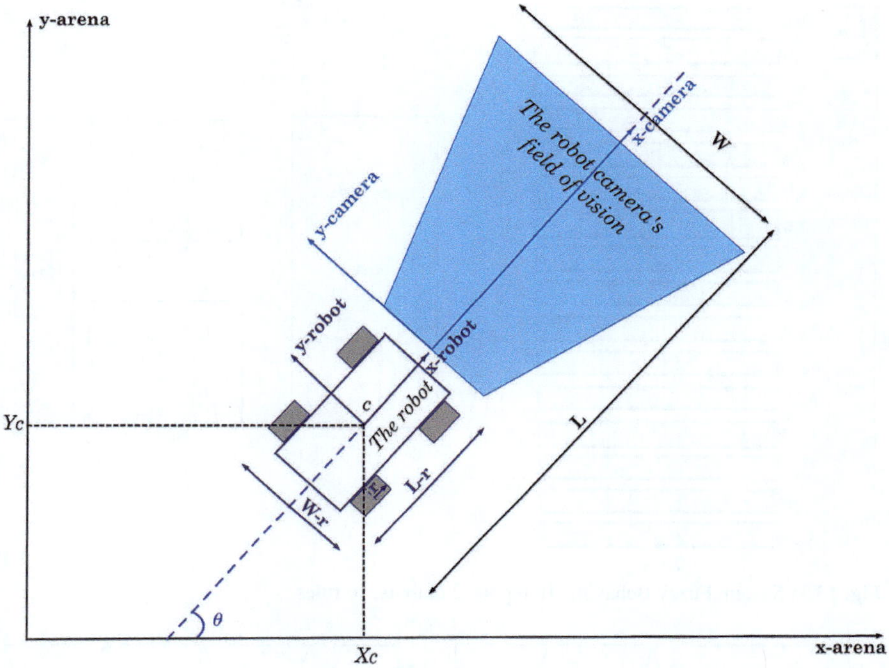

Fig. 13.2 The schematic diagram of the autonomous ground robot

In addition to sites and obstacles information, the estimation of the robot's current position ($X_c(t)$, $Y_c(t)$ and $\theta(t)$ as shown in Fig. 13.2) is communicated to the fuzzy behavior via the position estimator block.

In this chapter, a simple way is used to estimate pose. With knowledge of the wheel diameters, these values can be converted to translation and rotation values. This is easy to implement but the odometry is not representative of the exact distance travelled since errors in position quickly accumulate due to wheel slippage and due to the open-loop nature there is no way to correct these errors. In this work, the fuzzy behavior uses the information provided by the position estimator block in order to avoid revisiting the same areas twice. For this reason, basic wheel odometry is used as the only pose estimator in the robot system (it is not necessary to use a more sophisticated algorithm to know the exact position and orientation of the robot).

The concept of fuzzy behavior is similar to the conventional fuzzy logic control in that it performs an inference mapping from some input space to some output space. As shown in Fig. 13.3, the developed fuzzy behavior has 16 inputs, 2 outputs and 76 rules.

The first two inputs are the X,Y-coordinate of the nearest site detected in the video stream from the USB webcam. The used camera coordinates (x_camera, y_camera) is shown in Fig. 13.2 and the video stream image is 300 × 300. Figure 13.4 shows membership functions used to represent site position.

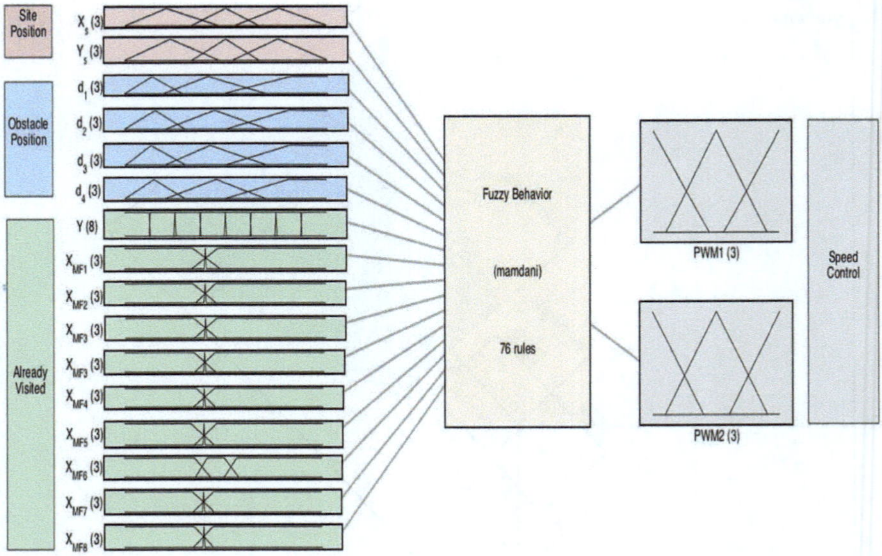

Fig. 13.3 System Fuzzy Behavior: 16 inputs, 2 outputs, 76 rules

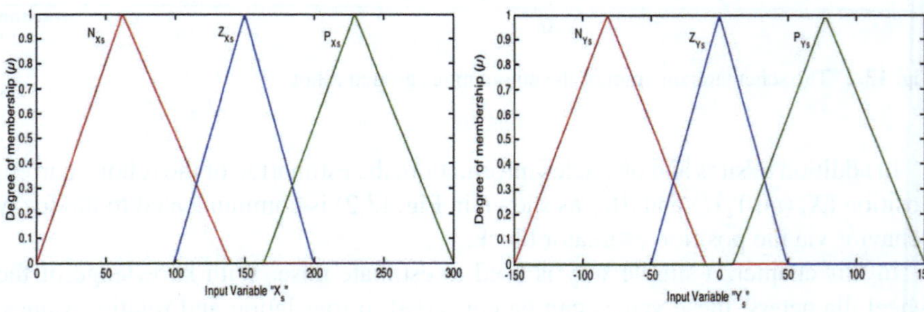

Fig. 13.4 Membership functions used to represent site position

The fuzzy variables used to control obstacle position are d_1, d_2, d_3 and d_4. Figure 13.5 shows membership functions used to represent distance to obstacle provided by the first ultrasonic sensor (associated with the input variable d_1). Note that the other three input variables (d_2, d_3 and d_4) have the same membership functions.

In order to avoid revisiting the same areas twice, the proposed reactive system has to keep track of which areas of the environment have already been explored. The challenge here resides in defining adaptive membership functions that change according to the robot movement.

The design of a fuzzy model that incorporates dynamical membership functions whose parameters are adjusted online is related to the idea of variable universe of discourse (VUD), which can tune online universes of discourse using a set of

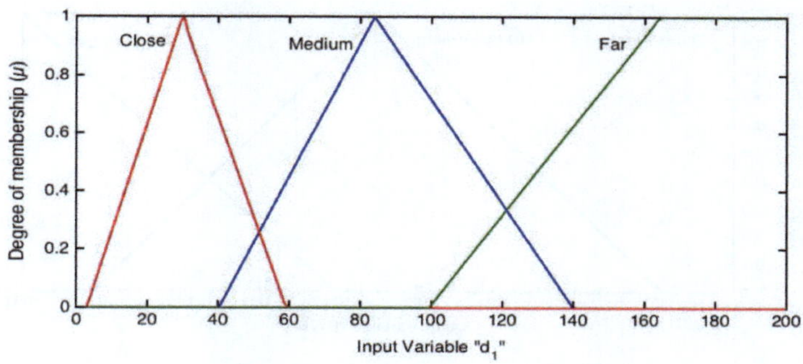

Fig. 13.5 Membership functions used to represent distance to obstacle

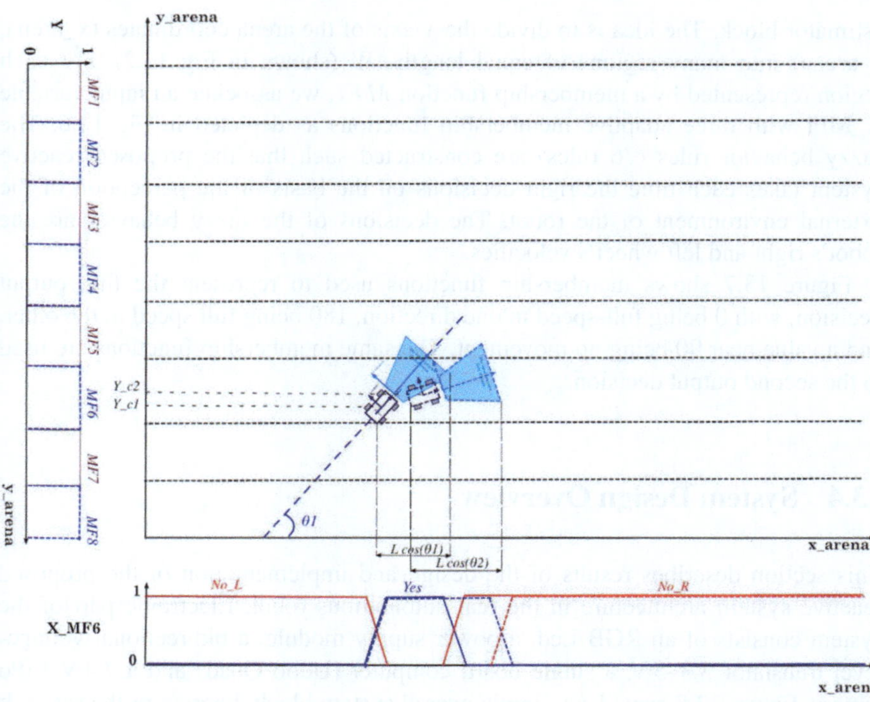

Fig. 13.6 Example of adaptive membership functions used to represent information provided by the position estimator block

nonlinear scale factors. The VUD fuzzy controllers have been applied in many fields of intelligent controlling which was applied to a nonlinear control system and obtained fine control effects (Long et al. 2010; Wang et al. 2017).

The developed fuzzy behavior uses nine input variables to represent information provided by the position estimator block. Figure 13.6 shows an example of adaptive membership functions used to represent information provided by the position

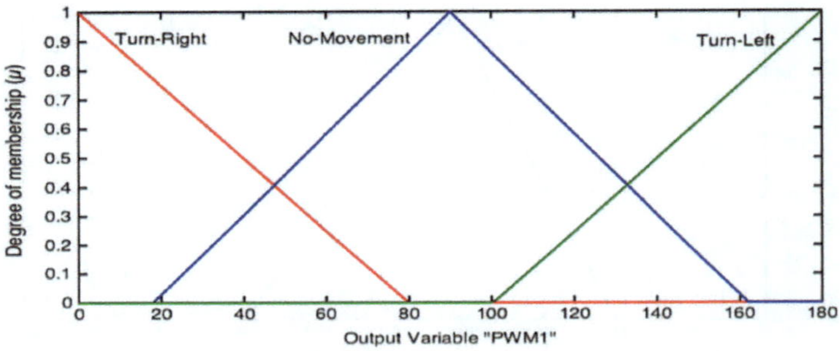

Fig. 13.7 Membership functions used to represent the output variable "PWM1"

estimator block. The idea is to divide the y axis of the arena coordinates (x_arena, y_arena) into many regions of equal lengths W (shown in Fig. 13.2). For each region represented by a membership function MFi, we associate an input variable X_MFi with three adaptive membership functions as depicted in Fig. 13.6. The fuzzy behavior rules (76 rules) are constructed such that the proposed reactive system takes each time the right decisions on the basis of the perception of the external environment of the robot. The decisions of the fuzzy behavior are the robot's right and left wheel's velocities.

Figure 13.7 shows membership functions used to represent the first output decision, with 0 being full-speed in one direction, 180 being full speed in the other, and a value near 90 being no movement. The same membership functions are used to the second output decision.

13.4 System Design Overview

This section describes results of the design and implementation of the proposed reactive system architecture in the real autonomous robot. Electronic part of the system consists of an RGB Led, a power supply module, a bidirectional voltage-level translator 3.3–5 V, a single board computer (Udoo Quad) and a 7.4 V LiPo battery. Figure 13.8 provides a simple overall system block diagram of the robot. It shows also how the various components connect together.

The main processing system is a powerful board computer (called Udoo Quad[1]), based on quad core ARM cortex-A9 CPU (Freescalei.MX6) with great performance both on Android and Linux OS, and a dedicated ARM processor (Atmel SAM3X8E) for the General Purpose Input/Output (GPIO) (UDOO 2013). Figure 13.9 shows the main components of the Udoo quad board.

The Freescale i.MX6 is used to perform complex operations such as the fuzzy behavior-based control scheme and the vision processing used to identify any sites

[1]UDOO website: http://www.udoo.org/

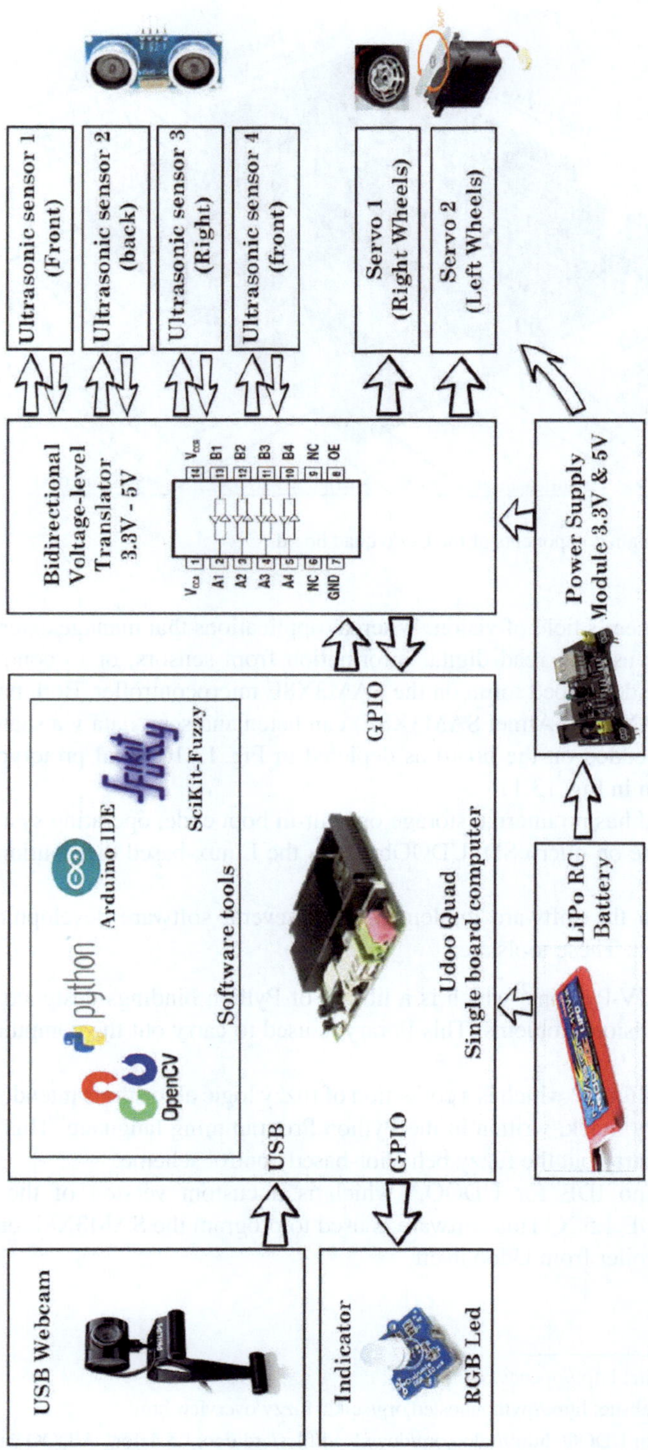

Fig. 13.8 Overall system block diagram

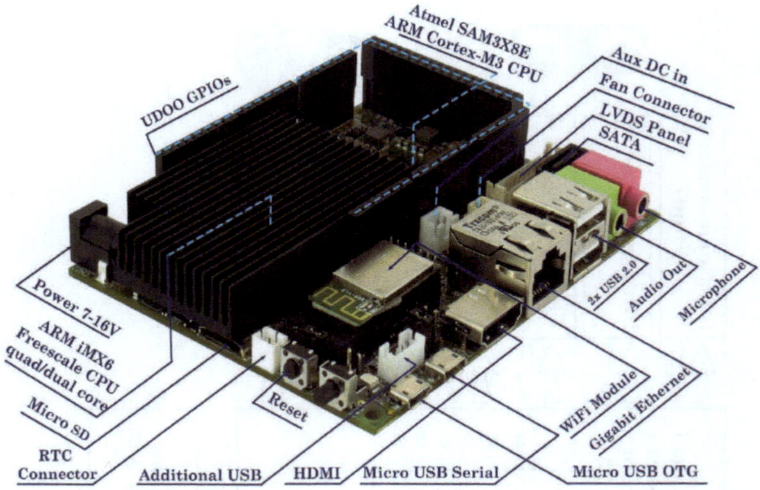

Fig. 13.9 The main components of the Udoo quad board

within the camera's field of vision. Whereas applications that manage external pins, which can be used to read digital information from sensors, or to control lights or motors, are developed torun on the SAM3X8E microcontroller. Both processors (Freescale i.MX6 and Atmel SAM3X8E) can listen and send data via shared serial line port embedded on the board as depicted in Fig. 13.10. Real prototype of the robot is shown in Fig. 13.11.

Udoo Quad has no internal storage or built-in boot code, operating system (OS) and storage are on micro SD. UDOObuntu is the Linux-based distribution used in this work.

To perform the software implementation, several software development tools have been used. These tools are:

- The OpenCV-Python,[2] which is a library of Python bindings designed to solve computer vision problems. This library is used to carry out the computer vision processing.
- The Scikit-fuzzy,[3] which is a collection of fuzzy logic algorithms intended for use in the SciPy Stack, written in the Python Programming language. This package is used to carry out the fuzzy behavior-based control scheme.
- The Arduino IDE for UDOO,[4] which is a custom version of the original Arduino IDE 1.5.X. This software is used to program the SAM3X8E embedded microcontroller from Udoo itself.

[2]OpenCV website: http://opencv.org

[3]Scikit-fuzzy website: http://pythonhosted.org/scikit-fuzzy/overview.html

[4]Arduino IDE for UDOO: http://udoo.org/download/files/arduino-1.5.4-for__UDOO.tar.gz

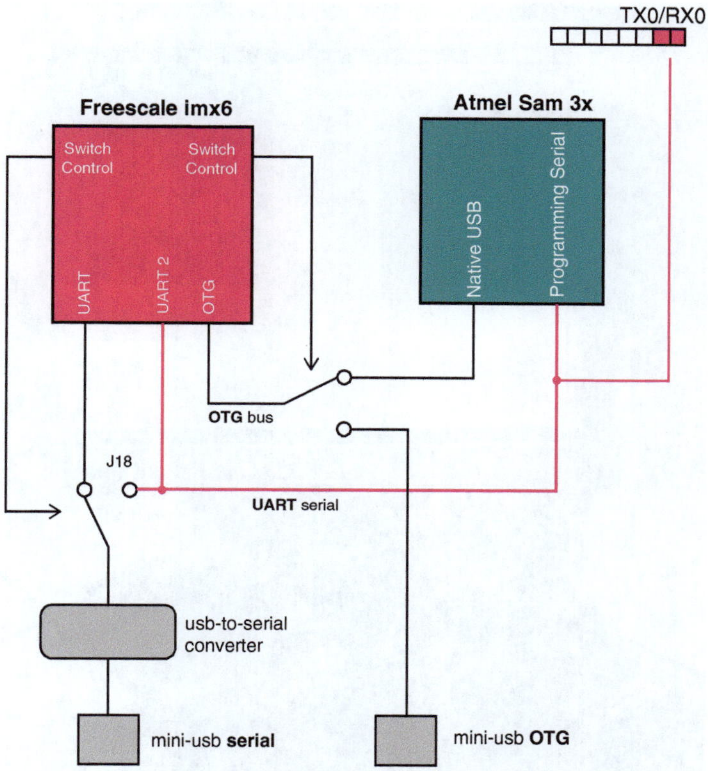

Fig. 13.10 Communication between I.MX6 and SAM3X (UDOO 2013)

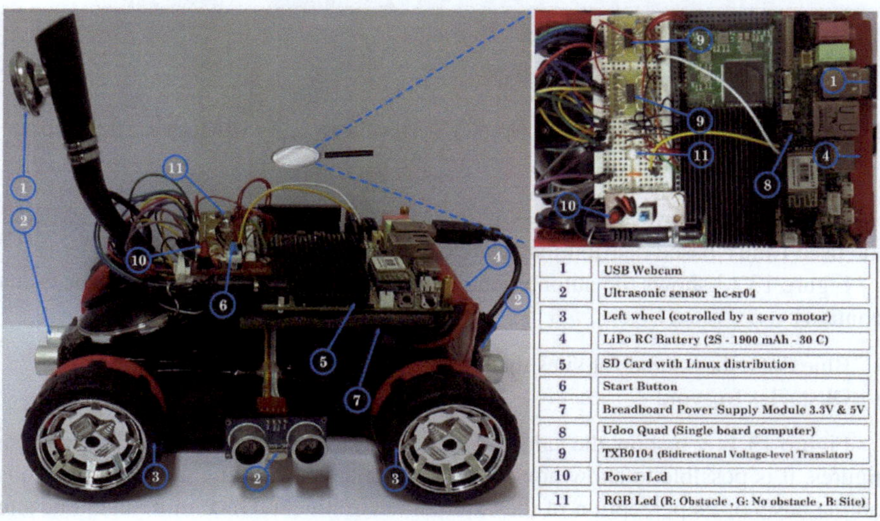

Fig. 13.11 Autonomous mobile robot prototype

Fig. 13.12 Example of code using the OpenCV-Python library under UDOO buntu distribution

Figure 13.12 shows an example of code using the OpenCV-Python library under the used Linux-based distribution (UDOObuntu).

Figure 13.13 shows the autonomous robot prototype in the real arena. The arena contains:

- Blue dots: they stand for sites that the robot has to locate.
- Obstacles to avoid

Figure 13.14a, b show the camera's field of vision obtained every 0.2 s. In these figures, it is possible to show how the real robot take each time the right decisions on the basis of its perception of the external environment. Once more than one site is detected in the camera's field of vision (Fig. 13.14b), the decisions of the fuzzy

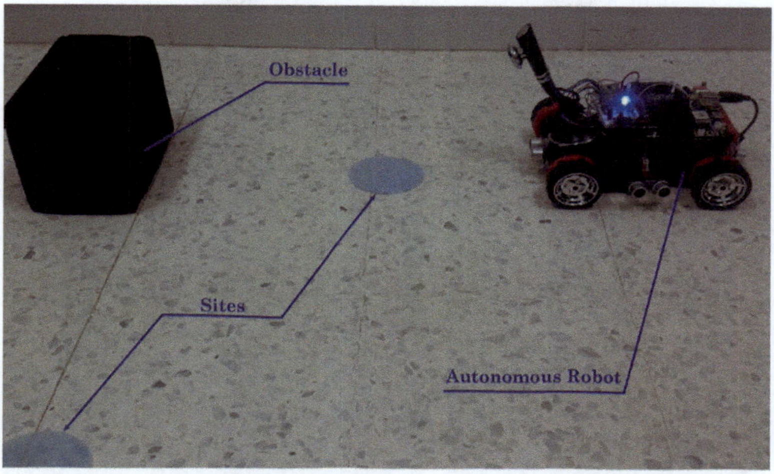

Fig. 13.13 The autonomous robot prototype in the real arena

behavior is to make the robot move toward the nearest site detected. This is done one the detected site is in the middle of the camera's field of view.

13.5 Conclusion

In this chapter, a novel reactive navigation system for an automated robot is presented. The proposed reactive system is based on a fuzzy behavior-based control scheme with an adaptive membership function. The goal of such fuzzy behavior is to increase a robot's knowledge of its workspace by selecting appropriate control actions (robot's right and left wheel's velocities).

To implement the proposed reactive system in the real autonomous robot we have adapted the technology available nowadays based on a powerful single board computer, called Udoo Quad. We have shown that the autonomous robot prototype equipped with a webcam and distance sensors is able to take each time the right decisions on the basis of its perception of the external environment, which will lead it to visit previously unseen areas and locate a number of sites, while avoiding the arena boundary and any obstacles it might encounter.

We are presently working towards implementing such approach using Robot Operating System (ROS) Platform.

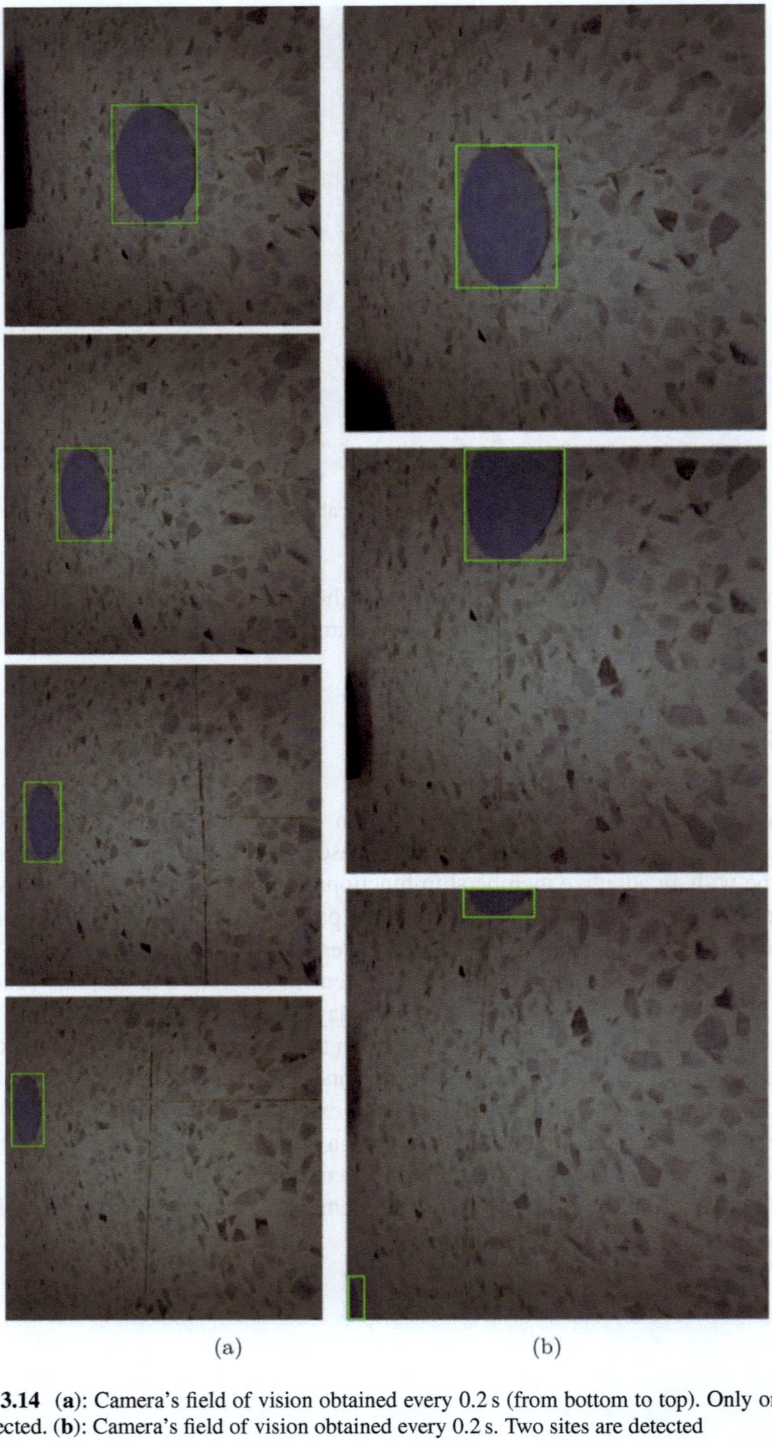

Fig. 13.14 (**a**): Camera's field of vision obtained every 0.2 s (from bottom to top). Only one site is detected. (**b**): Camera's field of vision obtained every 0.2 s. Two sites are detected

References

Arena, P., Di Giamberardino, P., Fortuna, L., La Gala, F., Monaco, S., Muscato, G., Rizzo, A., & Ronchini, R. (2004). Toward a mobile autonomous robotic system for Mars exploration. *Planetary and Space Science, 52*, 23–30.

Assis, L. S., Soares, A. S., Coelho, C. J., & Baalen, J. V. (2016). An evolutionary algorithm for autonomous robot navigation. *International Conference on Computational Science, 80*, 2261–2265.

Hashikawa, F., & Morioka, K. (2015). An assistance system for building intelligent spaces based on mapsharing among a mobile robot and distributed sensors. *International Journal on Smart Sensing and Intelligent Systems, 8*(1), 1–25.

Lin, H.-I., & Tzeng, H. Jr. (2014). Search strategy of a mobile robot for radiation sources in an unknown environment. In *International Conference on Advanced and Intelligent Systems* (pp. 56–60).

Long, Z., Liang, X., & Yang, L. (2010). Some approximation properties of adaptive fuzzy systems with variable universe of discourse. *Information Sciences, 180*(16), 2991–3005.

Mathers, N., Goktogen, A., Rankin, J., & Anderson, M. (2012). Robotic mission to Mars: Hands-on, minds-on, web-based learning. *Acta Astronautica, 80*, 124–131.

Murphy, R. (2000). *Introduction to AI robotics*. Cambridge: MIT Press.

Koren, Y., & Borenstein, J. (1991). Potential field methods and their inherent limitations for mobile robot navigation. In *IEEE International Conference on Robotics and Automation* (pp. 1398–1404).

Kunwar, F., & Benhabib, B. (2008). Advanced predictive guidance navigation for mobile robots: A novel strategy for rendezvous in dynamic settings. *International Journal on Smart Sensing and Intelligent Systems, 1*(4), 858–890.

Putney, J. S. (2006). *Reactive navigation of an autonomous ground vehicle using dynamic expanding zones*. Thesis, Faculty of the Virginia Polytechnic Institute and State University.

Siegwart, R., & Nourbakhsh, I. N. (2004). *Introduction to autonomous mobile robots*. Cambridge: MIT Press.

Siegwart, R., Nourbakhsh, I. R., & Scaramuzza, D. (2011). Introduction to autonomous mobile robots (2nd ed.). Cambridge: MIT Press.

Tang, S. H., Ang, C. K., Nakhaeinia, D., Karasfi, B., & Motlagh, O. (2013). A reactive collision avoidance approach for mobile robot in dynamic environments. *Journal of Automation and Control Engineering, 1*(1), 16–20.

UDOO Starting Manual. http://udoo.org/download/files/Documents/UDOO_Starting_Manual_beta0.4_11_28_2013.pdf.

Wang, R., Liu, Y. J., Yu, F. S., Wang, J. Y., & Yang, J. L. (2017). Adaptive variable universe of discourse fuzzy control for a class of nonlinear systems with unknown dead zones. *International Journal of Adaptive Control and Signal Processing, 31*(12), 1934–1951.

Xu, J., Qian, H., Ying, W., & Zhang, J. (2015). A deployment algorithm for mobile wireless sensor networks based on the electrostatic filed theory. *International Journal on Smart Sensing and Intelligent Systems, 8*(1), 516–537.

Zhang, Q., Yue, S. G., Yin, Q. J., & Zha, Y. B. (2013). Dynamic obstacle-avoiding path planning for robots based on modified potential field method. *Intelligent Computing Theories and Technology, 9*, 332–342.

Chapter 14
Area Coverage Algorithms for Networked Multi-robot Systems

Lamia Iftekhar, H. M. Nafid Rahman, and Imran Rahman

Abstract The area coverage problem in robotics alludes to the challenge of getting a network of mobile robots to efficiently cover or sense a spatial region of interest in an optimal way. A significant number of coverage algorithms exists, especially from works on sensor networks and related computer science fields, offering a wide range of coverage solutions. However, many of them are not readily applicable to robotic systems due to computational complexity of the theoretical algorithms as well as the physical limitations of the robots in terms of sensing and processing. In this chapter, we take a look at various area coverage algorithms that can be implemented upon a system of networked robots and try to identify the ideal characteristics of an area coverage algorithm whose target application is in mobile robotics. We propose a set of simple flocking-based algorithms which inherently captures the identified characteristics of a good coverage algorithm for a multi-robot system. The proposed algorithm is simulated under various conditions to demonstrate its feasibility. This chapter also gives the reader further direction towards open research problems on this topic.

Keywords Distributed control · Swarm robots · Sensor networks · Coverage optimization · Cooperative behavior

L. Iftekhar (✉)
Department of Electrical and Computer Engineering, North South University, Dhaka, Bangladesh
e-mail: lamia.iftekhar@northsouth.edu

H. M. Nafid Rahman
Ceridian HCM, Toronto, ON, Canada

I. Rahman
The Bradley Department of Electrical and Computer Engineering, Virginia Polytechnic Institute and State University, Blacksburg, VA, USA
e-mail: imran292@vt.edu

N. Derbel et al. (eds.), *New Developments and Advances in Robot Control*, Studies in Systems, Decision and Control 175,
https://doi.org/10.1007/978-981-13-2212-9_14

14.1 Introduction

Multi-robot systems have become the preferred framework in solving many different kinds of challenges in various fields of robotics. Instead of using a single, gigantic, prohibitively expensive, computationally complex and power-hungry robot, the same task can be completed with a group of simpler robots efficiently. On top of this, with the large-scale advent of distributed control and collaborative protocols, multi-robot systems have been yielding fascinating results, smartly utilizing comparatively inexpensive robots with limitations in both sensing and communication, limited battery life and low-power builds. One illustrative example can be found in the problem of coverage, where an area has to 'covered' or kept under surveillance all the time. Rather than having a lone patrolling robot (whose slightest malfunction may jeopardize the whole mission of coverage), a multi-robot system can perform the task in a more intelligent and robust manner, without compromising the overall performance, even if a few of the robots here or there are impaired.

The task of Area Coverage started out as a problem in robotics. Douglas W Gage in 1992 (Gage 1992) presented the problem of coverage in a systematic manner for a multi-robot (which he termed as 'many-robot') system and introduced new taxonomy for different aspects of the problem. Interestingly, computer and communication scientists captured the idea and applied it to Wireless Sensor Networks, making vast progress in devising algorithms for a network of sensors (nodes) to spread out over an area of interest, extracting and exchanging information efficiently. Recently, with the advancement in networked vehicular robotics as well as mobile robotics in general, a growing number of researchers have been working on getting robotic systems to perform coverage tasks with varied applications in military surveillance (Mahamuni 2016), structural surveillance (Adaldo et al. 2017), monitoring of forests and wildlife (Zheng et al. 2010), agriculture (Hameed 2014; Janani et al. 2016) indoor and outdoor cleaning (Xin et al. 2017; Yuming et al. 2011), clearing mines (Habib 2017), and many more (e.g. Chen et al. 2014 and Piciarelli et al. 2016). Although, robotics researchers build upon the algorithms from the large body of literature on wireless sensor networks, various challenges need to be addressed in new ways when nodes are replaced by mobile robots. In such cases, algorithms have to be customized or devised to suit the complexities associated with different kinds of robot dynamics, such as differential-drive robots, four-wheeled robots, fixed wing planes, quad-rotor models etc.

This chapter serves as a quick primer on latest works on coverage algorithms, especially designed for multi-robot systems (Sect. 14.2). It also applies modified flocking algorithms to achieve the area coverage task. We demonstrate that the proposed algorithms are simple compared to many existing ones. Yet these area coverage algorithms incorporate advantages, such as being able to cover an unstructured area as well as not waste coverage resources on obstacles and non-spaces, unnecessary. The description of the flocking-based area coverage algorithms is detailed in Sect. 14.3, after which numerical simulation results are presented in Sect. 14.4. The chapter closes with some guidance for researchers interested in delving into this interesting aspect of robotics by offering open problems.

14.2 Discussion on the Coverage Problem and Existing Solutions

We take a tour of the *recent* advances in the area coverage problem, focusing on the application of robotics, in this section. The coverage problem is an old one stemming from military applications three decades ago, and have gone through phases of different terminologies and popular solution strategies, but we focus on the recent works, preferably devised for robotics applications (Fig. 14.1).

14.2.1 What Is Coverage?

The task of coverage by a multi-robot system is the efficient autonomous distribution of the robots over an area or region of interest (ROI). Each robot's sensing capabilities extend over a limited spatial region around itself, and that region is said to be *covered* by the robot. In wireless sensor network (WSN) terminology, the sensors themselves are the discrete elements analogous to the robots, and are denoted as nodes. The objective in coverage tasks would then be to find the ideal placement of the nodes so that any *event* that might happen in the region of interest falls in the sensing region of at least one node and thus does not go undetected. An event could be an occurrence of interest which the sensors are attempting to detect, such as an intrusion by a foreign object. In this chapter, we use sensors, nodes and robots interchangeably (Fig. 14.2).

From literature, four types of coverage tasks can be extracted.

1. *Area coverage or Blanket coverage:* The robots are required to get distributed over the whole region of interest and ensure as much coverage of the entire area as possible. The area of interest is generally a convex area, bounded and connected. (E.g. Cheng and Savkin 2013, Nasimov and Matveev 2014, Santoso 2011)

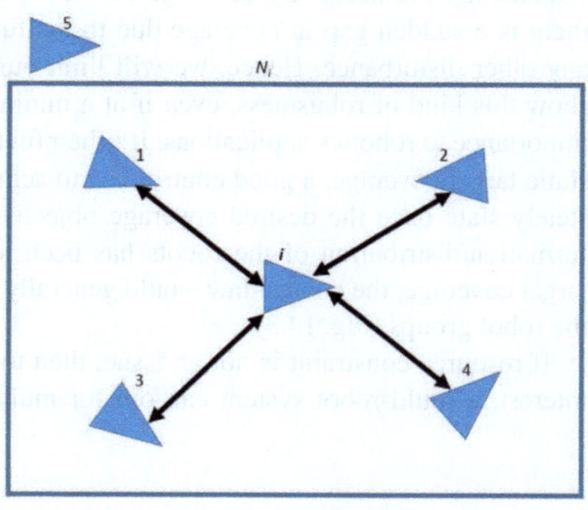

Fig. 14.1 An α-agent and its neighbors. The triangles represent α agents. Agent i has four neighbors because they fall within its sensing and communication range. Neighbors are assumed to have a mutual sensing and communication relationship. N_i denotes the set of neighbors of agent i. Agent 5 is not a neighbor of agent i Olfati-Saber (2006)

Fig. 14.2 An α-lattice and a quasi α-lattice. Note the triangular formations. Source: Olfati-Saber (2006)

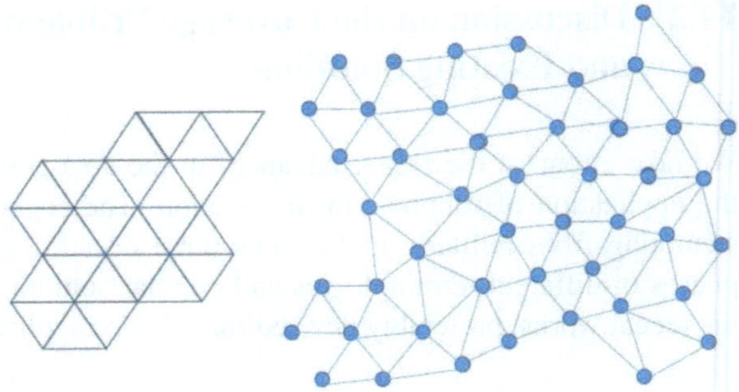

(a) α-lattice (b) quasi α-lattice

2. *Barrier coverage:* The robots are required to get distributed along the perimeter of an area of interest, or a slice of the perimeter, such as an entry point or exit point. (E.g. Tao and Wu 2015)
3. *Sweep coverage:* The robots are required to move in groups and continuously move around a region of interest, to maximize detection of any event. This is a preferred coverage protocol if the number of robots available is small compared to the size of the region of interest. (E.g. Banimelhem et al. 2017)
4. *Target coverage:* The robots are required to get distributed in a region of interest and always keep one or several target points under surveillance. The targets could be static or dynamic (in the latter case, the coverage algorithm would overlap with a target tracking algorithm). (E.g. Yang et al. 2015).

In the earliest of coverage problems, the sensors/nodes could be considered stationary and in that case, the placement of the sensors to achieve maximum coverage is calculated beforehand and the sensors are manually placed. However, in robotics, we are of course interested in dynamic nodes, that is, a group of mobile robots are deployed in the region of interest and they are expected to get themselves distributed in an automated fashion to achieve the desired coverage goals. This mobility and self-organization capability of the robots are beneficial, not only for establishing the necessary coverage of the region, but also to re-organize in case there is a sudden gap in coverage due to malfunction of a robot (node failure) or any other disturbance. Hence, we will limit our discussion to algorithms that can show this kind of robustness, even if at a minimum level, and thus be of practical importance to robotics applications. It is helpful to note that, for blanket, barrier and static target coverage, a good control law to achieve coverage will eventually reach steady state once the desired coverage objective has been achieved and the ideal formation/distribution of the robots has been identified. For sweep and dynamic target coverage, the control law would generally have to ensure perpetual motion of the robot groups (Fig. 14.3).

If resource constraint is not an issue, then to guarantee coverage of a region of interest, a multi-robot system can opt for multiple *layers of coverage*. A *k-layer*

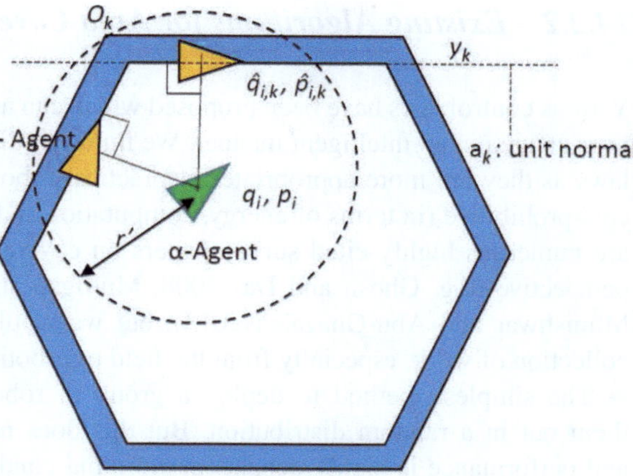

Fig. 14.3 Agent-based representation of obstacles: Hexagonal Wall. Source: Olfati-Saber (2006)

coverage of a region ensures that each point in the region is covered by at least *k* robots. The concept gains increased significance in barrier coverage problems where 1-barrier has a perimeter guarded by one layer of robots, whereas *k*-barrier is *k* layers of strings of robots – making it more difficult for an intruder to penetrate the barrier.

Sensing capability of a robot is normally assumed to be omnidirectional (isotropic sensing), where the sensor strength is maximal at its own position and fades out radially. Although this is a good enough model for some real-life sensors such as temperature sensor or circular laser scans, experimental robotics projects generally utilize sensors such as cone-shaped lasers, monocular cameras or the simple ultrasonic sensor, all of which are better modeled using directional sensor regions, an example of which can be found in Tao and Wu (2015).

In the coverage problem dealt within wireless sensor network community, two parameters of concern exist – adequate coverage and adequate connectivity. Coverage is associated with the sensing capability of a node, while connectivity is associated with the communication capability of the node. Connectivity denotes the state of robot being in communication range of another robot. In WSN, connectivity is of high importance, as generally information is passed from one node to the base (or sink node) through its connected nodes. In fact, to ensure connectivity is not compromised, many algorithms emphasize on having *k-connectivity*, which denotes having the state of one node lying within the communication range of at least *k* other nodes. In the coverage problems dealt within the robotics community, generally connectivity can be given less importance at point if we assume information picked by any robot in the system can be directly communicated to the base or the human-in-the-loop. Hence we can focus on achieving adequate coverage only.

We focus on area or blanket coverage algorithms in this chapter as it is an intriguing topic with a plethora of robotics applications, having much exciting work already done, yet with many open problems still present.

14.2.2 Existing Algorithms for Area Coverage

Various control laws have been proposed which can aid a multi-robot system spread over an area in an intelligent manner. We limit our discussion to distributed control laws as they are more appropriate for practical robotics application without being cost-prohibitive (in terms of energy, computation and consequently, money). There are numerous highly cited survey papers on coverage from the wireless network perspective (e.g. Ghosh and Das 2008, Mulligan and Ammari 2010, Fazli 2010, Munishwar and Abu-Ghazaleh 2013), but we would like to provide an updated collection of work, especially from the field of robotics.

The simplest method to deploy a group of robots over an area is to spread them out in a random distribution. But this does not guarantee proper coverage and performance is highly dependent on initial conditions. Some researchers have taken the next logical step and used probabilistic distributions spatially to control deployment of robots for area coverage (Chang et al. 2017; Fan et al. 2015; Prakash and Patidar 2014). For example, authors in Prakash and Patidar (2014) have utilized three distributions – Beta, Gamma and Fisher-Snedecor on large group of sensors, demonstrating that coverage of a region of interest is easily achieved (the amount of *uncovered area* gets very close to zero), the last distribution yielding the quickest results. To calculate the amount of uncovered area, their performance metric, the authors used two approaches, grid-based and disc-based. In grid-based approach, area coverage is achieved when each single cell of the entire grid has at least one sensor, whereas in disc-based approach, the coverage task is complete if only each and every point is covered by a sensor, making the latter metric a more difficult one to fulfill. A drawback to these probabilistic approaches would be the assumption that the distribution is made over a square or rectangular box overlay above the region of interest, and unless other explicit constraining laws are introduced, many of the sensors or robots get automatically placed outside the region of interest, wasting precious resources.

Partitioning a region of interest in Voronoi cells assigned to each robot and updating the robot's state using some distributed control law has been a popular approach (see references in Mulligan and Ammari 2010). Authors in Yang et al. (2014) took it a step further and divided the area of interest in Voronoi cells dynamically and applied modified bacterial foraging optimization to move the robots accordingly. An excellent example of Voronoi based application on a real multi-robot system can be found in the works of Lei Ma such as in Han et al. (2017), where multiplicatively-weighted Voronoi diagrams are used so that a group of heterogenous robots, with varying physical design and capabilities can perform area coverage utilizing each different robot's highest potential. This work as well as Han et al. (2017) and Zhang et al. (2016) implemented their area coverage algorithms on two wheeled differential robots as well as four wheeled car-like robots, in contrast to most works on multi-robot area coverage problems where the focus is mostly on algorithmic level, and hence on theoretical robots with particle dynamics.

Assuming a virtual geometric configuration akin to a lattice-like structure covering the region of interest and figuring out a control strategy to get the robots situated at the vertices is an underlying motivation for a huge body of grid-based and computational geometry-based approaches for area coverage. Earlier works (Cortes et al. 2005) devised protocols to make the robots or sensors form rectangular grids, e.g. authors in Wang et al. (2007) utilized grid structure over an area and divided sensors into clusters based on distance. Hexagonal and rhombus based topologies exist too but the equilateral triangle topology is the optimal pattern for minimum overlapping of sensing region and maximum connectivity with six neighbors (Gupta et al. 2016). A major positive feature of our coverage algorithm is that it automatically results in the robots forming triangular lattices without any explicit command.

The control laws used to achieve the above coverage formations are often force-based, capitalizing on attraction and replusion forces between neighboring robots. Nearest neighbor techniques are used in Santoso (2011) in which the algorithm was designed for airdrop deployment of robots which will perform a square-grid area coverage upon release. Compared to some of the results in the previous works, where a few robots in the group extends beyond the region of interest, in works such as Cheng and Savkin (2010) and Santoso (2011), the robots themselves manage to stay within the ROI, but the ones at the outer perimeter situate themselves at the border, wasting sensing resources. Our algorithm to be described later, attempts to keep not only the robots, but even their sensing regions within the region of interest, to make optimal use of all resources. In the past few years, consensus algorithms have been an increasingly popular choice to achieve robust and stable steady state for area coverage problems in finite time, such as algorithms in Cheng and Savkin (2010, 2013) and Nasimov and Matveev (2014), all of which achieve asymptotically optimal triangular blanket coverage.

Desired characteristics of an ideal area coverage protocol From the above discussion, we identify our objective as to devise simple rules to coordinate the motion of the robots that would result in intelligent and powerful control law to achieve the desired coverage. We also identify some desired characteristics of an ideal area coverage protocol for robotics applications: the final distribution should result in triangular lattice formation, achieve steady state quickly, minimize overlapping of sensing region (to cover more space using fewer robots and avoid redundancy and wastage of computational and sensing resources) and be implementable of real-life robot dynamics. Moreover, all robots should stay within the boundary of region of interest. Two more criteria may be added: it is preferred that the robots can work in a previously unknown environment (many of the prior algorithms only work when the sensors know of the boundaries beforehand) and eventually be able to complete the task of area coverage in the presence of obstacles efficiently.

Our flocking-based area coverage algorithm described in the remaining sections of this chapter is related to the consensus-based protocols and employs nearest neighbor rules, but using very simple laws. Compared to other consensus protocols, the parent algorithm here by Olfati-Saber (2006) is highly robust, not suffering from pitfalls such as defragmentation of an established structure over a period of time, or collapse of two neighboring robots into one singular space. It is also easily extendable to two-wheeled and four-wheeled robot dynamics (Olfati-Saber and Iftekhar 2012), and so is the proposed algorithm here, making our algorithm a perfect candidate to be implemented in real-life robots.

14.3 Flocking-Based Area Coverage Algorithms

In this section, a set of area coverage algorithms based on Olfati-Saber's flocking algorithms (Olfati-Saber 2006) are described. We start with the description of the multi-robot systems, assuming particle dynamics, to establish the terminology. The work in this chapter is easily extendable to two-wheeled robots with non-linear dynamics (Olfati-Saber and Iftekhar 2012).

14.3.1 Preliminaries

The Area Coverage algorithm we developed is heavily based on Reza Olfati's (Olfati-Saber 2006) Flocking Algorithms. The concept and rules of these particular flocking algorithms have never been applied to solve the problem of area coverage and hence the application of flocking presented in our work is a novel one. In this chapter, we present the background of flocking, and the notations and equations for the existing Flocking Algorithm explained in Olfati-Saber (2006).

Flocking is a form of collective behavior of a large number interacting agents, which includes a cooperative movement to stay in close proximity and maintain a coveted formation while moving. It is a behavioral trait that can be observed in multi-agent systems in nature, demonstrating a strongly adaptive capability to achieve a common objective. Examples of these agents include birds, fish, ants, bees, people, robots, etc. where each entity or members of flocks are referred as an α-agent in this chapter (following the terminology of Olfati-Saber 2006).

In 1986, Craig Reynolds proposed three heuristic rules for flocking which are Flock Centering, Obstacle Avoidance and Velocity Matching, also known as cohesion, separation and alignment respectively. Numerous studies have attempted to give these heuristic rules an algorithmic face, but most of them suffered from drawbacks and could not capture the smooth flocking behavior shown in nature. However, Olfati-Saber's flocking algorithms in Olfati-Saber (2006), finally provided a robust mathematical protocol to achieve the dynamic collective behavior researchers have been aiming for, without the common pitfalls of a formed flock falling apart over time (fragmentation) or the nodes inside the flock colliding with each other (collapse). They could also perform well in the presence of obstacles and successfully demonstrated splitting, rejoining and squeezing maneuvers.

Flocking has also been applied on achieving autonomous driving (Iftekhar 2012), in aerial vehicle formation (Crowther 2002) and target seeking (Lindhe et al. 2005). However it has not been used explicitly for area coverage yet, to the best of our knowledge. The unique lattice-formation capabilities of the highly scalable flocking algorithms in Olfati-Saber (2006) (that neither experience fragmentation nor have the nodes collapse upon each other) make them an excellent potential base to develop area coverage algorithms, which we do in this chapter.

We now provide some background and notations on Olfati-Saber's flocking algorithm (Olfati-Saber 2006). As we have mentioned before, the agents performing flocking behavior are denominated as α-agents.

A group of n α-agents moving in m-dimensional space are modeled by the following equations:

$$\begin{aligned} \dot{q}_i &= p_i \\ \dot{p}_i &= u_i \end{aligned} \tag{14.1}$$

where $q_i, p_i, u_i \in R^m$ represents the position, velocity and control input respectively of ith agent. The systematic configuration of all the α-agents in a system is denoted by the vector equation $q = col(q_1, ...q_n) \in R^{mn}$. An α-agent is assumed to have a sensing and communication radius of $r > 0$ and if any other α-agent enters that disc, they can interact with each other and are said to be neighbors. This two-way neighbor relationship for all the agents in a system can be represented by a graph where agents are the nodes $V = \{1, 2, ...n\}$ and the interaction between neighbors are represented by edges. The set of neighbors of α-agent i can thus be denoted by:

$$N_i(q)\{j : (i, j) \in \varepsilon(q)\} \tag{14.2}$$

where $\varepsilon(q) = \{(i, j) \in V \times V : \left\|q_j - q_i\right\| \leq r\}$ is a set of boundaries and $V = \{1, 2, ...n\}$. where $\|.\|$ is the Euclidean norm. A conformation q is said to form a *quasi* α*−lattice*, if each agent has at least two neighbors and the distances between neighboring agents are approximately equal to a desired distance d i.e.

$$\exists \epsilon, d > 0 : -\epsilon \leq \left\|q_j - q_i\right\| - d \leq \epsilon, \forall j \in N_i(q) \tag{14.3}$$

A quasi α-lattice is an α-lattice when $\epsilon = 0$. Moreover, the attraction or repulsion between the agents depends on the variation of $\left\|q_j - q_i\right\|$

Once a group of α-agents have created a quasi α-lattice, they can be considered to be demonstrating flocking behavior.

Olfati-Saber also introduces two more agents. The first one is a virtual γ agent, which incorporates the objective of the flock enabling them to move from one place to another without fragmentation. The second one is also a virtual agent, termed β-agent, which each α-agent creates on an obstacle, when the obstacle falls within the α-agent's sensing region, to maintain appropriate distance from it. Hence to achieve flocking according to Olfati-Saber (2006), each α-agent is subjected to the following control input:

$$u_i = u_i^{\alpha} + u_i^{\beta} + u_i^{\gamma} \tag{14.4}$$

where u_i^{α} refers to the (α, α) interaction term used to keep the agents in a lattice form, and u_j^{β} refers to the (α, β) interaction term used for obstacle avoidance, where the virtual β-agent is the nearest point on the obstacle from the α-agent.

Two scalable flocking protocols for α-agents have been used in u_i^{α}, a gradient-based one and a velocity consensus protocol.

$$u_i^{\alpha} = \sum_{j \in N_i} \phi_{\alpha}(\|q_j - q_i\|_{\sigma})n_{ij} + \sum_{j \in N_i} a_{ij}(q)(p_j - p_i) \tag{14.5}$$

The first algorithm, u_i^{α} represents all three features of Reynolds, i.e. the gradient-based term comprises flock centering and collision avoidance, and the consensus term is equivalent to velocity matching.

$$n_{ij} = (q_j - q_i)/\sqrt{1 + \epsilon \|q_j - q_i\|^2} \tag{14.6}$$

n_{ij} denotes an enclosed vector along the line connecting q_i and q_j, and

$$\|s\|_{\sigma} = \frac{1}{\epsilon}(\sqrt{1 + \epsilon \|s\|^2} - 1) \tag{14.7}$$

denotes the σ-norm of vector s and $\epsilon \in (0, 1)$. The group has a potential function defined as $V(q)$:

$$V(q) = \frac{1}{2} \sum_i \sum_{j \neq i} \psi_{\alpha}(\|q_j - q_i\|_{\sigma}) \tag{14.8}$$

where $\psi_{\alpha} = \int_{\|d\|}^{s} \phi_{\alpha}(h)dh$. The scalar function $\phi_{\alpha}(s)$ is defined as follows:

$$\phi_{\alpha}(s) = \rho_h(s/\|r\|_{\sigma})\phi(s - \|d\|_{\sigma}) \tag{14.9}$$

$$\phi(s) = \frac{1}{2}[(a + b)\sigma_1(s + c) + (a - b)]$$

where $\sigma_1(s) = \dfrac{s}{\sqrt{1 + s^2}}$. $\phi(s)$ depicts an uneven sigmoidal function having parameters that satisfy the conditions $0 < a \leq b$ and $c = |a - b|/\sqrt{4ab}$ to ensure $\phi(s) = 0$. $\rho_h(s)$, on the other hand, is a scalar function that varies in between the range of 0 and 1. The purpose of this function is to construct smooth potential functions with finite cut-offs and smooth adjacency matrices. In order to do that, the following are used:

$$\rho_h(s) = \begin{cases} 1 & \text{if } s \in [0, h] \\ \frac{1}{2}\left[1 + \cos\left(\pi \frac{s-h}{1-h}\right)\right] & \text{if } s \in [h, 1] \\ 0 & \text{otherwise} \end{cases}$$

where $h \in (0, 1)$.

Using the algorithm discussed above, one can achieve α-agents forming beautiful lattice structures.

14.3.2 Area Coverage Algorithm in an Unknown Environment

A swarm of α-agents instructed to move over an area while ensuring as much of the area as possible falls under the sensing region of at least one α-agent is called area coverage. In this chapter we used the concept of Flocking from Olfati-Saber (2006) to optimally cover a given area and minimize the elapsed time for the coverage without human intervention. The Area Coverage problem addressed here is particularly defined as follows: The multi-agent system finds stationary positions to assume so that the agents can optimally cover the given area using their sensing regions at given point of time. Once the agents reach a steady state/optimal position, they assume a stationary position.

We have slightly modified and implemented Olfati-Saber's (2006) flocking algorithm by using two agents only: α and β and ignoring the γ-agent. The algorithm allows α-agents to optimally cover the whole area, move in clusters and avoid any obstacles/boundaries which are encountered. In the new algorithm, each α-agent is subjected to the following control law:

$$u_i = u_i^\alpha + u_i^\beta \tag{14.10}$$

Here, the first term u_i^α is as explained in Sect. 14.3.1.

The second term of the algorithm is as follows:

$$u_i^\beta = \sum_{k \in N_i^\beta} \phi_\beta(\|\hat{q}_{i,k} - q_i\|_\sigma)\hat{n}_{i,k} + \sum_{k \in N_i^\beta} b_{i,k}(q)(\hat{p}_{i,k} - p_i) \tag{14.11}$$

where $b_{i,k}$ represents a *heterogeneous adjacency* between an α-agent at q_i and its neighboring β-agent at $\hat{q}_{i,k}$, and is denoted by:

$$b_{i,k}(q) = \rho_h(\|\hat{q}_{i,k} - q_i\|_\sigma / d_\beta) \tag{14.12}$$

In addition, the ϕ_β in Eq. (14.11), constitutes a *repulsive action function*, that is defined as $\phi_\beta(z) = \rho_h(z/d_\beta)(\sigma_1(z - d_\beta) - 1)$, and the vector $\hat{n}_{i,k}$ is given by $\hat{n}_{i,k} = \frac{\hat{q}_{i,k} - q_i}{\sqrt{1 + \epsilon \| \hat{q}_{i,k} - q_i \|^2}}$.

The obstacles in the territory are represented through virtual agents called β-agents, that move on the boundary of the obstacles and correspond the nearest distance between them and the α-agents. If the α-agents come in close contact with the β-agents, repulsion takes place, and hence, any collision with the obstacles is avoided.

Olfati-Saber (2006) mentions two different obstacles: hyperplane boundaries (walls) and spherical boundaries. To develop the algorithm for area coverage, we work with the m-dimensional space of an area, i.e. the hyperplane boundaries. We apply the concept of individual hyperplane boundaries to create walls which form an enclosed edge of the area to be covered by the α-agents. When any one of the walls is within the sensing range of an α-agent, a virtual β-agent is formed on that particular wall. This virtual β-agent then exerts a repulsive force on the α-agent, ensuring that the α-agent remains within the boundaries of the coverage area and ensures that there are no collisions between the α-agents and the walls. The position and velocity of a β-agent on a wall is denoted by $\hat{q}_{i,k}$ and $\hat{p}_{i,k}$ respectively, generated by an α-agent with state (q_i, p_i) on a given obstacle, such as O_k. Subsequently, the following applies:

- Consider an obstacle with a hyperplane boundary. Assume that is has a unit normal a_k, and passes through the point y_k. Then position and velocity of the β-agent is as follows:

$$\begin{aligned} \hat{q}_{i,k} &= Pq_i + (I - P)y_k \\ \hat{p}_{i,k} &= Pp_i \end{aligned} \tag{14.13}$$

  where $P = I - a_k a_k^T$ is a projection matrix.

The third part of Olfati-Saber's flocking algorithm, u_i^γ, which is used for the movement of agents from one point to another in a m-dimensional space, is omitted as our primary goal is optimal area coverage. Instead of specifying predetermined coordinates and directing the α-agents to move from point A to point B, we directed our focus specifically on optimal coverage of an area. We applied the concept of a multi-robot system getting deployed in the targeted area first, then apply flocking to move around in cluster and lattice-like form, eventually covering the entire territory.

14.3.3 Area Coverage Algorithm with Obstacle Avoidance

Here, we show multi-robot system with obstacle avoidance capabilities. In the last section, two different kinds of obstacles were mentioned, out of which we used the concept of hyperplane obstacles to set the basics of our area coverage algorithm. For

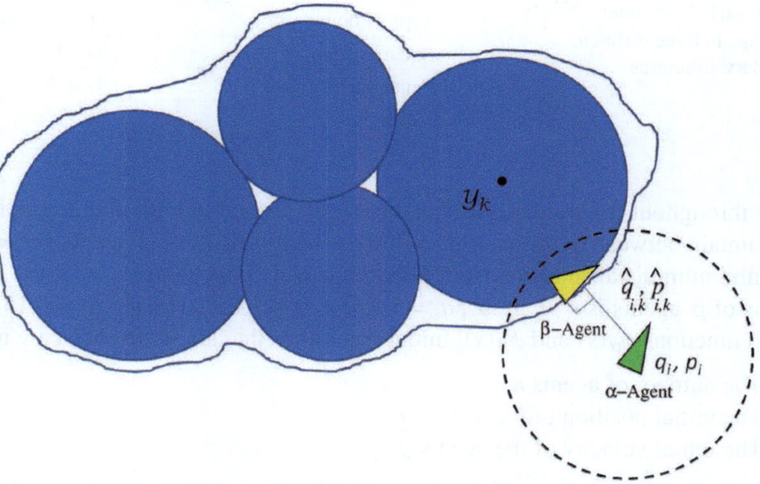

Fig. 14.4 Agent-based representation of obstacles: Approximate model of arbitrarily-shaped obstacle using spheres

this section, our objective is to upgrade this algorithm to make it capable of letting the robots in charge of the coverage task be able to avoid any sort of obstacles *within* the region of interest, while they themselves are getting distributed over the region. For that, we utilize the model of spherical obstacles – the second kind of obstacles from Olfati-Saber (2006).

To achieve the above, the α-agents now also keep an eye out for spherical obstacles by being able to project a new kind of β-agents which is defined below.

- Consider a spherical obstacle centered at y_k and having a radius R_k. μ Then the position and velocity of the β-agent are as follows:

$$\begin{aligned}\hat{q}_{i,k} &= \mu q_i + (1-\mu) y_k \\ \hat{p}_{i,k} &= \mu P p_i\end{aligned} \tag{14.14}$$

Here $P = I - a_k a_k^T$, $\mu = R_k / \|q_i - y_k\|$ and $a_k = (q_i - y_k)/\|q_i - y_k\|$

The strength of such spherical obstacles is that, multiple small versions of them can be used to make an approximate model of many kinds of arbitrarily shaped obstacles (Fig. 14.4).

14.4 Simulation Results

In this section, we present simulation results for Area Coverage by a multi-robot system, where each robot is modeled as an α-agent, for three different environments: square, pentagonal and hexagonal boundaries. The following parameters remain

Table 14.1 Total final coverage in three different boundary structures

Type of boundary	Final covered area
Square	98.29%
Pentagon	99.03%
Hexagon	96.73%

fixed throughout the simulations that have been carried out: the minimum distance to maintain between α-agents is $d = 7$, the sensing radius of α-agents is $r = 1.2d$, the minimum distance to maintain between α and β agents is $d' = 8$, the sensing radius of β-agents is $r' = 1.5d'$, $a = 5$ and $b = 4.5$ for $\phi(s)$ and $h = 0.15$ for the bump functions $\phi_\alpha(s)$ and $\phi_\beta(s)$. Initial conditions that are set in each case include:

i. The number of agents n
ii. The initial position of the agents q
iii. The initial velocity of the agents p

In the case of a square boundary, initial position coordinates are chosen at random from the box $[0, 21]^2$ and velocity coordinates are chosen randomly within the box $[0, 3]^2$. For a pentagonal boundary the initial conditions are set such that initial position coordinates are chosen at random within the box $[0, n]^2$ and velocity coordinates are chosen at random from the box $[0, 5]^2$. In the event of the hexagonal boundaries the initial position coordinates are also chosen at random from the box $[0, n]^2$ and velocity coordinates are chosen at random within the box $[0, 5]^2$.

The following figure shows steady state snapshots from different region of interests with illustrative boundaries:

Table 14.1 projects an approximate percentage area coverage for the three aforementioned boundaries. Coverage is calculated manually using geometric analysis of the two-dimensional space. It is seen that the area coverage of the agents in any given area is at least 96%, thereby making it a highly efficient area coverage algorithm.

Figure 14.5 gives a series of snapshots show the simulation results for the pentagonal boundaries.

To demonstrate the performance of our area coverage algorithm in the presence of obstacles (as described in Sect. 14.3.3), we run another set of simulations with different kinds of convex obstacles (Fig. 14.6). Here we present the snapshots of a single circular obstacle in an area of interest with hexagonal boundary.[1]

It is worth mentioning that once steady state is reached the lattice formation covering the ROI becomes fully still and no oscillation is observed no matter how long the simulation is run, demonstrating stability of the system.

[1]Step size for the simulation in Fig. 14.7 was different than the one used in obstacle-free environment simulations.

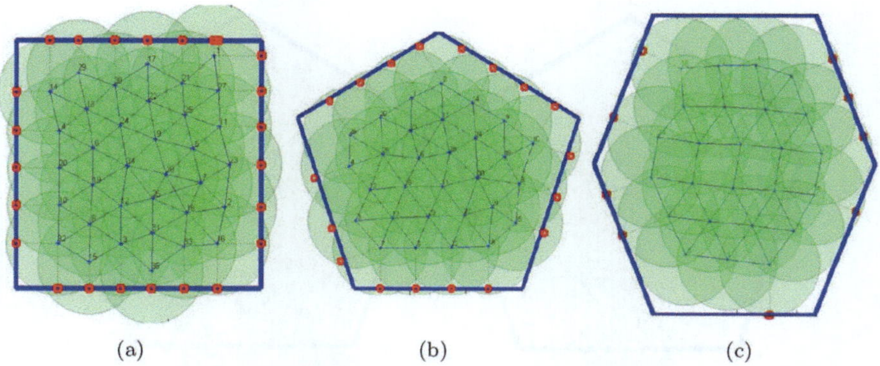

Fig. 14.5 Final coverage by multi-robot systems applying the proposed algorithm in Sect. 14.3.2 in different environments. The boundary structure is not known to the robots. The green disks are the robots' sensing regions. (**a**) Square ROI. (**b**) Pentagonal ROI. (**c**) Hexagonal ROI

14.5 Conclusion

In this chapter, we have explored briefly the task of area coverage by a group of mobile robots and demonstrated the performance of some cooperative area coverage algorithms. Our flocking-based algorithms are decentralized control laws which enable the mobile robots to autonomously form a lattice-like coverage over a region of interest. The region of interest could be of any arbitrary convex shape and the robots do not need prior knowledge of the boundaries. Implementation of the algorithm results in quick distribution of robots in triangular lattices, ensuring maximum coverage and minimal sensing region overlap, as well as attempting to limit extension of robot sensing region beyond the boundary of the area of interest, thus offering more intelligent resource usage. Most significantly with very minimal computational addition, the robots can also avoid any arbitrarily shaped obstacles within the region of interest and distribute themselves in optimal configurations in the presence of any number of obstacles, once again, not wasting any sensing resource by being too close to the obstacles or other non-spaces.

We now provide some future directions and discuss some aspects off the area coverage algorithms that still require more research on.

1. **Heterogeneous sensing regions**: We have only used a specific circular sensing region with a fixed radius for all the α-agents in this chapter. Future work involves implementing various sensing regions for α-agents of different shapes and sizes, thereby giving more freedom to explore and increase the efficiency of the α-agents during area coverage. This would enable a multi-robot group to utilize different kinds of robots available with varying capabilities. For more practical applications, algorithms that can handle robots directional sensing regions should be focused upon, which would also reduce the issue of overlapping caused by

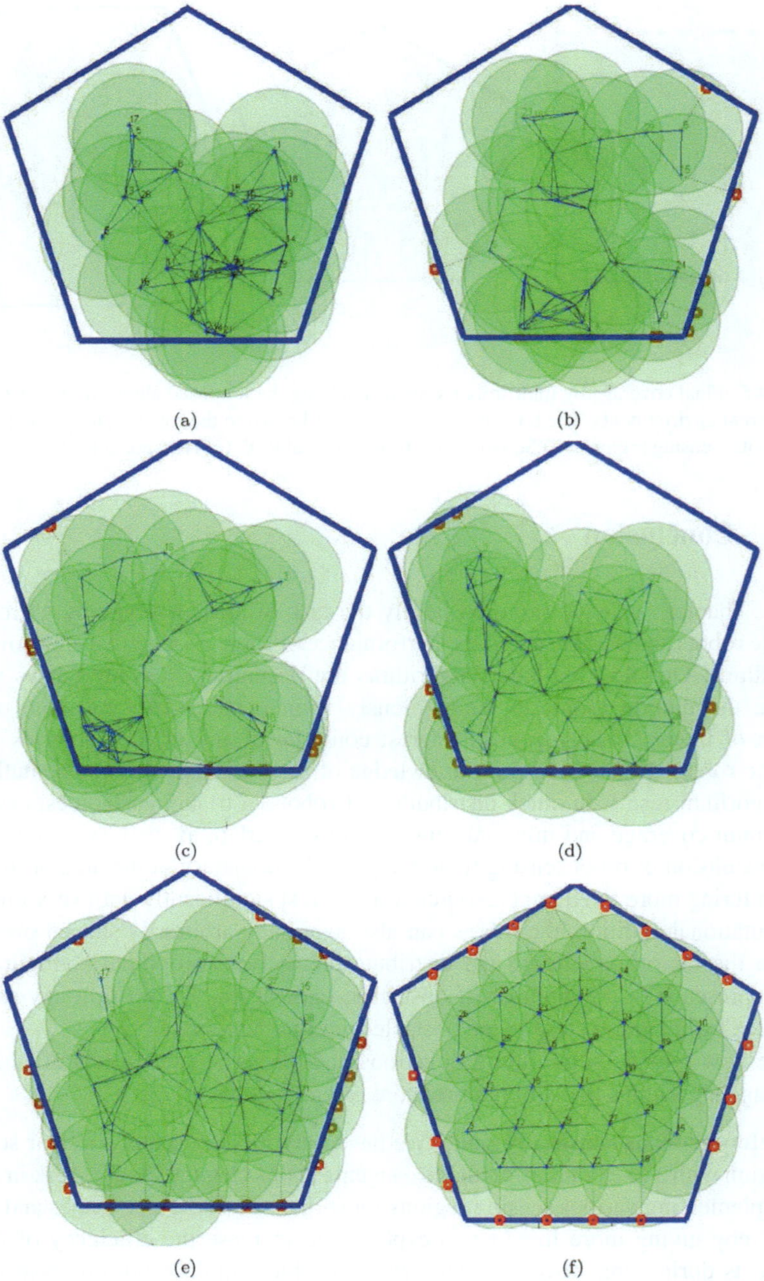

Fig. 14.6 A snapshot of area coverage by 30 robots in a pentagonal region. The robots have no prior knowledge of the boundary. They are initially distributed randomly within the region of interest. They spread out, detect boundary and distribute themselves quite uniformly over the region of interest. (**a**) t = 0.0 (sec) (**b**) t = 4.03 (sec) (**c**) t = 8.91 (sec) (**d**) t = 11.12 (sec) (**e**) t = 14.08 (sec) (**f**) t = 18.02 (sec)

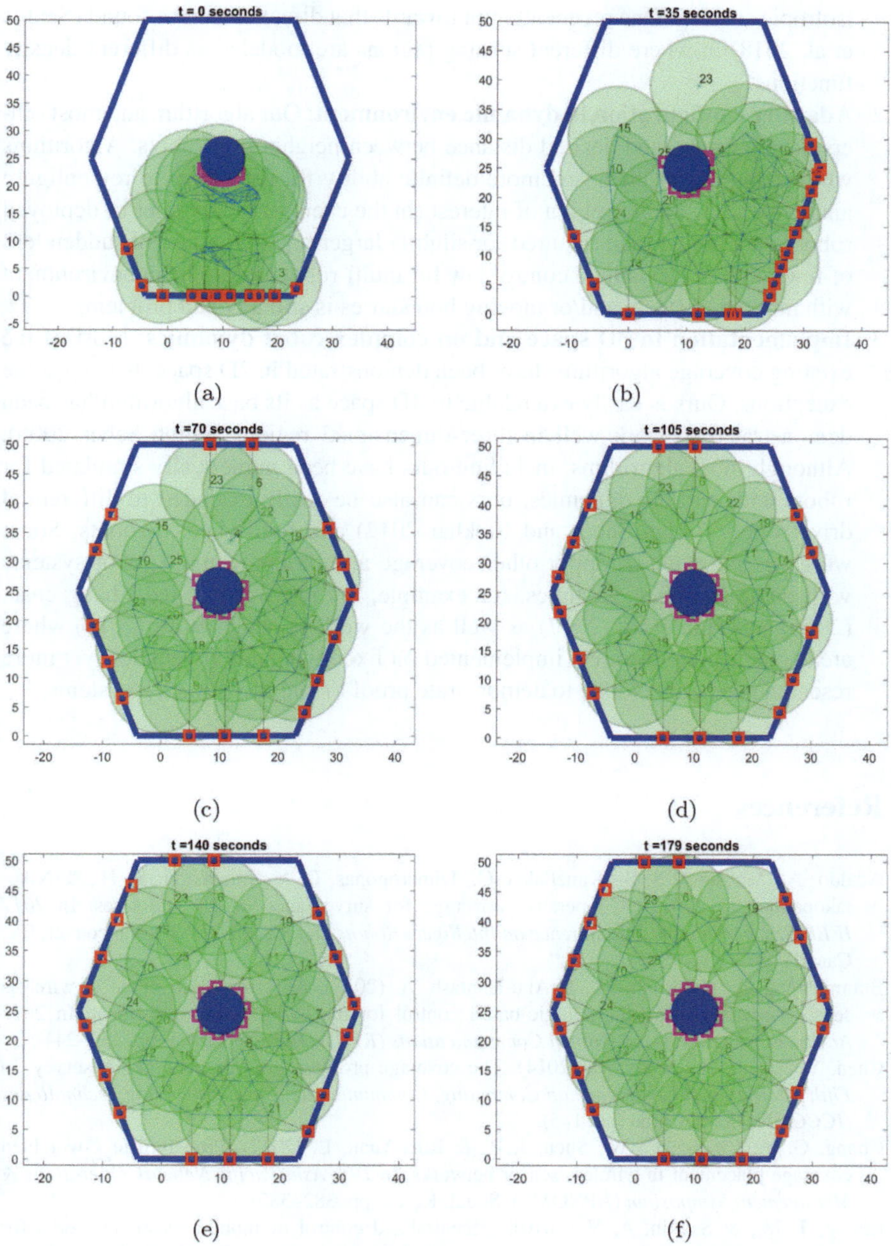

Fig. 14.7 A snapshot of area coverage by 30 robots in a hexagonal region with a large obstacle. The robots have no prior knowledge of the boundary or the obstacle. They are initially distributed randomly within the region of interest. They spread out, detect boundary and obstacle and distribute themselves quite uniformly over the region of interest. They maintain adequate distance from the boundary and obstacle and thus do not waste sensing region over non-spaces or beyond region of interest. **(a)** t = 0 (sec) **(b)** t = 35 (sec) **(c)** t = 70 (sec) **(d)** t = 105 (sec) **(e)** t = 140 (sec) **(f)** t = 179 (sec)

isotropic sensing. One recent attempt towards that direction can be found (Santos et al. 2018) in where different sensing regions are modeled as different density functions.

2. **Adaptive configuration in dynamic environment**: Our algorithm and most others have a predefined desired distance between neighboring robots. Algorithms which can give stronger and more definite ability for the robots to re-configure and adapt to a smaller region of interest (in the event of the number of deployed robot being higher than required) or slightly larger region (in case of sudden loss of few robots). Devising a control law for multi-robot system in an environment with moving obstacle and/or moving boundaries is still an open problem.
3. **Implementation in 3D space and on complex robot dynamics**: Most of the existing coverage algorithms have been demonstrated in 2D space, barring some exceptions. Ours is easily extendable to 3D space as its base algorithm has been demonstrated to work well in three-dimensional regions (Olfati-Saber 2006). Although most algorithms, including ours have been numerically simulated for robots with particle dynamics, ours can also be easily extended to differential drive robots (Olfati-Saber and Iftekhar 2012) and four-wheeled robots. Some works do exist which apply other coverage algorithms to multi-robot systems with complex robot dynamics, for example, the previously cited Zhang et al. (2016) and Han et al. (2017) as well as the work in Varga et al. (2015) where area coverage algorithm is implemented on fixed wing aerial vehicles, yet more research needs to be done to demonstrate proof of stability for such systems.

References

Adaldo, A., Mansouri, S. S., Kanellakis, C., Dimarogonas, D. V., Johansson, K. H., & Nikolakopoulos, G. (2017). Cooperative coverage for surveillance of 3D structures. In *2017 IEEE/RSJ International Conference on Intelligent Robots and Systems (IROS)*, Vancouver, BC, Canada (pp. 1838–1845).

Banimelhem, O., Naserllah, M., & Abu-Hantash, A. (2017). An efficient coverage in wireless sensor networks using fuzzy logic-based control for the mobile node movement. In *2017 Advances in Wireless and Optical Communications (RTUWO)*, Riga, Latvia (pp. 239–244).

Chen, Y., Zhang, H., & Xu, M. (2014). The coverage problem in UAV network: A survey. In *Fifth International Conference on Computing, Communications and Networking Technologies (ICCCNT)*, Hefei, China (pp. 1–5).

Chang, G. Y., Charng, C. W., Sheu, J. P., & Ruei-Yuan, L. (2017). Probabilistic k-weighted coverage placement in wireless sensor networks. In *19th Asia-Pacific Network Operations & Management Symposium* (APNOMS), Seoul, Korea (pp. 382–385).

Cheng, T. M., & Savkin, A. V. (2010). Decentralized control of mobile sensor networks for asymptotically optimal self-deployment in blanket coverage. In *49th IEEE Conference on Decision and Control (CDC)*, Atlanta, Georgia, USA (pp. 1257–1262).

Cheng, T. M., & Savkin, A. V. (2013). Decentralized control of mobile sensor networks for asymptotically optimal blanket coverage between two boundaries. *IEEE Transactions on Industrial Informatics, 9*(1), 365–376.

Cortes, J., Martinez, S., & Bullo, F. (2005). Spatially-distributed coverage optimization and control with limited-range interactions. *ESAIM: Control, Optimisation and Calculus of Variations,*

11(4), 691–719.

Crowther, B. (2002). Rule-based guidance for flight vehicle flocking. *Proceedings of the Institute of Mechanical Engineers, Part G: Journal of Aerospace Engineering, 218*(2), 111–124.

Fan, X., Yang, J., & Wang, C. (2015). An effective construction scheme of probabilistic k-barrier coverage in WSN. In *IEEE International Conference on Cyber Technology in Automation, Control, and Intelligent Systems (CYBER)*, Shenyang, China (pp. 671–676).

Fazli, P. (2010). On multi-robot area coverage. In *9th International Conference on Autonomous Agents and Multiagent Systems*. AAMAS'10 (Vol. 1, pp. 1669–1670). International Foundation for Autonomous Agents and Multiagent Systems, Toronto, Canada.

Gage, D. W. (1992). *Command control for many-robot systems*. Naval Command Control and Ocean Surveillance Center RDT and E Div, San Diego, CA, Technical Report

Ghosh, A., & Das, S. K. (2008). Coverage and connectivity issues in wireless sensor networks: A survey. *Pervasive and Mobile Computing, 4*(3), 303–334. [Online]. Available: http://www.sciencedirect.com/science/article/pii/S1574119208000187.

Habib, M. K. (2017). Coordinated multi robotic system for demining activities. In *The First International Conference on Landmine: Detection, Clearance and Legislations (LDCL2017)*, Faculty of Engineering, Lebanese University, Campus of Hadath, Beirut, Lebanon (pp. 1–5).

Hameed, I. (2014). Intelligent coverage path planning for agricultural robots and autonomous machines on three-dimensional terrain. *Journal of Intelligent and Robotic Systems, 74*(3–4), 965–983.

Han, X., Ma, L., & Feng, K. (2017). Distributed coverage control of networked heterogeneous robots. In *6th Data Driven Control and Learning Systems (DDCLS)*, Chongqing, China (pp. 286–291).

Gupta, P., Prakash, V., & Suman, P. (2016). Noticeable key points and issues of sensor deployment for large area wireless sensor network: A survey. In *2016 International Conference System Modeling Advancement in Research Trends (SMART)*, Moradabad, India (pp. 164–169).

Iftekhar, L., & Olfati-Saber, R. (2012). Cooperative autonomous driving for vehicular networks. In *9th International Conference on Informatics in Control, Automation and Robotics ICINCO*, Rome, Italy (Vol. 2, pp. 345–352). SciTePress.

Janani, A.-R., Alboul, L., & Penders, J. (2016). Multi robot cooperative area coverage, case study: Spraying. In *Towards Autonomous Robotics Systems: 17th Annual Conference TAROS 2016*, Sheffield, UK (Proceedings of the lecture notes in computer science, Vol. 2716, pp. 165–176). Springer.

Lindhe, M., Ogren, P., & Johansson K. H. (2005). Flocking with obstacle avoidance: A new distributed coordination algorithm based on Voronoi partitions. *In IEEE International Conference on Robotics and Automation*, Barcelona, Spain (Vol. 2, pp. 1785–1790).

Mahamuni, C. V. (2016). A military surveillance system based on wireless sensor networks with extended coverage life. In *2016 International Conference on Global Trends in Signal Processing, Information Computing and Communication (ICGTSPICC)*, Jalgaon, India (pp. 375–381).

Mulligan, R., & Ammari, H. M. (2010). Coverage in wireless sensor networks: A survey. *Network Protocols & Algorithms, 2*(2), 27–53. [Online]. Available: https://doi.org/10.5296/npa.v2i2.276.

Munishwar, V. P., & Abu-Ghazaleh, N. B. (2013). Coverage algorithms for visual sensor networks. *ACM Transactions on Sensor Networks, 9*(4), 1–45.

Nasimov, M., & Matveev, A. (2014). Suboptimal decentralized blanket coverage control of mobile autonomous sensor networks. In *6th International Congress on Ultra Modern Telecommunications and Control Systems and Workshops (ICUMT)*, St. Petersburg, Russia (pp. 366–371).

Olfati-Saber, R. (2006). Flocking for multi-agent dynamic systems: Algorithms and theory. *IEEE Transactions On Automatic Control, 51*(3), 401–420.

Olfati-Saber, R., & Iftekhar, L. (2012). Flocking for networks of mobile robots with nonlinear dynamics. In *9th International Conference on Informatics in Control, Automation and Robotics ICINCO*, Rome, Italy (Vol. 2, pp. 353–359). SciTePress.

Piciarelli, C., Esterle, L., Khan, A., Rinner, B., & Foresti, G. L. (2016). Dynamic reconfiguration in camera networks: A short survey. *IEEE Transactions on Circuits and Systems for Video Technology, 26*(5), 965–977.

Prakash, S. K. L. V. S., & Patidar, N. (2014). Area based coverage for three deployment distributions and their connectivity in wireless sensor networks. In *4th International Conference on Communication Systems and Network Technologies*, Bhopal, India (pp. 128–132).

Santos, M., Diaz-Mercado, Y., & Egerstedt, M. (2018). Coverage control for multirobot teams with heterogeneous sensing capabilities. *IEEE Robotics and Automation Letters, 3*(2), 919–925.

Santoso, F. (2011). A decentralised self-dispatch algorithm for square-grid blanket coverage intrusion detection systems in wireless sensor networks. In *2011 IEEE Vehicular Technology Conference (VTC Fall)*, San Francisco, CA, USA (pp. 1–5).

Tao, D., & Wu, T. Y. (2015). A survey on barrier coverage problem in directional sensor networks. *IEEE Sensors Journal, 15*(2), 876–885.

Varga, M., Basiri, M., Heitz, G., & Floreano, D. (2015). Distributed formation control of fixed wing micro aerial vehicles for area coverage. In *2015 IEEE/RSJ International Conference on Intelligent Robots and Systems (IROS)*, Hamburg, Germany (pp. 669–674).

Wang, Y. H., Yu, C. Y., & Fu, P. F. (2007). A coverage and connectivity method to cluster topology in wireless sensor networks. In *21st International Conference on Advanced Information Networking and Applications Workshops, AINAW'07*, Niagara Falls, ON, Canada (Vol. 1, pp. 97–102).

Xin, B., Gao, G. Q., Ding, Y. L., Zhu, Y. G., & Fang, H. (2017). Distributed multi-robot motion planning for cooperative multi-area coverage. In *13th IEEE International Conference on Control & Automation (ICCA)*, Ohrid, Macedonia (pp. 361–366).

Yang, B., Ding, Y., & Hao, K. (2014). Target searching and trapping for swarm robots with modified bacterial foraging optimization algorithm. In *11th World Congress on Intelligent Control and Automation*, Shenyang, China (pp. 1348–1353).

Yang, B., Ding, Y., & Hao, K. (2015). Area coverage searching for swarm robots using dynamic voronoi-based method. In *34th Chinese Control Conference (CCC)*, Hangzhou, China (pp. 6090–6094).

Yuming, L., Ruchun, W., Zhenli, Z., & Junlin, Z. (2011). Dynamic coverage path planning and obstacle avoidance for cleaning robot based on behavior. In *International Conference on Electric Information and Control Engineering*, Wuhan, China (pp. 4952–4956).

Zhang, J., Zhou, P., & Ma, L. (2016). Coverage control of multiple heterogeneous mobile robots with nonholonomic constraints. In *35th Chinese Control Conference (CCC)*, Chengdu, China (pp. 6272–6277).

Zheng, X., Koenig, S., Kempe, D., & Jain, S. (2010). Multirobot forest coverage for weighted and unweighted terrain. *IEEE Transactions on Robotics, 26*(6), 1018–1031.

Chapter 15
Backstepping-Based Nonlinear RISE Feedback Control for an Underactuated Quadrotor UAV Without Linear Velocity Measurements

Jawhar Ghommam, Luis F. Luque-Vega, and Maarouf Saad

Abstract In this paper, a robust three dimensional output feedback control problem is proposed for a 6-degrees-of-freedom model of a quadrotor unmanned aerial vehicle (UAV) to track a bounded and sufficiently smooth reference trajectory in the presence of slowly varying force disturbances. Due to the underactuation structure of the UAV, a nonlinear output feedback controller based on the robust integral of the sign error signal (RISE) mechanism is first designed for the translational dynamics to ensure position reference tracking without velocity measurement. The angular velocity is then regarded as intermediate control signal for the rotational dynamics to fulfill the task of attitude angle reference tracking. The torque input is designed taking full advantage of the smooth exact differentiator that circumvents derivatives computation of virtual controls, the backstepping technique is then judiciously modified to allow the use of the RISE control technique to compensate for the external disturbances. The proposed controller yields semi-global asymptotic stability tracking despite the added disturbances in the dynamics. Simulation results are shown to demonstrate the proposed approach.

Keywords RISE-backstepping · Exact differentiator · Underactuated quadrotor · Output feedback

J. Ghommam (✉)
Department of Electrical and Computer Engineering, College of Engineering, Sultan Quaboos University, Muscat, Oman
e-mail: jawher@squ.edu.om

L. F. Luque-Vega
CIIDETEC-UVM, Universidad del Valle de Mexico, Mexico City, Mexico
e-mail: luis.luque@uvmnet.edu

M. Saad
Department of Electrical Engineering, École de Technologie Superieure, Montreal, QC, Canada
e-mail: maarouf.saad@etsmtl.ca

N. Derbel et al. (eds.), *New Developments and Advances in Robot Control*, Studies in Systems, Decision and Control 175,
https://doi.org/10.1007/978-981-13-2212-9_15

15.1 Introduction

The last few decades have witnessed quick trends in the field of unmanned aerial vehicle (UAV), with increasingly affordable devices endowing these types of vehicles with versatile capabilities to perform many types of tasks in multitude applications. Most of the tasks envisioned to be within the reach of the UAVs, involve search and rescue, security, civil engineering inspection and weather observation (Ilcev and Sibiya 2015; Magsino et al. 2016). In particular, quadrotor type UAVs have recently captured the attention of the control community since their characteristics being highly nonlinear dynamics poses an extremely good challenge to control system designers (Castillo et al. 2004; Cunha et al. 2009; Huang et al. 2010; Kim and Shim 2003). Moreover, the quadrotor type UAVs are steadily becoming instrumental tool for real life missions, ranging from surveillance to entertainment.

Given the difficulties encountered with underactuation and strong nonlinear coupling in the quadrotor dynamics, some investigations have explored various control solutions for this type of aircraft vehicle. Earlier attempts to achieve acceptable autonomous flight control have been developed based on linear control design and they basically involve PID controllers (Garcia et al. 2012; Salih et al. 2010), LQR controllers (Rinaldi et al. 2013) and H_∞ controllers (Ortiz et al. 2016). Of special significance, the input/output linearization approach has been used in Ciulkin et al. (2015) and Garcia et al. (2012) to design controllers for the stabilization and trajectory tracking of the quadrotor. Of particular note, the quadrotor aircraft often operates in three dimensional space. This limits the control design proposed in the aforementioned work which restricts the control of the quadrotor in a vertical plane only. Just recently the authors in Do (2015) and Ghommam et al. (2015), proposed new schemes that give ability to the designer of controlling the quadrotor in six degrees of freedom. In particular, the authors in Kendoul (2009), Wang and Jia (2014) and Van Den Eijnden (2017) proposed a cascade design procedure for the position control of a rotary aircraft UAV. In Bertrand et al. (2011) the authors proposed a hierarchical design procedure for an UAV vehicle based on the singular perturbation theory. Similar approach on using the hierarchical design procedure has been presented in Yao (2017) and Zou (2017).

In almost all the aforementioned works, the control designs require most often instantaneous measurements of the linear velocity of the quadrotor UAV. As there is no sensor which directly measures the linear velocity of the quadrotor in the frame body, those controllers require some kind of observer relying on the global positioning sensor. The global position tracking problem of VTOL UAVs without linear-velocity measurement has been addressed in Abdessameud and Tayebi (2010), where the authors exploited the natural cascaded structure of the UAV to design their controller based on thrust and desired attitude extraction algorithm. In Lee et al. (2007), the authors proposed a an adaptive output feedback scheme based on an exponential observer to estimate the velocities for a tracking control of quadrotor UAV using only linear and angular positions. In Madani and Benallegue

(2007) and Benallegue et al. (2008), sliding-mode observers were used to estimate the effect of external perturbations using measurement of positions and yaw angle. Although the nonlinear controllers realized along with the observer design to estimate the linear velocity of the quadrotor can achieve the desired performance of the system, but the resulting performance in the presence of structural uncertainties and external disturbances can only achieve ultimately boundedness and not asymptotic stability tracking.

In this paper, a nonlinear backstepping controller is designed concurrently with a filter that estimates the linear velocity of the quadrotor vehicle. The robust integral of the sign of the error (RISE) control structure is introduced in the design to compensate for uncertain disturbances affecting the translational dynamics. The resulting desired translational commands are then utilised as attitude commands in the backstepping approach, in that the virtual force command is backstepped through an exact differentiator to the rotational dynamics. The backstepping technique is then judiciously modified to allow the use of the (RISE) control structure in order to compensate for disturbances. A Lyapunov based stability analysis is used to prove that the proposed controller guarantees semiglobal asymptotic tracking. The novelty of this approach is that the trajectory tracking controller is designed based on the imbrication of the RISE feedback technique and makes use of the position measurement only for the stability of the translational dynamics. Information about the derivatives of the force commands being used as virtual command to the rotational dynamics are estimated through exact differentiator through the super twisting algorithm (Levant 2005).

The remainder of this paper is as follows: Sect. 15.2 provides mathematical preliminaries. Section 15.3 presents the problem statement of this paper. Section 15.4 details the main control development, Sect. 15.5 performs some numerical simulation to validate the proposed approach and finally, Sect. 15.6, concludes the paper. The stability analysis is provided in the appendix of the paper.

15.2 Preliminaries

In the following, $\mathbb{R}^{m \times n}$ denotes the $m \times n$ Euclidean space, $|x|$ denotes the absolute value of a scalar x, and $\|\boldsymbol{x}\|$ denotes the Euclidean norm of a vector $\boldsymbol{x} \in \mathbb{R}^n$. The infinity norm of a vector $\boldsymbol{x}$ is denoted $\|\boldsymbol{x}\|_\infty$. $\text{Sgn}(\boldsymbol{x})$ denotes the extension of the signum function to a vector $\boldsymbol{x} = [x_1, x_2, \ldots, x_n]^\top$, is such that $\text{Sgn}(\boldsymbol{x}) = [\text{sign}(x_1), \text{sign}(x_2), \ldots, \text{sign}(x_n)]^\top \in \mathbb{R}^n$.

In this section, useful definition and lemma are presented, which are used to the design of the control law and also for the purpose of the stability analysis of the three dimensional motion of the underactuated quadrotor system.

Lemma 1 (Levant 2005) *Consider the first order sliding mode differentiator designed as follows*

$$\dot{\varrho}_0 = \rho_0 = -\epsilon_0|\varrho_0 - f(t)|^{\frac{1}{2}} sign(\varrho_0 - f(t)) + \varrho_1$$
$$\dot{\varrho}_1 = -\epsilon_1 sign(\varrho_1 - \rho_0) \tag{15.1}$$

where ϱ_0, ϱ_1 and ρ_0 are the states of the first-order differentiator, ϵ_0 and ϵ_1 are the design parameters of the differentiator (15.1) *and $f(t)$ is a known scalar function. The output of the differentiator ρ_0 can approximate the term $\frac{d}{dt}f(t)$ to an arbitrary accuracy if $|\varrho_0 - f(t_0)|$ and $|\rho_0 - \dot{f}(t_0)|$ are bounded.*

Definition 1 (Filippov Solution Filippov 1964:) The vector function $\boldsymbol{\vartheta}$ is said to be a solution of $\dot{\boldsymbol{\vartheta}} = \Upsilon(\boldsymbol{\vartheta})$ on the interval $[t_0, t_1]$ with discontinuity on the right hand side such that Υ is Lebesgue measurable and essentially locally bounded function. If $\boldsymbol{\vartheta}$ is absolutely continuous on $[t_0, t_1]$ and for almost all $t \in [t_0, t_1]$, $\dot{\boldsymbol{\vartheta}} \in \Psi[\Upsilon](\boldsymbol{\vartheta})$ where $\Psi[\Upsilon](\boldsymbol{\vartheta}) \triangleq \bigcap_{\varepsilon>0} \bigcap_{\mu N=0} \overline{co}\Upsilon\left(\Psi(\boldsymbol{\vartheta}, \varepsilon_0) - N, t\right)$ in which $\bigcap_{\mu N=0}$ denotes the intersection overall sets N of Lebesgue measure zero, $\overline{co}$ denotes the convex closure and $\Psi(\boldsymbol{\vartheta}, \varepsilon)$ is an open ball of radius ε_0 centred at $\boldsymbol{\vartheta}$.

Lemma 2 (Xian et al. 2004) *Let the auxiliary function $P(t) \in \mathbb{R}$ be defined as follows:*

$$P(t) = -\int_0^t \boldsymbol{r}^\top(\mathbf{M}_d - K_1 \text{Sng}(\boldsymbol{x}(\mu)))d\mu + \sum_{i=1}^{n} K_1|x_i(0)| - \mathbf{M}_d^\top(t_0)\boldsymbol{r}(t_0) \tag{15.2}$$

where the subscript $i = 1, \ldots, n$ denotes the ith element of the vector $\boldsymbol{x}$. If the gain K_1 is chosen such that the following hold:

$$K_1 > \|\mathbf{M}_d\|_\infty + \|\dot{\mathbf{M}}_d\|_\infty \tag{15.3}$$

then $P(t) \geq 0$ and $\boldsymbol{r}^\top\mathbf{M}_d - \sum_{i=1}^n K_1|x_i(0)| - \boldsymbol{r}^\top\dot{\mathbf{M}}_d \leq 0$.

15.3 Problem Statement

15.3.1 Motion Equations of a Quadrotor

The quadrotor helicopter studied in this paper is shown in Fig. 15.1. Due to reactional torques, the quadrotor's rotors are divided by pairs. One pair (1,3) turns in the same clockwise direction, while the pair (2,4) turns in the counter-clockwise direction. The flight mechanism of the quadrotor can be explained as follows: The vertical motion of the vehicle is produced by generating a lift force through collectively varying the rotor speeds with the same velocity, this would enable the take-off/landing of the quadrotor. The yaw angle is controlled by the difference of speed of each motors pair. While rotation around the y-axis (pitch) will cause the quadrotor to move along the x-axis direction. Consider a body fixed reference frame

Fig. 15.1 Quadrotor model

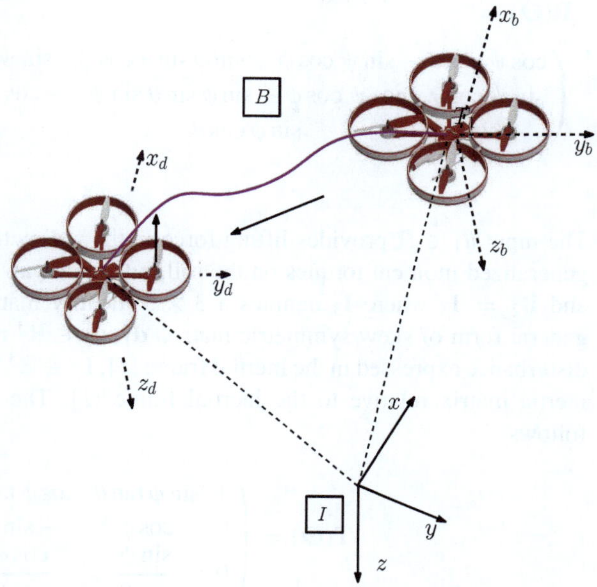

$\{B\}$ attached to the vehicle's center of mass which moves relative to a fixed inertial frame $\{I\}$. The kinematic and dynamic equations of motion for the quadrotor in the body-fixed reference frame can be described as follows (Ding et al. 2017):

$$
\begin{aligned}
\dot{\boldsymbol{p}} &= \mathbf{R}\boldsymbol{v} \\
m\dot{\boldsymbol{v}} &= -m\mathbf{S}(\omega)\boldsymbol{v} + \mathbf{N}_1(\boldsymbol{v}) + \mathbf{G} + \mathbf{B}_1 u_1 + \mathbf{d}_1 \\
\dot{\boldsymbol{\Theta}} &= \mathbf{T}(\boldsymbol{\Theta})\boldsymbol{\omega} \\
\mathbf{I}_f\dot{\boldsymbol{\omega}} &= -\mathbf{S}(\omega)\mathbf{I}_f\boldsymbol{\omega} - \mathbf{N}_2(\boldsymbol{\omega}) + \mathbf{B}_2 \boldsymbol{u}_2 + \mathbf{d}_2
\end{aligned}
\tag{15.4}
$$

where $\boldsymbol{p} = [x, y, z]^\top$ represents the relative position of the center of mass of the quadrotor with respect to an inertial frame and $\boldsymbol{\Theta} = [\theta, \phi, \psi]^\top$ represents the three Euler angles (roll, pitch, yaw) that define the orientation of the quadrotor in space. $\boldsymbol{v} = [v_x, v_y, v_z]^\top \in \mathbb{R}^3$ and $\boldsymbol{\omega} = [\omega_x, \omega_y, \omega_z]^\top \in \mathbb{R}^3$ denote the linear and angular velocities expressed in the body-fixed frame $\{B\}$ respectively. $\mathbf{N}_1(.)$ and $\mathbf{N}_2(.) \in \mathbb{R}^3$ denote the nonlinear aerodynamic damping interactions. $\mathbf{G} \in \mathbb{R}^3$ is the gravity vector defined as $\mathbf{G} = mg\mathbf{R}^\top \mathbf{e}_z$, where m is the mass of the quadrotor, $g \in \mathbb{R}$ denotes the gravitational acceleration and $\mathbf{e}_z = [0, 0, 1]^\top$ is the unit vector in the inertial frame. The rotation matrix $\mathbf{R}(\boldsymbol{\Theta})$ is given as below:

$$\mathbf{R}(\boldsymbol{\Theta}) = \begin{pmatrix} \cos\psi\cos\theta & -\sin\psi\cos\phi + \sin\phi\sin\theta\cos\psi & \sin\psi\sin\phi + \sin\theta\cos\psi\cos\phi \\ \sin\psi\cos\theta & \cos\psi\cos\phi + \sin\phi\sin\theta\sin\psi & -\cos\psi\sin\phi + \sin\theta\sin\psi\cos\phi \\ -\sin\theta & \sin\phi\cos\theta & \cos\phi\cos\theta \end{pmatrix} \quad (15.5)$$

The input $u_1 \in \mathbb{R}$ provides lifting force in the z-direction, $\boldsymbol{u}_2 \in \mathbb{R}^3$ represents the generalized moment torques on the roll, pitch and yaw directions. $\mathbf{B}_1 = [0, 0, 1]^\top$ and $\mathbf{B}_2 = \mathbf{I}_3$ where $\mathbf{I}_3$ denotes a 3×3 identity matrix. $\mathbf{S}(.) \in SO_3$ denotes a general form of skew symmetric matrix. $\mathbf{d}_1, \mathbf{d}_2 \in \mathbb{R}^3$ represent the external forces disturbance expressed in the inertial frame $\{I\}$. $\mathbf{I}_f \in \mathbb{R}^{3\times 3}$ denotes a positive definite inertia matrix relative to the inertial frame $\{I\}$. The matrix $\mathbf{T}(\boldsymbol{\Theta})$ is defined as follows

$$\mathbf{T}(\boldsymbol{\Theta}) = \begin{pmatrix} 1 & \sin\phi\tan\theta & \cos\phi\tan\theta \\ 0 & \cos\phi & -\sin\phi \\ 0 & \dfrac{\sin\phi}{\cos\theta} & \dfrac{\cos\phi}{\cos\theta} \end{pmatrix} \quad (15.6)$$

The transformation matrix $\mathbf{T}(\boldsymbol{\Theta})$ is nonsingular if the pitch angle satisfies $\theta \neq (2k-1)\frac{\pi}{2}$.

To facilitate the control design, the following assumptions are made:

Assumption 1 *The reference trajectory $\boldsymbol{p}_d(t) = [x_d(t), y_d(t), z_d(t)]^\top$ is sufficiently smooth, twice differentiable and satisfies $\|\dot{\boldsymbol{p}}_d\|_\infty \leq \ell_{1d}$, $\|\ddot{\boldsymbol{p}}_d\|_\infty \leq \ell_{2d}$ where ℓ_{1d} and ℓ_{2d} are positive constants.*

Assumption 2 *To ensure that the angular transformation $\mathbf{T}(\boldsymbol{\Theta})$ is nonsingular, we assume that $\theta \neq (2k-1)\frac{\pi}{2}$ and $\phi \neq (2k-1)\frac{\pi}{2}$, where $k \in \mathbb{N}$ and that $\|\mathbf{T}(\boldsymbol{\Theta})\|_\infty$ is bounded (i.e., there exists a positive constant ε_1 such that $\|\mathbf{T}(\boldsymbol{\Theta})\|_\infty \leq \varepsilon_1$).*

Assumption 3 *We assume that the quadrotor translational velocity signal $\boldsymbol{v}(t)$ is not measurable and thus cannot be available for state feedback, whereas the angular velocity is measurable via Inertial system.*

Assumption 4 *The mass m and the UAV inertia matrix $\mathbf{I}_f$ are assumed to be known. The disturbances $\mathbf{d}_1$ and $\mathbf{d}_2$ are time varying and their first two time derivatives (i.e., $d_i, \dot{d}_i, \ddot{d}_i$, i=1,2) exist and are bounded by known constants. (i.e, there exists $\epsilon_{i1}, \epsilon_{i2}$ and ϵ_{i3} such that $\|d_i\| \leq \epsilon_{i1}$, $\|\dot{d}_i\| \leq \epsilon_{i2}$ and $\|\ddot{d}_i\| \leq \epsilon_{i3}$).*

Remark 1 The assumption on the unavailability of the quadrotor's linear velocity measurement stems from the fact that most commercially available quadrotor are often equipped with inertial measurement unit (IMU) which is a set of sensors consisting of accelerometer, magnetometer and gyroscope from which the attitude can be measured with acceptable accuracy. However, there is no specific sensor that is capable of measuring the body-fixed velocity $\boldsymbol{v}$. This justifies Assumption 3.

15.3.2 Control Objective

The aim of the control design is to force the quadrotor AUV to track a desired a three dimensional trajectory $\boldsymbol{p}_d = [x_d, y_d, z_d]^\top$ with a desired reference yaw angle $\psi_d(t)$. Specifically for any initial conditions $\boldsymbol{p}(t_0)$, $\boldsymbol{\Theta}(t_0)$ and $\boldsymbol{\omega}(t_0)$ and under Assumptions 1, 2, 3 and 4, design the thrust force u_1 and the torque input $\boldsymbol{u}_2$ such that

$$\lim_{t\to\infty} \mathbf{e}_p(t) = 0, \quad \lim_{t\to\infty} \psi(t) - \psi_d(t) = 0 \tag{15.7}$$

while keeping all other states of the quadrotor bounded for all initial conditions $\boldsymbol{p}(t_0)$, $\boldsymbol{\Theta}(t_0)$ and $\boldsymbol{\omega}(t_0)$, where $\mathbf{e}_p(t)$ is the position tracking error expressed in the body-fixed frame to be defined later.

15.4 Control Development

To address the positioning tracking problem, we start by defining the error variable, henceforth called position tracking error $\mathbf{e}_p$, which is expressed in the body-fixed frame and it is given as follows

$$\mathbf{e}_p = \mathbf{R}^\top(\boldsymbol{p} - \boldsymbol{p}_d) \tag{15.8}$$

Note that the tracking error coordinates (15.8) is expressed with respect to a body-fixed frame $\{B\}$ rather than the inertial frame {I}, this is advantageous as it avoids the dependence on the choice of the inertial reference frame. The position tracking error rate $\dot{\mathbf{e}}_p(t)$ is obtained by tacking the time derivative of (15.8) as follows:

$$\begin{aligned}\dot{\mathbf{e}}_p(t) &= \dot{\mathbf{R}}^\top(\boldsymbol{p} - \boldsymbol{p}_d) + \mathbf{R}^\top\dot{\boldsymbol{p}} - \mathbf{R}^\top\dot{\boldsymbol{p}}_d \\ &= -\mathbf{S}(\boldsymbol{\omega})\mathbf{R}^\top(\boldsymbol{p} - \boldsymbol{p}_d) + \boldsymbol{v} - \mathbf{R}^\top\dot{\boldsymbol{p}}_d \\ &= -\mathbf{S}(\boldsymbol{\omega})\mathbf{e}_p + \frac{1}{m}(m\boldsymbol{v} - \mathbf{R}^\top\dot{\boldsymbol{p}}_d) + (\frac{1}{m} - 1)\mathbf{R}^\top\dot{\boldsymbol{p}}_d\end{aligned} \tag{15.9}$$

The translational linear velocity tracking error can now be defined from Eq. (15.9) as $\mathbf{e}_v \in \mathbb{R}^3$ to be defined in the body-fixed fame as

$$\mathbf{e}_v = m\boldsymbol{v} - \mathbf{R}^\top\dot{\boldsymbol{p}}_d - \boldsymbol{\delta} \tag{15.10}$$

where $\boldsymbol{\delta} \in \mathbb{R}^3$ is a scalar vector defines as $\boldsymbol{\delta} = [0, 0, \delta]^\top$ with $\delta \in \mathbb{R}$ is a design constant. The time derivative of the velocity tracking error along the solutions of the second equation of the quadrotor dynamics in (15.4) yields

$$\dot{\mathbf{e}}_{\upsilon} = -m\mathbf{S}(\omega)\boldsymbol{\upsilon} + \mathbf{N}_1(\boldsymbol{\upsilon}) + \mathbf{G} + \mathbf{B}_1 u_1 + \mathbf{d}_1 + \mathbf{S}(\omega)\mathbf{R}^\top \dot{\boldsymbol{p}}_d - \mathbf{R}^\top \ddot{\boldsymbol{p}}_d$$
$$= -\mathbf{S}(\omega)\mathbf{e}_{\upsilon} - \mathbf{S}(\omega)\boldsymbol{\delta} + \mathbf{N}_1(\boldsymbol{\upsilon}) + \mathbf{G} - \mathbf{R}^\top \ddot{\boldsymbol{p}}_d + \mathbf{B}_1 u_1 + \mathbf{d}_1 \qquad (15.11)$$

The yaw angle tracking error as defined in (15.7) is given by $e_\psi = \psi - \psi_d \in \mathbb{R}$, its rate error is obtained by taking its time derivative as follows

$$\dot{e}_\psi = \dot{\psi} - \dot{\psi}_d = T_z(\boldsymbol{\Theta})\boldsymbol{\omega} - \omega_{zd} \qquad (15.12)$$

where $T_z(\boldsymbol{\Theta}) \in \mathbb{R}^{3\times 1}$ is the third row of $\mathbf{T}(\boldsymbol{\Theta})$ defined in (15.6) and $\omega_{zd} = \dot{\psi}_d$ is the desired yaw rate angle expressed in the body-fixed frame. Based on the tracking errors (15.9), (15.11), (15.12) and the last equation of the quadrotor's dynamics (15.4), the control scheme construction is proposed based on the backstepping approach and the nonlinear robust control technique of Cai et al. (2006). The technique unfolds in two steps.

Step 1 Due to the unmeasured state $\upsilon(t)$ in (15.10), the control design cannot contains terms involving the tracking error $\mathbf{e}_\upsilon$. To overcome such obstruction, the velocity tracking error is estimated through a second order filter designed as follows:

$$\mathbf{r}_f = \mathbf{q} - (k_2 + 1)\int_0^t \mathbf{e}_\upsilon(\lambda)d\lambda \qquad (15.13)$$
$$\dot{\mathbf{q}} = -\mathbf{r}_f - (k_2 + 1)(\mathbf{e}_p + \mathbf{r}_f + \boldsymbol{\delta}) + \mathbf{e}_p - \mathbf{e}_f \qquad (15.14)$$
$$\dot{\mathbf{e}}_f = -\mathbf{e}_f + \mathbf{r}_f \qquad (15.15)$$

where $\mathbf{r}_f, \mathbf{e}_f \in \mathbb{R}^3$ are filtered signals, $\mathbf{q}(t) \in \mathbb{R}^3$ is an auxiliary variable useful for the implementation of the filter and k_2 is a positive constant to be determined later. The initial conditions for the filter (15.13), (15.14) and (15.15) are as follows:

$$\mathbf{e}_f(t_0) = 0 \qquad (15.16)$$
$$\mathbf{q}(t_0) = (k_2 + 1)t_0\mathbf{e}_\upsilon(t_0) \qquad (15.17)$$

Remark 2 Note from (15.13), that the second right hand term $\int_0^t \mathbf{e}_\upsilon(\lambda)d\lambda$ is completely known and can be evaluated directly by the measured signals $\boldsymbol{p}$, $\boldsymbol{\Theta}$ and $\boldsymbol{p}_d$.

In order to further proceed with the development of the control input for the translational dynamics, new error states are defined. Let the signal $\boldsymbol{\eta} \in \mathbb{R}^3$ be defined as follows:

$$\boldsymbol{\eta} = \mathbf{e}_\upsilon + \mathbf{e}_p + \mathbf{r}_f + \boldsymbol{\delta} \qquad (15.18)$$

Take the time derivative of (15.13) using, (15.14) and (15.18), we obtain

$$
\begin{aligned}
\dot{\mathbf{r}}_f &= \dot{\mathbf{q}} - (k_2+1)\mathbf{e}_v \\
&= -\mathbf{r}_f - (k_2+1)(\mathbf{e}_p + \mathbf{r}_f + \boldsymbol{\delta}) + \mathbf{e}_p - \mathbf{e}_f - (k_2+1)(\boldsymbol{\eta} - \mathbf{e}_p - \mathbf{r}_f - \boldsymbol{\delta}) \\
&= -\mathbf{r}_f - (k_2+1)\boldsymbol{\eta} + \mathbf{e}_p - \mathbf{e}_f
\end{aligned}
\tag{15.19}
$$

Take the time derivative of η and using (15.19), we obtain the following:

$$
\begin{aligned}
\dot{\boldsymbol{\eta}} &= \dot{\mathbf{e}}_v + \dot{\mathbf{e}}_p + \dot{\mathbf{r}}_f \\
&= \dot{\mathbf{e}}_v + \dot{\mathbf{e}}_p - (k_2+1)\boldsymbol{\eta} - \mathbf{r}_f + \mathbf{e}_p - \mathbf{e}_f \\
&= -\mathbf{S}(\omega)\boldsymbol{\eta} + \mathbf{S}(\omega)\mathbf{r}_f - \mathbf{S}(\omega)\boldsymbol{\delta} + \frac{1}{m}\mathbf{e}_v + \frac{1}{m}\boldsymbol{\delta} + (\frac{1}{m} - 1)\mathbf{R}^\top \dot{p}_d - \mathbf{R}^\top \ddot{p}_d \\
&\quad + \mathbf{N}_1(\boldsymbol{v}) + \mathbf{G} + \mathbf{d}_1 + \mathbf{S}(\omega)\boldsymbol{\delta} + \mathbf{B}_1 u_1 - (k_2+1)\boldsymbol{\eta} - \mathbf{r}_f + \mathbf{e}_p - \mathbf{e}_f
\end{aligned}
\tag{15.20}
$$

Rearranging the terms in (15.20) yields

$$
\dot{\boldsymbol{\eta}} = -(k_2+1)\boldsymbol{\eta} - \mathbf{S}(\omega)\boldsymbol{\eta} + \widetilde{\mathbf{M}} + \mathbf{M}_d + \mathbf{d}_1 + \Big[-\mathbf{S}(\boldsymbol{\delta})\boldsymbol{\omega} + \mathbf{B}_1 u_1\Big] - \mathbf{e}_p \tag{15.21}
$$

where the auxiliary functions $\widetilde{\mathbf{M}} \in \mathbb{R}^3$ and $\mathbf{M}_d \in \mathbb{R}^3$ are defined as

$$
\widetilde{\mathbf{M}} = \mathbf{S}(\omega)\mathbf{r}_f + \mathbf{S}(\omega)\boldsymbol{\delta} + \frac{1}{m}\mathbf{e}_v + \mathbf{N}_1(\boldsymbol{v}) - \mathbf{M}_d - (\mathbf{r}_f + \mathbf{e}_f - 2\mathbf{e}_p) \tag{15.22}
$$

$$
\mathbf{M}_d = (\frac{1}{m} - 1)\mathbf{R}^\top \dot{p}_d - \mathbf{R}^\top \ddot{p}_d + \mathbf{G} + \frac{1}{m}\boldsymbol{\delta}\mathbf{d}_1 \tag{15.23}
$$

The filtered position tracking error (15.21) can now be combined with the yaw tracking error (15.12) to create a composite tracking error $\chi(t) \in \mathbb{R}^4$ such that $\chi = [\boldsymbol{\eta}^\top, \mathbf{e}_\psi]^\top$. The dynamics of the composite tracking error is therefore obtained as follows:

$$
\dot{\chi} := \begin{bmatrix} \dot{\boldsymbol{\eta}} \\ \dot{\mathbf{e}}_\psi \end{bmatrix} = \begin{bmatrix} -(k_2+1)\boldsymbol{\eta} - \mathbf{S}(\omega)\boldsymbol{\eta} + \widetilde{\mathbf{M}} + \mathbf{M}_d - \mathbf{e}_p \\ -\omega_{zd} \end{bmatrix} + \underbrace{\begin{bmatrix} -\mathbf{S}(\boldsymbol{\delta}) & \mathbf{B}_1 \\ T_z(\boldsymbol{\Theta}) & 0 \end{bmatrix}}_{\mathbf{B}_C} \underbrace{\begin{bmatrix} \boldsymbol{\omega} \\ u_1 \end{bmatrix}}_{\zeta} \tag{15.24}
$$

Note from (15.24), two components of the vector $\zeta := [\boldsymbol{\omega}^\top, u_1]^\top$ can have influence on the translational dynamics. The first component is u_1 which is viewed as a physical input to the system and can be directly specified in the control design, while the second component is $\boldsymbol{\omega}$ which can only be indirectly specified through the design of the torque input u_2. The backstepping technique is therefore pursued to control the translation dynamics via the angular velocity $\boldsymbol{\omega}$. To this end, we regard $\boldsymbol{\omega}$ as a desired angular velocity to force the filtered state χ to converge asymptotically to zero. To this purpose, we define the backstepping variable error z as follows:

$$
z = \boldsymbol{\omega} - \mathbf{B}_z \overline{\mathbf{u}}_1 \tag{15.25}
$$

where $\overline{\mathbf{u}}_1 = [\boldsymbol{\alpha}_\omega^\top, u_1]^\top \in \mathbb{R}^4$ and the matrix $\mathbf{B}_z = [I_3, \mathbf{0}_{3\times 1}]$. From (15.24) and (15.25), it can be shown that:

$$\mathbf{B}_C\boldsymbol{\zeta} = \mathbf{B}_C\overline{\mathbf{u}}_1 + \mathbf{B}_C\begin{bmatrix} z \\ 0 \end{bmatrix} \tag{15.26}$$

Then according to (15.26), the system dynamics (15.24) can be written as follows:

$$\dot{\boldsymbol{\chi}} := \begin{bmatrix} \dot{\boldsymbol{\eta}} \\ \dot{\mathbf{e}}_\psi \end{bmatrix} = \begin{bmatrix} -(k_2+1)\boldsymbol{\eta} - \mathbf{S}(\omega)\boldsymbol{\eta} + \widetilde{\mathbf{M}} + \mathbf{M}_d - \mathbf{e}_p \\ -\omega_{zd} \end{bmatrix} + \mathbf{B}_C\overline{\mathbf{u}}_1 + \mathbf{B}_C\begin{bmatrix} z \\ 0 \end{bmatrix} \tag{15.27}$$

From (15.27), the control input $\overline{\mathbf{u}}_1$ can be designed using the robust mechanism presented in Xian et al. (2004) as follows:

$$\overline{\mathbf{u}}_1 = \underbrace{\begin{bmatrix} 0 & -\frac{1}{\delta} & 0 & 0 \\ \frac{1}{\delta} & 0 & 0 & 0 \\ -\frac{1}{\delta}\frac{\sin\phi}{\cos\phi} & 0 & 0 & \frac{\cos\theta}{\cos\phi} \\ 0 & 0 & 1 & 0 \end{bmatrix}}_{B_C^{-1}} \begin{bmatrix} -K_1\mathrm{Sgn}(\mathbf{e}_p + \mathbf{e}_f) + (k_2+1)\mathbf{r}_f \\ \omega_{zd} \end{bmatrix} \tag{15.28}$$

Notice that according to Assumption 2, the matrix B_C is invertible and well defined, this is due to the assumption that the pitch and the roll angles never reach singularity $\pm\frac{\pi}{2}$. Clearly from (15.28), it is seen that the last term of this equation quantifies an error term that can be controlled through the input to the rotational dynamics. Substituting (15.28) into (15.27) to obtain the following closed-loop translational system

$$\begin{bmatrix} \dot{\boldsymbol{\eta}} \\ \dot{\mathbf{e}}_\psi \end{bmatrix} = \begin{bmatrix} (k_2+1)(\mathbf{r}_f - \boldsymbol{\eta}) - \mathbf{S}(\omega)\boldsymbol{\eta} + \widetilde{\mathbf{M}} + \mathbf{M}_d - K_1\mathrm{Sgn}(\mathbf{e}_p + \mathbf{e}_f) - \mathbf{e}_p \\ 0 \end{bmatrix} + \mathbf{B}_C\begin{bmatrix} z \\ 0 \end{bmatrix} \tag{15.29}$$

Remark 3 Note the importance of the scalar δ introduced in the filtering tracking error in (15.14) as well as in (15.18), was to ensure that the matrix B_C is nonsingular, which based on Assumption 2 the matrix B_C is invertible. The control input $\overline{\mathbf{u}}_1$ is therefore well defined.

Step 2 The control input $\overline{\mathbf{u}}_1$ was designed to control the motion of the quadrotor along the three dimensional axes. In order to control the quadrotor's attitude, this signal must be injected as a reference to be tracked by the rotational dynamics. The design proceeds by modifying the standard backstepping procedure so that the robustification mechanism of Cai et al. (2006) can be used in a simpler manner. To do that, consider the following augmented system:

$$\mathbf{r}_a = \frac{1}{k_3}\dot{z} + z \tag{15.30}$$

where k_3 is a positive design gain. Note that the augmented system (15.30) is discontinuous and involves the time derivative of the control signal $\overline{\mathbf{u}}_1$. Since $\overline{\mathbf{u}}_1$ contains signum function, its derivative is hard enough to figure out. To proceed with the design an exact differentiator is needed to estimate the first two derivatives of the control input $\overline{\mathbf{u}}_1 = [\overline{u}_{1i}]_{i=1,\ldots,4}^{\top}$. The following finite time estimator is proposed:

$$\hat{\dot{\overline{\mathbf{u}}}}_1 \triangleq \begin{cases} \dot{\varrho}_{0i} = \rho_{0i} = \varrho_{1i} - \epsilon_{0i}\sqrt{|\varrho_{0i} - \overline{u}_{1i}|}\,\text{sign}(\varrho_{0i} - \overline{u}_{1i}) \\ \dot{\varrho}_{1i} = -\epsilon_{1i}\,\text{sign}(\varrho_{1i} - \rho_{0i}), \qquad i = 1, \ldots, 4 \end{cases} \tag{15.31}$$

$$\hat{\ddot{\overline{\mathbf{u}}}}_1 \triangleq \begin{cases} \dot{\varrho}_{2i} = \rho_{1i} = \varrho_{3i} - \epsilon_{2i}\sqrt{|\varrho_{2i} - \hat{\dot{\overline{u}}}_{1i}|}\,\text{sign}(\varrho_{2i} - \hat{\dot{\overline{u}}}_{1i}) \\ \dot{\varrho}_{3i} = -\epsilon_{3i}\,\text{sign}(\varrho_{3i} - \rho_{1i}), \qquad i = 1, \ldots, 4 \end{cases} \tag{15.32}$$

where $\varrho_{0i}, \varrho_{1i}, \varrho_{2i}, \varrho_{3i}, \rho_{0i}$ and ρ_{1i} are the states of (15.31) and (15.32) respectively. $\epsilon_{0i}, \epsilon_{1i}, \epsilon_{2i}$ and ϵ_{3i} are positive design constants. According to (15.32) and Lemma 1, we have

$$\hat{\ddot{\overline{\mathbf{u}}}}_1 = \boldsymbol{\rho}_1 + \boldsymbol{\varpi} \tag{15.33}$$

where $\boldsymbol{\rho}_1 = [\rho_{11}, \rho_{12}, \ldots, \rho_{14}]^{\top}$, $\boldsymbol{\varpi}$ is the estimation error of the first order sliding mode differentiator (15.32). From Lemma 1, we have for each component of $\boldsymbol{\varpi}$, the following bound $|\varpi_i| \leq \varsigma_i$, with $\varsigma_i > 0$, $i = 1, 2, \ldots, 4$.

Remark 4 Note that if the filtered error $\mathbf{r}_a$ of the state z converges asymptotically to zero then the state tracking error z also does. The control design will be responsible of ensuring boundedness of the filtered error as well as its asymptotic convergence to zero.

Taking the time derivative of $\mathbf{r}_a(t)$, multiplying by the inertia matrix $\mathbf{I}_f$ and using the expression in (15.25), yields

$$\begin{aligned} \mathbf{I}_f\dot{\mathbf{r}}_a &= \frac{\mathbf{I}_f}{k_3}\ddot{z} + \mathbf{I}_f\dot{z} \\ &= \frac{\mathbf{I}_f}{k_3}\ddot{z} + k_3\mathbf{I}_f(\mathbf{r}_a - z) \\ &= \frac{1}{k_3}\big(\dot{\Phi} + \mathbf{B}_2\dot{\mathbf{u}}_2 + \mathbf{D} - \mathbf{I}_f\mathbf{B}_z\hat{\ddot{\overline{\mathbf{u}}}}_1\big) + k_3\mathbf{I}_f(\mathbf{r}_a - z) \end{aligned} \tag{15.34}$$

where Φ denotes the nonlinear part of the attitude dynamics in the last equation of (15.4), such that $\Phi(t) = -\mathbf{S}(\omega)\mathbf{I}_f\omega - \mathbf{N}_2(\omega)$ and $\mathbf{D}$ is a disturbance compound given as $\mathbf{D} = \dot{\mathbf{d}}_2 + \mathbf{B}_z\boldsymbol{\varpi}$. Note that the differentiation of the backstepping variable z involves the differentiation of the virtual control $\overline{\mathbf{u}}_1$ which is a hectic operation as it contains signum function. It therefore has been replaced by its estimate using the finite time estimator (15.32).

Based on (15.34), we design $\dot{\mathbf{u}}_2$ as follows:

$$\dot{\mathbf{u}}_2 = \mathbf{B}_2^{-1}\Big(- \dot{\Phi} + \mathbf{I}_f \mathbf{B}_z \ddot{\hat{\bar{\mathbf{u}}}}_1 - K_2 \mathrm{Sng}(z) - k_3(k_3 \mathbf{I}_f + \mathbf{I}_3)\mathbf{r}_a\Big) \tag{15.35}$$

where $k_3 > 0$ and K_2 is a positive gain control to be determined later. The first term (15.35) is the nonlinear term to be canceled out form the attitude dynamics, the second term is due to the time derivative of the virtual control law in the previous step and third and the forth terms are used to compensate for the unknowns and crossing terms during the Laypunov stability analysis. From (15.35), the actual control law can be calculated according to the following:

$$\begin{aligned}\mathbf{u}_2 = \mathbf{B}_2^{-1}\Big(&- \Phi(t) + \Phi(t_0) + \mathbf{I}_f \mathbf{B}_z\big(\hat{\bar{\mathbf{u}}}_1(t) - \hat{\bar{\mathbf{u}}}_1(t_0)\big) - (k_3 \mathbf{I}_f + \mathbf{I}_3)(z(t) - z(t_0)) \\ &- \int_{t_0}^{t} k_3(k_3 \mathbf{I}_f + \mathbf{I}_3) z(\mu) - K_2 \mathrm{Sng}(z(\mu)) d\mu\Big)\end{aligned} \tag{15.36}$$

The control design is now completed, the main result of this paper can be summarized in the following theorem:

Theorem 1 *Consider the translational and rotational dynamics of the quadrotor in closed loop system. the control input* (15.28) *and* (15.36) *ensure that all closed-loop signals of the quadrotor system are bounded. In particular* $\mathbf{e}_p, \mathbf{e}_f, \mathbf{r}_f, e_\psi, \boldsymbol{\eta}, z$ *and* $\mathbf{r}_a$ *converge asymptotically to zero if the gain controllers* K_1 *and* K_2 *are chosen to satisfy the following:*

$$K_1 > \|\mathbf{M}_d\|_\infty + \|\dot{\mathbf{M}}_d\|_\infty, \quad K_2 > \|\mathbf{D}\|_\infty + \|\dot{\mathbf{D}}\|_\infty \tag{15.37}$$

and the positive constant k_2 *is chosen to satisfy*

$$k_2 > \frac{\varsigma}{4(\varsigma + \gamma)} \varphi \left(\sqrt{\frac{\max\{1, \frac{m}{2}\}}{\min\{\frac{1}{2}, \frac{m}{2}\}}} \|\mathbf{y}(t_0)\| \right)^2 \tag{15.38}$$

where an explicit expression for $\|\mathbf{y}(t_0)\|$ *is given as*

$$\|\mathbf{y}(t_0)\| = \sqrt{\|\mathbf{e}_p(0)\|^2 + e_\psi(0)^2 + \|\boldsymbol{\eta}\|^2 + z(t_0) + \mathbf{r}_a(t_0) + P(t_0) + Q(t_0)} \tag{15.39}$$

In particular the velocity tracking error $m\boldsymbol{v} - \mathbf{R}^\top \dot{\boldsymbol{p}}_d$ *converges asymptotically to zero.*

Remark 5 The challenge of the proposed control for the aerial underactuated quadrotor, is first to systematically incorporate the RISE feedback technique into a backstepping design procedure which is made quite hard for most of underactuated systems. The second challenge is to achieve asymptotic convergence of the closed-loop system in the presence of a disturbance. On the other hand, a major disadvantage of the scheme is that the controller being designed has discontinuous feedback due to the presence of signum functions. However, we emphasize that although it is discontinuous control design, the proposed controller doesn't suffer from limitations revealed by other relevant control schemes such as the sliding mode techniques characterized by their infinite bandwidth and chatter. An alternative way to convert the proposed controller into a continuous feedback is to implement the sigmoid function in place of the signum function. Nevertheless, incorporating the sigmoid function in the implementation of the proposed controller would imply different stability results (the reader is referred to Xian and Zhang 2017 for more details).

15.5 Numerical Simulations

In this section, we illustrate the effectiveness of the proposed tracking controller through a numerical simulation and assess the robustness property by adding disturbances into the position and attitude loop. Simulations are carried out by means of the "Matlab/Simulink" software program. For simplicity of the simulation, the aerodynamics interaction terms $\mathbf{N}_1(.)$ and $\mathbf{N}_2(.)$ are neglected, hence the physical parameters of the quadrotor UAV are taken as: $m = 1.1\,\text{kg}$, $d = 0.2125\,\text{m}$, $I_{rx} = I_{ry} = 0.013\,\text{kg/m}^2$, $I_{rz} = 0.024\,\text{kg/m}^2$, $b = 1.41871 \times 10^{-5}$, $c = 1.84 \times 10^{-6}$, and $g = 9.81\,\text{m/s}^2$. The reference trajectories are taken $\mathbf{p}_d = [10\sin(0.1t), 10\cos(0.1t), 0.2t]\,\text{m}$ and the desired yaw angle $\psi_d(t) = 0.1t$ rad. The initial position and angle of the quadrotor at the center of mass are selected as follows: $\boldsymbol{p}(t_0) = [2, 1, 5]^\top$, $\Theta(t_0) = [\frac{\pi}{9}, 0, 0]^\top$ and the constant vector design $\boldsymbol{\delta}$ is chosen as $[0, 0, -1]^\top$. The control parameters for the filter and the controller are chosen to be $K_1 = 30$, $k_2 = 15$, $k_3 = 5$, $K_2 = 20$, the gain parameters for the differentiator are chosen as $\varepsilon_{0i} = \varepsilon_{2i} = 10$ and $\varepsilon_{1i} = \varepsilon_{3i} = 10$, for $i = 1, \ldots, 4$. We also consider that a normally distributed noise signal with zero mean and variance of 0.1 is introduced into the motion equations of the quadrotor.

Figure 15.2 depicts the 3D trajectory tracking of the quadrotor with respect to the desired trajectory, which clearly shows the accomplishment of the control objective stated in (15.7). Figures 15.3 and 15.4 illustrates the asymptotic stability of the trajectory tracking $\boldsymbol{p} - \boldsymbol{p_d}$ and the asymptotic convergence of the backstepping error tracking variable $\boldsymbol{z} = [z_1, z_2, z_3]^\top$ respectively. These three figures illustrate the validity of the robust position and attitude loops tracking results. Figure 15.5 illustrates the time variation of the resulting pitch and roll angles. Further, Fig. 15.6 illustrates the generated forces by the four motors of the quadrotor.

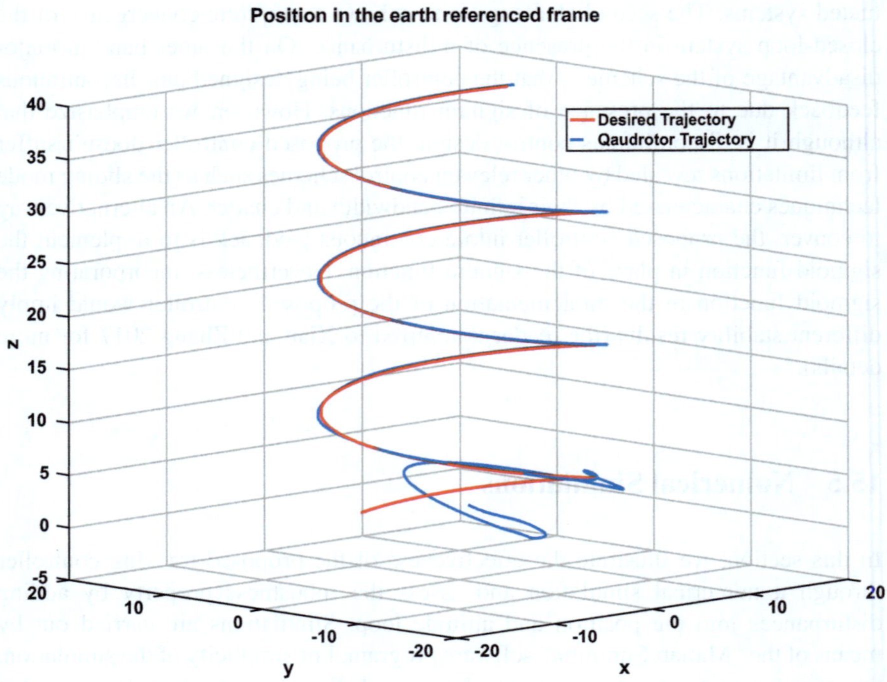

Fig. 15.2 Figure of 3D trajectory tracking

15.6 Conclusion

This paper presented a nonlinear backstepping control design to achieve trajectory tracking for quadrotor despite the presence of unknown external time varying perturbation. The RISE technique has been employed within the backstepping procedure in order to overcome the uncertainties in the system. During the position tracking loop control design and due to the unavailability of the linear velocity measurement a filter design is introduced to estimate the linear velocities in the first step of the design. The integrator backstepping in combination with the RISE feedback are then considered to overcome difficulties due to underactuation. It has been demonstrated that, despite the presence of external unknown disturbances the resulting closed-loop tracking error system is semi-globally asymptotically stable. Future work will improve the proposed technique by considering parametric model uncertainties affecting the system.

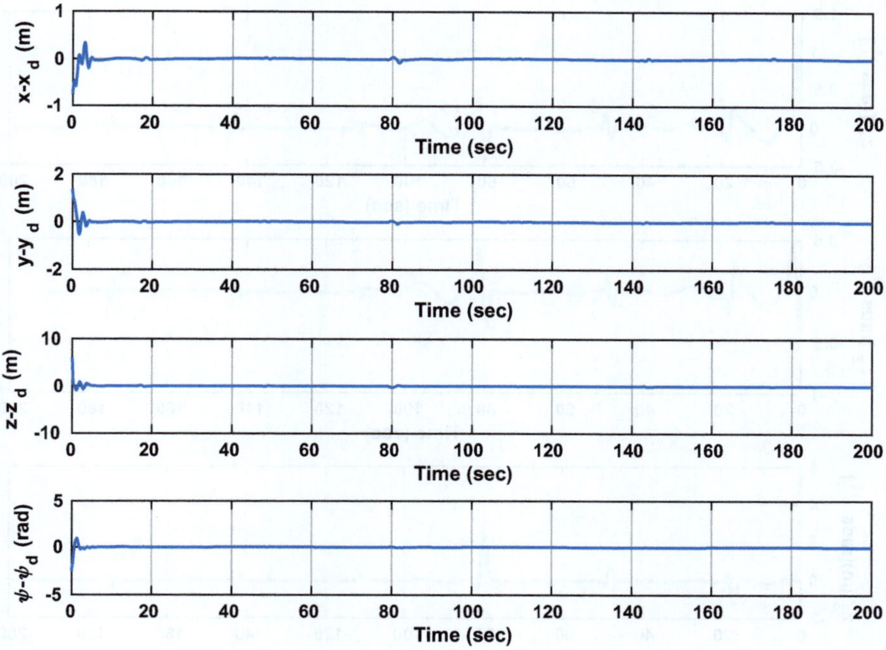

Fig. 15.3 Position and orientation tracking error variables

Appendix

For the clarity of the stability proof, we first write the closed loop-system by substituting the control input (15.35) into the dynamics (15.34), yields

$$\begin{bmatrix} \dot{\boldsymbol{\eta}} \\ \dot{\mathbf{e}}_{\psi} \end{bmatrix} = \begin{bmatrix} (k_2+1)(\mathbf{r}_f - \boldsymbol{\eta}) - \mathbf{S}(\omega)\boldsymbol{\eta} + \widetilde{\mathbf{M}} + \mathbf{M}_d - K_1 \mathrm{Sgn}(\mathbf{e}_p + \mathbf{e}_f) - \mathbf{e}_p \\ 0 \end{bmatrix} + \overline{\mathbf{B}}_C z \tag{15.40}$$

$$\mathbf{I}_f \dot{\mathbf{r}}_a = \frac{1}{k_3}\Big(\mathbf{D} - K_2 \mathrm{Sng}(z) - k_3(k_3 \mathbf{I}_f + \mathbf{I}_3)\mathbf{r}_a\Big) + \mathbf{I}_f(\mathbf{r}_a - z) \tag{15.41}$$

where we have used the following relationship:

$$\mathbf{B}_C \begin{bmatrix} z \\ 0 \end{bmatrix} = \begin{bmatrix} -\mathbf{S}(\boldsymbol{\delta})z \\ T_z(\boldsymbol{\Theta})z \end{bmatrix} = \underbrace{\begin{bmatrix} -\mathbf{S}(\boldsymbol{\delta}) \\ T_z(\boldsymbol{\Theta}) \end{bmatrix}}_{\overline{\mathbf{B}}_C} z$$

For the closed-loop system (15.41) and (15.40), we consider the Lyapunov function candidate function V which is Lipschitz positive definite function defined as:

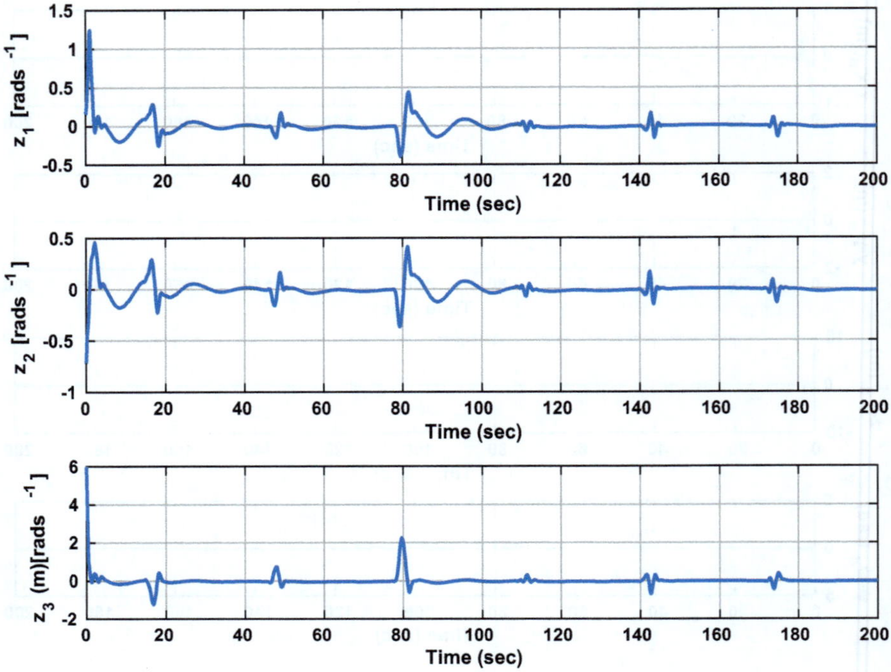

Fig. 15.4 Backstepping variable time evolution **z**

$$V(\boldsymbol{y}, t) = V_1 + V_2 + P + Q \tag{15.42}$$

where we define the functions V_1 and V_2 as follows

$$V_1 = \frac{m}{2}\mathbf{e}_p^\top \mathbf{e}_p + \frac{1}{2}\mathbf{e}_f^\top \mathbf{e}_f + \frac{1}{2}\mathbf{r}_f^\top \mathbf{r}_f + \frac{1}{2}\boldsymbol{\chi}^\top \boldsymbol{\chi} \tag{15.43}$$

$$V_2 = \frac{1}{2}z^\top \mathbf{I}_f z + \frac{1}{2}\mathbf{r}_a^\top \mathbf{I}_f \mathbf{r}_a \tag{15.44}$$

Let $\boldsymbol{y} \in R^{20}$ be defined as

$$\boldsymbol{y} = [\boldsymbol{X}^\top, \sqrt{P(t)}, \sqrt{Q(t)}]^\top \tag{15.45}$$

$$\boldsymbol{X} = [\mathbf{e}_p^\top, \mathbf{e}_f^\top, \mathbf{r}_f^\top, \boldsymbol{\eta}^\top, e_\psi, \mathbf{z}^\top, \mathbf{r}_a^\top]$$

and the auxiliary function $P \in \mathbb{R}$ and $Q \in \mathbb{R}$ are defined as the Filippov solutions (Filippov 1964) to the following differential equations respectively:

$$\dot{P} = -\boldsymbol{\eta}^\top \Big(\mathbf{M}_d - K_1 \mathrm{Sng}(\mathbf{e}_p + \mathbf{e}_f)\Big) \tag{15.46}$$

$$\dot{Q} = \frac{\mathbf{r}_a^\top}{k_3}\Big(\mathbf{D} - K_2 \mathrm{Sgn}(z)\Big) \tag{15.47}$$

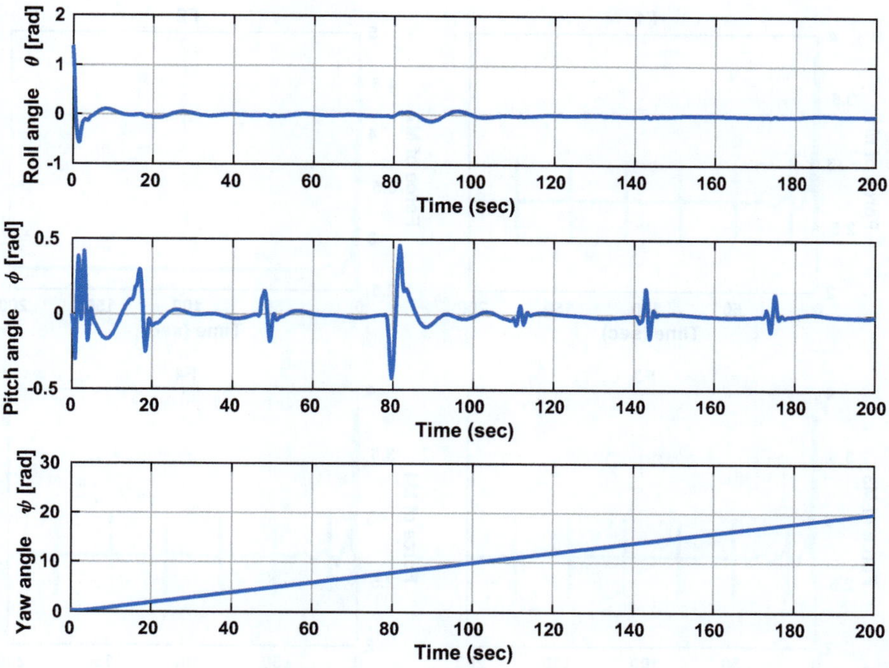

Fig. 15.5 Attitude angles

with initial conditions $P(t_0)$ and $Q(t_0)$ are given as follows

$$P(t_0) = \sum_{i=1}^{3} K_1 |e_{pi}(0)| - \mathbf{e}_p^\top(0) M_d(0) \tag{15.48}$$

$$Q(t_0) = \frac{1}{k_3}\Big(\sum_{i=1}^{3} K_2 |z_i(0)| - z(0)^\top \mathbf{D}(0)\Big) \tag{15.49}$$

with the subscript $i = 1, \ldots, n$ denotes the i-th element of the vectors $\mathbf{e}_p(t)$ and $z(t)$. The gain control K_1 and K_2 are chosen according to the sufficient condition (15.3) underlined in Lemma 2.

It can be verified that the Lyapunov function candidate defined in (15.42) can be bounded as :

$$U_1(\boldsymbol{y}) \leq V(\boldsymbol{y}, t) \leq U_2(\boldsymbol{y}) \tag{15.50}$$

where $U_1 : \mathbb{R}^{20} \to \mathbb{R}$ and $U_2 : \mathbb{R}^{20} \to R$ are positive definite functions defined as $U_1(\boldsymbol{y}) = \min\{\frac{1}{2}, \frac{m}{2}\}\|\boldsymbol{y}\|^2$ and $U_2(\boldsymbol{y}) = \max\{1, \frac{m}{2}\}\|\boldsymbol{y}\|^2$. Under Filippov's framework, strong stability of the closed-loop system will be established in the following.

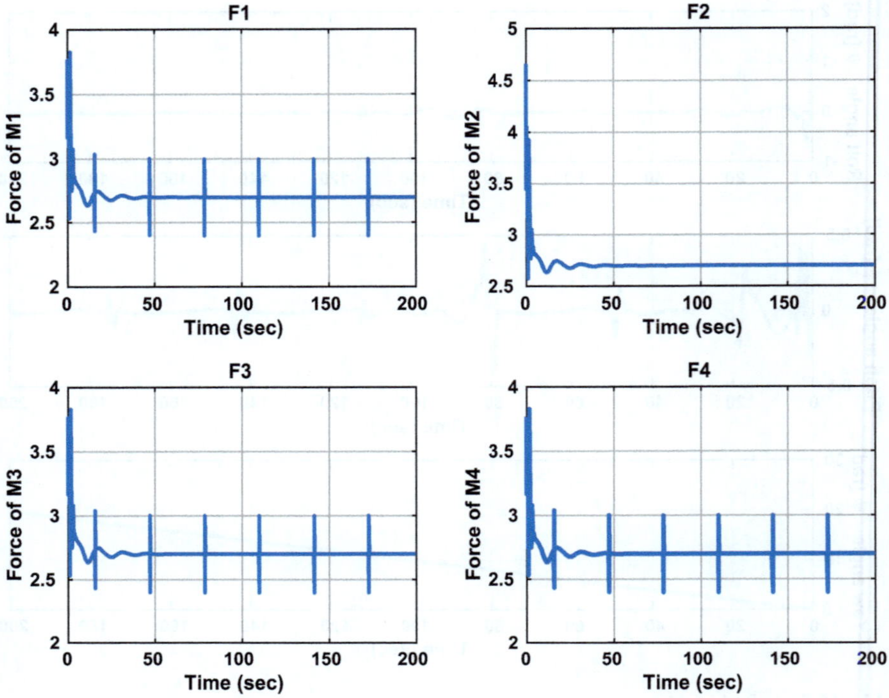

Fig. 15.6 Torques for Motors $M_1, \ldots, M_4$

The time derivative of (15.42) along the Filippov trajectories exists almost everywhere (a.e), for almost all $t \in [t_0, t]$, one can get:

$$\dot{V} \overset{\text{a.e}}{\in} \dot{\tilde{V}} = \bigcap_{\xi \in \partial V(y,t)} \xi^\top \Psi[\dot{\mathbf{e}}_p^\top, \dot{\mathbf{e}}_f^\top, \dot{\mathbf{r}}_f^\top, \dot{\boldsymbol{\eta}}^\top, \dot{e}_\psi, \dot{\mathbf{z}}^\top, \dot{\mathbf{r}}_a^\top, \frac{1}{2}P^{\frac{1}{2}}\dot{P}, \frac{1}{2}Q^{\frac{1}{2}}\dot{Q}, 1] \tag{15.51}$$

where ∂V is the generalized gradient of V (Clarke 1990). Since V is continuously differentiable with respect to $\boldsymbol{y}$, then one has

$$\dot{\tilde{V}} \subset \nabla V^\top \Psi[\dot{\mathbf{e}}_p^\top, \dot{\mathbf{e}}_f^\top, \dot{\mathbf{r}}_f^\top, \dot{\boldsymbol{\eta}}^\top, \dot{e}_\psi, \frac{1}{2}P^{\frac{1}{2}}\dot{P}, \frac{1}{2}Q^{\frac{1}{2}}\dot{Q}, 1] \tag{15.52}$$

where $\nabla V = \Psi[\mathbf{e}_p^\top, \mathbf{e}_f^\top, \mathbf{r}_f^\top, \boldsymbol{\eta}^\top, e_\psi, \mathbf{z}^\top\mathbf{I}_f, \mathbf{r}_a^\top\mathbf{I}_f, 2P^{-\frac{1}{2}}, 2Q^{-\frac{1}{2}}]^\top$. Substituting (15.9), (15.15), (15.24), (15.40) and (15.41) into (15.52) and using the calculus for $\Psi[.]$ from Paden and Sastry (1987), yields:

$$\begin{aligned}\dot{\tilde{V}} \subset\ & \mathbf{e}_p^\top(\boldsymbol{\eta} - \mathbf{e}_p - \mathbf{r}_f - \boldsymbol{\delta}) + (1-m)\mathbf{e}_p^\top\mathbf{R}^\top\dot{\mathbf{p}}_d - \mathbf{e}_f^\top\mathbf{e}_f + \mathbf{e}_f^\top\mathbf{r}_f - \mathbf{r}_f^\top\mathbf{r}_f \\ & -(k_2+1)\mathbf{r}^\top\boldsymbol{\eta} + \mathbf{r}_f^\top\mathbf{e}_p - \mathbf{r}_f^\top\mathbf{e}_f - (k_2+1)\boldsymbol{\eta}^\top\boldsymbol{\eta} + \boldsymbol{\eta}^\top\widetilde{\mathbf{M}}\end{aligned}$$

$$
\begin{aligned}
&+\boldsymbol{\eta}^{\top}\Big(\mathbf{M}_d - K_1\Psi[\mathrm{Sng}(\mathbf{e}_p+\mathbf{e}_f)]\Big) - \boldsymbol{\eta}^{\top}\mathbf{e}_p + (k_2+1)\boldsymbol{\eta}^{\top}\mathbf{r}_f\\
&+\boldsymbol{\eta}^{\top}\overline{\mathbf{B}}_C z - \Big(\mathbf{M}_d - K_1\Psi[\mathrm{Sng}(\mathbf{e}_p+\mathbf{e}_f)]\Big) - k_3 z^{\top}\mathbf{I}_f z - \mathbf{r}_a^{\top}\mathbf{r}_a\\
&+\frac{1}{k_3}\mathbf{r}_a^{\top}\Big(\mathbf{D} - K_2\Psi[\mathrm{Sng}(z)]\Big) - \frac{1}{k_3}\mathbf{r}_a^{\top}\Big(\mathbf{D} - K_2\Psi[\mathrm{Sng}(z)]\Big) \qquad (15.53)
\end{aligned}
$$

where $\Psi[\mathrm{Sng}(\boldsymbol{x})] = \mathrm{SGN}(\boldsymbol{x})$ such that $\forall \boldsymbol{x} = [x_1, x_2, x_3]^{\top}$, the set valued map $\mathrm{SNG}(x_i) = 1$ if $x_i > 0$, $[-1,\ 1]$ if $x_i = 0$ and -1 if $x_i < 0$ for $i = 1, 2, 3$. Using the fact that the set in (15.53) reduces to a scalar equality since the right hand side is continuous a.e, the following upper bound can be obtained on $\dot{V}$:

$$
\begin{aligned}
\dot{\tilde{V}} \overset{\text{a.e}}{\le} &-\|\mathbf{e}_p\|^2 - \|\mathbf{e}_f\|^2 - \|\mathbf{r}_f\|^2 - (k_2+1)\|\boldsymbol{\eta}\|^2 - k_3\|\mathbf{I}_f\|\|z\|^2 - \|\mathbf{r}_a\|^2 + \Big\|\boldsymbol{\eta}^{\top}\widetilde{\mathbf{M}}\Big\|\\
&+\Big\| - \mathbf{e}_p^{\top}\big(\boldsymbol{\delta} + (m-1)\mathbf{R}^{\top}\dot{\mathbf{p}}_d\big)\Big\| \qquad (15.54)
\end{aligned}
$$

Denote by $\gamma = \|\boldsymbol{\delta}\| + (m+1)\|\dot{\mathbf{p}}_d\|$, using the Mean Value Theorem (De Queiroz et al. 1997), the function $\widetilde{\mathbf{M}}$ can be upper bounded as $\|\widetilde{\mathbf{M}}\| \le \|\varphi(\|\boldsymbol{X}\|)\|\|\boldsymbol{X}\|$, then using the young's inequality yields:

$$
\| - \mathbf{e}_p^{\top}\big(\boldsymbol{\delta} + (m-1)\mathbf{R}^{\top}\dot{\mathbf{p}}_d\big)\| \le \gamma\|\mathbf{e}_p\| \le \frac{\gamma}{\|\boldsymbol{X}\|}\|\boldsymbol{X}\|^2 \qquad (15.55)
$$

$$
\|\widetilde{\mathbf{M}}\|\|\boldsymbol{\eta}\| \le \frac{\|\varphi(\|\boldsymbol{X}\|)\|^2}{4k_2} + k_2\|\boldsymbol{\eta}\|^2 \qquad (15.56)
$$

substituting (15.55) and (15.56) into (15.54), we obtain

$$
\begin{aligned}
\dot{\tilde{V}} &\overset{\text{a.e}}{\le} -\|\mathbf{e}_p\|^2 - \|\mathbf{e}_f\|^2 - \|\mathbf{r}_f\|^2 - \|\boldsymbol{\eta}\|^2 - k_3\|\mathbf{I}_f\|\|z\|^2 - \|\mathbf{r}_a\|^2 + \frac{\|\varphi(\|\boldsymbol{X}\|)\|^2}{4k_2}\\
&\quad + \frac{\gamma}{\|\boldsymbol{X}\|}\|\boldsymbol{X}\|^2\\
&\overset{\text{a.e}}{\le} -\|\boldsymbol{X}\|^2 + \frac{\|\varphi(\|\boldsymbol{X}\|)\|^2}{4k_2} + \frac{\gamma}{\|\boldsymbol{X}\|}\|\boldsymbol{X}\|^2 \qquad (15.57)
\end{aligned}
$$

It follows through the choice of ς as any given positive constant such that $\|\boldsymbol{X}\|^2 \ge \varsigma^2$, the expression in (15.57) can be upper bounded as:

$$
\dot{\tilde{V}} \overset{\text{a.e}}{\le} -\left(1 - \frac{\gamma}{\varsigma} - \frac{\|\varphi(\|\boldsymbol{X}\|)\|^2}{4k_2}\right)\|\boldsymbol{X}\|^2 \qquad (15.58)
$$

From (15.58), it can be concluded that for some positive constant C_0, we have:

$$
\dot{\tilde{V}} \overset{\text{a.e}}{\le} -C_0\|\boldsymbol{X}\|^2, \quad \text{for } k_2 > \frac{\|\varphi(\|\boldsymbol{X}\|)\|^2}{4(1-\frac{\gamma}{\varsigma})}, \text{ and } \varsigma < \gamma \qquad (15.59)
$$

It can be concluded from inequalities (15.50) and (15.59) that $V(\boldsymbol{y}, t) \in L_\infty$, therefore $\mathbf{e}_p, \mathbf{e}_f, \mathbf{r}_f, \eta, z, \mathbf{r}_a, P$ and $Q \in \mathcal{L}_\infty$. From (15.13), (15.14) and (15.15), a straightforward conclusion can be drawn to show that $\dot{\mathbf{e}}_f, \mathbf{q}, \dot{\mathbf{q}} \in \mathcal{L}_\infty$. From the fact that the desired trajectory is sufficiently smooth then using (15.8) we can conclude that $\mathbf{p}$ and $\dot{\mathbf{p}} \in \mathcal{L}_\infty$. From (15.28) and (15.36) we know that $\bar{\mathbf{u}}_1$ and $\mathbf{u}_2 \in \mathcal{L}_\infty$. Given these boundedness statements, it is clear from that the time derivative of $U(\boldsymbol{X}) = \gamma\|\boldsymbol{X}\|^2$ is such that $\dot{U}(\boldsymbol{X}) \in \mathcal{L}_\infty$, which also implies that $U(\boldsymbol{X})$ is uniformly continuous. Define the region $\mathcal{S}_\mathcal{D}$ as follows:

$$\mathcal{S}_\mathcal{D} = \left\{ \boldsymbol{y} \in \mathcal{D} \mid \quad U_2(\boldsymbol{y}) < \min\{\frac{1}{2}, \frac{m}{2}\}\varphi^{-1}\left(2\sqrt{\frac{k_2(\varsigma - \gamma)}{\varsigma}}\right)^2 \right\} \tag{15.60}$$

The region of attraction in $\mathcal{S}_\mathcal{D}$ can be made arbitrary large by appropriately tuning the control gain k_2. As a matter of fact, the region of attraction is calculated based on the set given in (15.60) as follows:

$$\|\boldsymbol{y}(t_0)\| \le \sqrt{\frac{\min\{\frac{1}{2}, \frac{m}{2}\}}{\max\{1, \frac{m}{2}\}}\varphi^{-1}\left(2\sqrt{\frac{k_2(\varsigma - \gamma)}{\varsigma}}\right)} \tag{15.61}$$

which can be rearranged as in (15.38). Based on the definition of the vector state $\boldsymbol{y}$ in (15.45) and the initial conditions defined in (15.16) and (15.17) and (15.48) and (15.49), an explicit expression for $\|\boldsymbol{y}(\boldsymbol{t_0})\|$ can be derived as in (15.39).

From (15.59) it can be concluded that $C_0\|\boldsymbol{X}\|^2 \to 0$ as time goes to infinity, $\forall \boldsymbol{y} \in \mathcal{S}_\mathcal{D}$, it is then obvious to show form the definition of $\boldsymbol{X}$ that $\mathbf{e}_p, \mathbf{e}_f, \mathbf{r}_f, e_\psi, \eta, z, \mathbf{r}_a \to 0$ as $t \to \infty$, $\forall \boldsymbol{y} \in \mathcal{S}_\mathcal{D}$. This further implies that $\mathbf{e}_v \to [0, 0, -\delta]^\top$ as $t \to \infty$, consequently the velocity tracking error $m\boldsymbol{v} - \mathbf{R}^\top \dot{\boldsymbol{p}}_d \to 0$ as $t \to \infty$. This completes the proof.

References

Abdessameud, A., & Tayebi, A. (2010). Global trajectory tracking control of VTOL-UAVs without linear velocity measurements. *Automatica, 46*(6), 1053–1059.

Benallegue, A., Mokhtari, A., & Fridman, L. (2008). High-order sliding-mode observer for a quadrotor UAV. *International Journal of Robust and Nonlinear Control, 18*(4–5), 427–440.

Bertrand, S., Guénard, N., Hamel, T., Piet-Lahanier, H., & Eck, L. (2011). A hierarchical controller for miniature VTOL UAVs: Design and stability analysis using singular perturbation theory. *Control Engineering Practice, 19*(10), 1099–1108.

Cai, Z., de Queiroz, M. S., & Dawson, D. M. (2006). Robust adaptive asymptotic tracking of nonlinear systems with additive disturbance. *IEEE Transactions on Automatic Control, 51*(3), 524–529.

Castillo, P., Dzul, A., & Lozano, R. (2004). Real-time stabilization and tracking of a four-rotor mini rotorcraft. *IEEE Transactions on Control Systems Technology, 12*(4), 510–516.

Ciulkin, M., Pawluszewicz, E., Kaparin, V., & Kotta, U. (2015). Input-output linearization by dynamic output feedback on homogeneous time scales. In *20th International Conference on Methods and Models in Automation and Robotics (MMAR)*, Międzyzdroje, Poland (pp. 477–482).

Clarke, F. H. (1990). *Optimization and nonsmooth analysis*. Philadelphia: SIAM.

Cunha, R., Cabecinhas, D., & Silvestre, C. (2009). Nonlinear trajectory tracking control of a quadrotor vehicle. In *Proceedings of ECC 2009 – 10th European Control Conference*, Budapest, Hungary.

De Queiroz, M., Hu, J., Dawson, D., Burg, T., & Donepudi, S. (1997). Adaptive position/force control of robot manipulators without velocity measurements: Theory and experimentation. *IEEE Transactions on Systems, Man, and Cybernetics, 27-B*(5), 796–809.

Ding, X., Wang, X., Yu, Y., & Zha, C. (2017). Dynamics modeling and trajectory tracking control of a quadrotor unmanned aerial vehicle. *Journal of Dynamic Systems, Measurement, and Control, 139*, 1–11.

Do, K. D. (2015). Path-tracking control of stochastic quadrotor aircraft in three-dimensional space. *Journal of Dynamic Systems, Measurement, and Control, 137*(10), 1–11.

Filippov, A. (1964). Differential equations with discontinuous right-hand side. *American Mathematical Society Translations, 42*(2), 199–231.

Garcia, R., Rubio, F., & Ortega, M. (2012). Robust PID control of the quadrotor helicopter. *IFAC Proceedings Volumes, 45*(3), 229–234.

Ghommam, J., Charland, G., & Saad, M. (2015). Three-dimensional constrained tracking control via exact differentiation estimator of a quadrotor helicopter. *Asian Journal of Control, 17*(3), 1093–1103.

Huang, M., Xian, B., Diao, C., Yang, K., & Feng, Y. (2010). Adaptive tracking control of underactuated quadrotor unmanned aerial vehicles via backstepping. In *Proceedings of the IEEE American control Conference (ACC)*, Baltimore, MD, USA (pp. 2076–2081).

Ilcev, D. S., & Sibiya, S. S. (2015). Weather observation via stratospheric platform stations. In *2015 IST-Africa Conference*, Lilongwe, Malawi (pp. 1–12).

Kendoul, F. (2009). Nonlinear hierarchical flight controller for unmanned rotorcraft: Design, stability, and experiments. *Journal of Guidance Control and Dynamics, 32*(6), 1954–1958.

Kim, H. J., & Shim, D. H. (2003). A flight control system for aerial robots: Algorithms and experiments. *Control Engineering Practice, 11*(2), 1389–1400.

Lee, D. B., Burg, T. C., Xian, B., & Dawson, D. M. (2007). Output feedback tracking control of an underactuated quad-rotor UAV. In *Proceedings of the 2007 American Control Conference*, New York, USA (pp. 1775–1780).

Levant, A. (2005). Quasi-continuous high-order sliding-mode controllers. *IEEE Transactions on Automatic Control, 50*(11), 1812–1816.

Madani, T., & Benallegue, A. (2007). Sliding mode observer and backstepping control for a quadrotor unmanned aerial vehicles. In *2007 American Control Conference*, New York, USA (pp. 5887–5892).

Magsino, E. R., Chua, J. R. B., Chua, L. S., de Guzman, C. M., & Gepaya, J. V. L. (2016). A rapid screening algorithm using a quadrotor for crack detection on bridges. In *2016 IEEE Region 10 Conference (TENCON)*, Singapore (pp. 1829–1833).

Ortiz, J. P., Minchala, L. I., & Reinoso, M. J. (2016). Nonlinear robust h-infinity pid controller for the multivariable system quadrotor. *IEEE Latin America Transactions, 14*(3), 1176–1183.

Paden, B., & Sastry, S. (1987). A calculus for computing filippov's differential inclusion with application to the variable structure control of robot manipulators. *IEEE Transactions on Circuits System, 34*(1), 73–82.

Rinaldi, F., Chiesa, S., & Quagliotti, F. (2013). Linear quadratic control for quadrotors UAVs dynamics and formation flight. *Journal of Intelligent and Robotic Systems, 70*(1–4), 203–220.

Salih, A. L., Moghavvemi, M., Mohamed, H. A. F., & Gaeid, K. S. (2010). Modelling and PID controller design for a quadrotor unmanned air vehicle. In *2010 IEEE International Conference on Automation, Quality and Testing, Robotics (AQTR)*, Cluj-Napoca, Romania (Vol. 1, pp. 1–5).

Van Den Eijnden, S. (2017). *Cascade based tracking control of quadrotors*. Master's thesis, Eindhoven University of Technology.

Wang, L., & Jia, H. (2014). The trajectory tracking problem of quadrotor UAV: Global stability analysis and control design based on the cascade theory. *Asian Journal of Control, 16*(2), 574–588.

Xian, B., Dawson, D. M., de Queiroz, M. S., & Chen, J. (2004). A continuous asymptotic tracking control strategy for uncertain nonlinear systems. *IEEE Transactions on Automatic Control, 49*(7), 1206–1211.

Xian, B., & Zhang, Y. (2017). A new smooth robust control design for uncertain nonlinear systems with non-vanishing disturbances. *International Journal of Control, 89*(6), 1285–1302.

Yao, Z. (2017). Nonlinear robust adaptive hierarchical sliding mode control approach for quadrotors. *International Journal of Robust and Nonlinear Control, 27*(6), 925–941.

Zou, Y. (2017). Nonlinear hierarchical control for quad-rotors with rotation matrix. *International Journal of Control, 90*(7), 1308–1318.

Chapter 16
On the Tip–Path-Plane Flapping Angles Estimation for Small–Scale Flybarless Helicopters at Near–Hover

Mohammad K. Al–Sharman and Mamoun F. Abdel–Hafez

Abstract Observing the Tip-path-plane (TPP) flapping motion is essential for characterizing the helicopter's main rotor dynamics. The high-order harmonics of the rotor blade flapping periodic motion are considered the major sources of helicopter vibration. However, incorporating precise information of the flapping dynamics with the overall helicopter dynamic model describes better the main rotor motion. This enables designers to innovate blade designs that have less dynamic vibration. Moreover, the flapping states are crucial for analysing the main rotor axial force and moment components. However, obtaining measurements for the flapping states is not directly possible. This mandates researchers in the field to exclude the flapping dynamics, despite being essential, and use some algebraic expressions and numerical approximations instead. This chapter addresses the problem of designing a model-based estimation algorithm for the unobservable flapping angles while a near-hover flight is being carried out. A Kalman state estimation algorithm is designed to provide accurate flapping estimates for the Maxi Joker 3 flapping angles. The presented estimator has succeeded in obtaining accurate longitudinal and lateral flapping angles estimates with small root mean square error of estimation of 0.3770° and 0.2464°, respectively. Besides several simulation tests, a real outdoor near-hover flight was performed to validate the proposed estimation method.

Keywords Small-scale helicopter · Flapping angles estimation · State estimation · Near–hover dynamics

---

M. K. Al–Sharman (✉)
Department of Electrical and Computer Engineering, University of Waterloo, ON, Canada
e-mail: mkalsharman@uwaterloo.ca

M. F. Abdel–Hafez
Mechanical Engineering Department, American University of Sharjah, Sharjah, UAE
e-mail: mabdelhafez@aus.edu

N. Derbel et al. (eds.), *New Developments and Advances in Robot Control*, Studies in Systems, Decision and Control 175,
https://doi.org/10.1007/978-981-13-2212-9_16

16.1 Introduction

Unmanned miniature aerial vehicles should be in a high degree of agility and maneuverability (Al-Sharman et al. 2015, 2018a; Marantos et al. 2017). Agile helicopter platforms have been deemed candidates in a range of civil and military applications (Al-Sharman et al. 2018b; Alshoubaki et al. 2015). Nonetheless, helicopters are highly nonlinear unstable dynamic systems; their dynamic model suffers from a high number of states, subsystems cross coupling, model uncertainties and noise associated with measurements. In addition, the unmeasured flapping dynamic states makes the problem of helicopter's states estimation and control more sophisticated.

The blade flapping motion of the full-scaled helicopter's main rotor has been extensively studied (Chen 1987, 1990; Johnson 1981). A correlation between the measured and the calculated values of the lateral flapping angle for the helicopter's rotor model was presented in Johnson (1981). A discrepancy was discovered between the calculated flapping angle and the one obtained from static inflow models. This discrepancy is because of the nonuniform distribution of the induced velocity by the rotor blades vortices. However, this discrepancy was reduced by combining the high-order aerodynamic models to the existed rotor model. In Chen (1987), a thorough formulation of the flap lag equations of motion for an articulated rotor was proposed. Recently, researches have addressed the problem of observing the flap-lead-lag motion for full-scaled helicopter models (Allred et al. 2017; Zappa et al. 2016). A real-time contactless instrumentation system was designed to observe the TPP pitch flap lag motion for an articulated motor (Zappa et al. 2016). The system consists of a set of laser and other vision sensors installed on the rotor head of an Agusta Westland AW139 helicopter. While in Allred et al. (2017), a set of gyros and MEMS accelerometers was evenly placed to monitor the flap, lead-lag and pitch motion of the main rotor.

Small-scale helicopters have noticeably different flapping dynamics from that of the full-scaled helicopters. Practically, full-scaled helicopters have flapping hinges; nonetheless, these hinges do not exist in the small-scale ones. In addition, some small-scale helicopters have a mechanical stabilizing flybar attached to the rotor head to slow down the rapid dynamic response of the helicopter. Compared to the full-scaled helicopters, the dynamic response of small-scale helicopters is remarkably faster and more sensitive to disturbances (Zhu and Zuo 2017). However, a robust control and effective sensor fusion design is required to navigate them precisely (Al-Sharman 2015; Alshoubaki et al. 2015). Some research efforts have focused on minimizing the state space model of the small-scale helicopter by replacing the flapping dynamics with static algebraic approximation to reduce the complexity of the control design (Gavrilets et al. 2002; Zhu and Zuo 2017). In Gavrilets et al. (2002), a robust control design for an autonomous aerobatic flight for a small-scale helicopter was introduced. Instead of implementing a flapping angles state estimator, an approximation of the flapping angles was proposed for them to maintain the robustness of the aerobatic flight controller. In Zhu and Zuo

(2017), the flapping dynamic states were approximated by algebraic expressions, and a Bell-Hiller stabilizing bar of the small-scale helicopter was used to slow down the model dynamics and provide some damping for the rotor dynamics. However, in this study, a helicopter with no mechanical stabilizer (flybarless) is used to perform an autonomous near-hover flight. While designing near-hover controllers, the helicopter dynamic states, including the flapping states, are linearized at near hover point. These linearized states are precisely estimated and controlled by a set of linear estimators and controllers (Al-Sharman 2016; Liu et al. 2013).

It is discerned that the existing research effort in the field of helicopter's navigation and control removed the need to estimate the flapping states of small-scale helicopters by excluding their dynamics or using static algebraic expressions. However, to improve the accuracy of the helicopter's maneuverability, an accurate knowledge of each vehicle's state is needed. Therefore, an accurate estimation of the flapping angles states is required for a precise hovering performance of the miniature flybarless helicopter. The flapping angles estimation problem is considered difficult due to the fact that no sensor can be directly attached to the main rotor of the small-scale helicopter to provide the estimator with flapping angles measurements. In this case, the accuracy of the flapping angles estimation at near hover is only based on the accuracy of the open loop near-hover dynamics of the helicopter. Therefore, this chapter addresses the problem of small-scale flapping angles estimation and proposes an implementation of a model-based estimator for these angles at near hover conditions.

The present chapter is structured as follows. Section 16.2 describes the dynamic model of the small-scale helicopter. In Sect. 16.3, the flapping angles state estimation algorithm is presented. Section 16.4 illustrates the flapping angles estimation results of the simulation tests. In Sect. 16.5, the experimental validation of the proposed estimation method is illustrated. Finally, the proposed work is concluded in Sect. 16.6.

16.2 Small-Scale Flybarless Helicopter Modeling

The dynamic model of the small-scale flybarless helicopter is a highly nonlinear unstable dynamic model. This is due to the high number of states along with its subsystems cross-coupling, model uncertainties and the noise associated with measurements. In addition, the immeasurable main rotor flapping dynamic states makes the problem of helicopter's navigation and control more sophisticated. Therefore, incorporating accurate helicopter's state estimates of the unobservable flapping angles into the near-hover control design would boost the robustness of the near-hover flight performance. This study uses the nonlinear dynamic model of the flybarless Maxi Joker 3 helicopter for designing a model-based flapping angles estimators (Al-Sharman et al. 2014; Alshoubaki et al. 2015) (See Fig. 16.1).

Fig. 16.1 Maxi Joker 3 helicopter

16.2.1 Reference Frames

Two frames are commonly used to describe the helicopter motion and orientation: the inertial earth frame (EF) and the body frame (BF). Both frames are used to describe the position and attitude of the helicopter. The inertial earth frame is positioned at a stationary point on earth, where the ground station is located; whereas, the origin of the body frame is in the center of gravity of the vehicle. The body frame is vital to describe the equations of motion and characterize the aerodynamic forces and torques.

16.2.2 Rotation Matrix

The rotation matrix illustrated in Eq. (16.1) is used to transfer vectors from the body frame to the earth frame and vice versa. The orthonormal rotation matrix R_b^e is used to change the representation of the helicopter forces and moments from frame to another. The matrix is formulated by performing successive roll, pitch and yaw rotations about the x, y, and z-axes, respectively.

$$R_b^e = \begin{bmatrix} \cos\theta \sin\psi & \cos\psi \sin\theta \sin\phi - \sin\psi \cos\phi & \cos\psi \sin\theta \cos\phi + \sin\psi \sin\phi \\ \cos\theta \sin\psi & \sin\psi \sin\theta \sin\phi + \cos\psi \cos\phi & \sin\psi \sin\theta \cos\phi - \cos\psi \sin\phi \\ -\sin\theta & \cos\theta \sin\phi & \cos\theta \cos\phi \end{bmatrix} \tag{16.1}$$

where we denote by $c(.) = \cos(.)$ and $s(.) = \sin(.)$.

16.2.3 Rigid Body Equations

The helicopter rigid body equations describe the position and the attitude in the body frame of the helicopter. Equations (16.2), (16.3) and (16.4) represent the rotational dynamics of the helicopter while Eqs. (16.5), (16.6) and (16.7) represent the helicopter's translational dynamics. The body rates around the X_b, Y_b and Z_b axes are p, q and r as shown in Fig. 16.2, respectively. Equations (16.2), (16.3), (16.4), (16.5), (16.6) and (16.7) represent rigid body equations of the helicopter (Alshoubaki et al. 2015):

$$\dot{p} = \frac{L}{J_x} - \frac{J_z - J_y}{J_x} qr \tag{16.2}$$

$$\dot{q} = \frac{M}{J_y} - \frac{J_x - J_z}{J_y} pr \tag{16.3}$$

$$\dot{r} = \frac{N}{J_z} - \frac{J_y - J_x}{J_z} pq \tag{16.4}$$

$$\dot{u} = rv - qw + \frac{F_{Xb}}{m} \tag{16.5}$$

$$\dot{v} = -ru + pw + \frac{F_{Yb}}{m} \tag{16.6}$$

$$\dot{w} = qu - pv + \frac{F_{zb}}{m} \tag{16.7}$$

where: p, q and r denote the angular rates around the X_B, Y_B and Z_B axes, respectively; u, v and w denote the velocities in the body frame along the X_B, Y_B and Z_B axes, respectively; J_X, J_Y and J_Z are the moments of inertia about the X_B, Y_B and Z_B axes, respectively; m is the mass of the helicopter; F_{Xb}, F_{Yb} and F_{Zb} define the force components in the body frame along the X_B, Y_B and Z_B axes, respectively. The body moments about the X_B, Y_B and Z_B axes are denoted by L, M and N.

16.2.4 Euler Angles Dynamics

Euler angles characterize the orientation of the helicopter with respect to earth frame. These angles are formulated based on the 6 DOF vehicle model. Equations (16.11), (16.12) and (16.13) define the Euler angles as:

$$\dot{\phi} = p + \big(q \sin\phi + r \cos\phi\big) \tan\theta \tag{16.8}$$

$$\dot{\theta} = q \cos\theta - r \sin\phi \tag{16.9}$$

$$\dot{\psi} = \frac{1}{\cos\phi} (p \sin\phi + r \cos\phi) \tag{16.10}$$

16.2.5 Main Rotor Thrust Model

The thrust generated by the helicopter rotor is needed to sustain the vehicle in the air. This thrust is governed by the collective command as given in Eq. (16.11) below (Leishman 2000; Padfield 1996):

$$T_{MR} = \left[\theta_0\left(\frac{1}{3} + \frac{\mu_{ar}^2}{2}\right) - \left(\frac{\mu_z + \lambda_i}{2}\right)\right]\frac{as}{2}\rho(\Omega R_{mr})^2 A_d \tag{16.11}$$

where: μ_{ar} represents the advanced ratio, which is the ratio between the rotor tip speed and the helicopter translational velocity; λ_i is the rotor inflow; μ_z defines the ratio of vertical velocity to the rotor tip speed; a and s denote the rotor lift curve slope and the rotor solidity, respectively. R_{mr} is the radius of the main rotor, and A_d is the rotor disk area.

16.2.6 Main Rotor Flapping Dynamics Model

The problem of modeling the flapping angles has been investigated in numerous published works (Perdomo and Wei 2017; Taamallah 2011). In Taamallah (2011), the blade pitch-lag-flap angles of a small-scale flybarless helicopter rotor were expanded using the Fourier series as shown in Eq. (16.12). In Perdomo and Wei (2017), a numerical solution was proposed for computing the high order coefficients of equation (16.12) in order to estimate the high order harmonics of the blade flapping periodic motion. Estimating these harmonics offers better understanding of innovating blade designs that have lesser rotor dynamic vibration.

$$\beta_{bl}(\varphi_{bl}) = \beta_0 + \beta_{1c}\cos\varphi_{bl} + \beta_{1s}\sin\varphi_{bl} \tag{16.12}$$

where: φ_{bl} is defined by the blade's azimuthal angular position, β_0 describes the TPP coning angle of the rotor, and β_{1c} describes the TPP pitch angle with respect to the X-axis of the body frame. Whereas, β_{1s} describes the TPP roll angle with respect to the Y-axis.

The flapping dynamic model of the small-scale helicopter is presented in Gavrilets et al. (2001). As shown in Eq. (16.13), the flapping angles commands, β_{1c}^c and β_{1s}^c , are proportionate to the longitudinal and latitudinal input commands, respectively.

Through the system identification process (Gavrilets et al. 2001), the main rotor time constants τ_{fc} and τ_{fs} are determined. At near-hover, τ_{fc}, τ_{fs} , $A_{\delta_{lon}}$ and $B_{\delta_{lat}}$ have the values of 0.113 s, 0.101 s, 4.2 and 4.2, respectively.

$$\begin{bmatrix}\dot\beta_{1c}\\ \dot\beta_{1s}\end{bmatrix} = \begin{bmatrix}-\frac{1}{\tau_{fc}} & 0\\ 0 & -\frac{1}{\tau_{fs}}\end{bmatrix}\begin{bmatrix}\beta_{1c}\\ \beta_{1s}\end{bmatrix} + \begin{bmatrix}0 & -1 & \frac{A_{\delta_{lon}}}{\tau_{fc}} & 0\\ -1 & 0 & 0 & \frac{B_{\delta_{lat}}}{\tau_{fs}}\end{bmatrix}\begin{pmatrix}p\\ q\\ \beta_{1c}^c\\ \beta_{1s}^c\end{pmatrix} \tag{16.13}$$

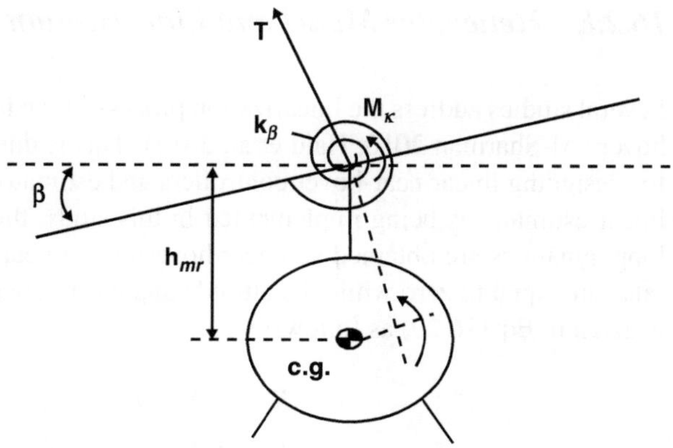

Fig. 16.2 Rotor lateral flapping moment acting on the fuselage (Gavrilets 2003)

16.2.7 Rotor Forces and Moments

The vector of the force acting on the main rotor of the helicopter consists of three components along the axes X, Y and Z. Assuming that the generated main rotor thrust is normal to the rotor disc, the main rotor forces are expressed as a function of the flapping angles and the main rotor thrust. These forces are described in Eqs. (16.14), (16.15) and (16.16) as follows:

$$f_{x,MR} = -T_{MR} \sin \beta_{1c} \cos \beta_{1c} \tag{16.14}$$

$$f_{y,MR} = T_{MR} \sin \beta_{1s} \cos \beta_{1s} \tag{16.15}$$

$$f_{z,MR} = -T_{MR} \cos \beta_{1s} \cos \beta_{1c} \tag{16.16}$$

The rotor flapping motion produces the dominant moments of the main rotor. Figure 16.2 displays the generated moment from the lateral TPP flapping angle. A linear torsional spring with K_β stiffness coefficient is used to express the restraint in the blade attachment (Gavrilets 2003).

The rotor has three axial components of moments that are described in the bodyframe. These components are given in Eqs. (16.17), (16.18) and (16.19) as follows:

$$L_{MR} = f_{y,MR} h_{mr} - f_{z,MR} y_{mr} + K_\beta \beta_{1s} - L_b \tau_{fs} p \tag{16.17}$$

$$M_{MR} = f_{x,MR} h_{mr} - f_{z,MR} y_{mr} + K_\beta \beta_{1c} - M_a \tau_{fc} q \tag{16.18}$$

$$N_{MR} = -\rho(\Omega R)^2 \frac{bcR^2}{s} C_Q^{MR} \tag{16.19}$$

where: h_{mr} is the distance in the z-axis from CG to the main rotor. l_{mr} is the distance in the x-axis from CG to the main rotor. y_{mr} is the y-axis distance from CG to the main rotor.

16.2.8 Helicopter Model and Linearization

Several studies address the linearization process of the helicopter's dynamics at near hover (Al-Sharman 2016; Hald et al. 2015). This is due to its importance and need for designing linear near-hover controllers and estimators (Al-Sharman 2016). As a linear estimator is being implemented in this study, the linearized helicopter open loop dynamics are obtained at a near hovering. At near hovering point, the attitude rates are equal to zero while the attitude angles are constant. The linearized system is given in Eq. (16.20) as follows:

$$\dot{X} = AX + BU + w \qquad (16.20)$$

$$Y = CX + DU + v$$

where X, U and Y represent the state space vector, the system's input vector and the output vector, respectively. The state-space matrices are defined in the A, B, C and D matrices. w and v are the dynamics and the measurement Gaussian white noise processes, respectively. As shown in Eq. (16.21), the helicopter's state vector is reduced to have the attitude, attitude rates and the main rotor flapping angles.

$$X = \left[\phi\ \theta\ \psi\ \dot{\phi}\ \dot{\theta}\ \dot{\psi}\ \beta_{1c}\ \beta_{1s}\right]^T \qquad (16.21)$$

The input vector is given as:

$$U = \begin{bmatrix} u_{lon} \\ u_{lat} \\ u_{col} \\ u_{ped} \end{bmatrix} \qquad (16.22)$$

where: u_{lon} is the longitudinal stick input u_{lat} is the lateral stick input, the collective lever input is denoted by u_{col}, and u_{ped} is the rudder pedal input. The dynamics matrix, A, is computed to be:

$$A = \begin{bmatrix} 0\,0\,0\,1\,0\,0 & 0 & 0 \\ 0\,0\,0\,0\,1\,0 & 0 & 0 \\ 0\,0\,0\,0\,0\,1 & 0 & 0 \\ 0\,0\,0\,0\,0\,0 & -36.63 & \frac{K_\beta + \frac{\pi + K_m h_{mr}}{86}}{J_x} \\ 0\,0\,0\,0\,0\,0 & \frac{K_\beta + \frac{\pi + K_m h_{mr}}{86}}{J_y} & 14.29 \\ 0\,0\,0\,0\,0\,0 & 0 & 0 \\ 0\,0\,0\,0\,0\,0 & -T_a & 0 \\ 0\,0\,0\,0\,0\,0 & 0 & -T_b \end{bmatrix} \qquad (16.23)$$

The input matrix B is shown in Eq. (16.24). The C matrix is a 6 × 8 unity diagonal matrix while the D matrix is a 6 × 4 zero matrix. The Maxi Joker 3 parameters, including the parameters of Eqs. (16.23) and (16.24), can be found in Al-Sharman (2016).

$$B = \begin{pmatrix} 0 & 0 & 0 & 0 \\ 0 & 0 & 0 & 0 \\ 0 & 0 & 0 & 0 \\ 0 & 0 & 0 & \frac{K_t l_{tr}}{J_x} \\ 0 & 0 & 0 & 0 \\ 0 & 0 & 0 & \frac{K_t l_{tr}}{J_z} \\ T_a & 0 & 0 & 0 \\ 0 & T_b & 0 & 0 \end{pmatrix} \quad (16.24)$$

16.3 Flapping Angles Estimation at Near–Hover

Precise state estimation technique is needed to ensure noise reduction and accurate navigation (Abdel-Hafez and Speyer 2008; Abdel-Hafez 2010). This section proposes the flapping angles state estimator design under near-hover conditions using a Kalman filter. The Kalman state estimation cycle is described, and the simulation results are presented. The results demonstrate accurate flapping states estimation with small RMSEE values.

16.3.1 Kalman Filter Design

Kalman filter is deemed an optimal minimum mean square error state estimator. Therefore, this study uses the Kalman state estimation algorithm to obtain optimal estimates for the TPP flapping angles. This linear model-based estimation algorithm uses the linearized near-hover dynamic model described in Sect. 16.2. It relies on the accurate open loop dynamics of the system along with the measurements of the onboard sensors to estimate the helicopter's states (Al-Sharman 2016). The Kalman state estimation cycle is described in the set of Eq. (16.25), (16.26), (16.27), (16.28), (16.29), (16.30) and (16.31) below:

1. State and measurement prediction

$$\hat{x}_{k+1|K} = A_K \hat{x}_{k|K} + B_K U_k \quad (16.25)$$

$$P_{K+1|k} = A_K P_{K|k} A_K^{\top} + Q_k \quad (16.26)$$

2. State update

$$v_{K+1} = z_{K+1} - H_{K+1}\hat{x}_{k+1|K} \tag{16.27}$$

$$S_{K+1} = R_{K+1} + H_{K+1}P_{K+1|k}H_{K+1}^{\top} \tag{16.28}$$

$$W_{K+1} = P_{K+1|k}H_{K+1}^{\top}S_{K+1}^{-1} \tag{16.29}$$

$$\hat{x}_{k+1|K+1} = \hat{x}_{k+1|K} + W_{K+1}v_{K+1} \tag{16.30}$$

$$P_{K+1|K+1} = P_{K+1|k} - W_{K+1}S_{K+1}W_{K+1}^{\top} \tag{16.31}$$

By assuming the prior knowledge of the matrices A_K, B_K, H_K, R_K and Q_K noise matrices, the Kalman filter begins its recursive state estimation cycle. Equation (16.28) represents the innovation sequence which provides the Kalman filter with the additional information as new measurement becomes available. Equations (16.28), (16.29), (16.30) and (16.31) incorporate the measurement updates into a priori estimate to enhance the posterior estimate and covariance.

The variances of the diagonal measurement noise covariance matrix and process noise covariance matrix are precisely chosen to avoid divergence of the estimator. As no measurements are available for the flapping angles, R_K contains only the attitude and the body rates variances.

16.4 Simulation Results

This section presents the simulation results of the proposed flapping angles state estimation method. Since there are no measurements for the flapping angles, their estimation is still challenging unless an accurate dynamic helicopter model is identified. The IMU measurements of body angles and the body rates were obtained by adding some measurement white noise sequences to the truth signals of the Maxi Joker 3 helicopter dynamic states. A near-hover simulated test was conducted to demonstrate the robustness of the proposed estimator. As shown in Figs. 16.3, 16.4 and 16.5, the Maxi Joker 3 is controlled to change its orientation by small degrees to maintain the near-hover dynamics. It can be seen that the added noise has been filtered out, and the attitude states are accurately estimated. The attitude estimation problem for the flybarless is thoroughly investigated in Al-Sharman (2016).

Results illustrated in Figs. 16.6, 16.7 and 16.8 exhibit the TPP flapping angles estimation of the Maxi Joker 3. Figure 16.6 shows the real and estimated flapping angles. The significant coincidence between the actual and the estimated flapping angles can be seen. Figure 16.7 shows the estimation result of the longitudinal flapping angle. The estimated longitudinal flapping angle coincides well with the truth angle. It is estimated with RMSEE of 0.3770°. The latitudinal angle is also estimated accurately as shown in Fig. 16.8. The proposed estimator has succeeded in estimating the latitudinal flapping angle with RMSEE of 0.2464°. The flapping angles estimation errors are illustrated in Fig. 16.9. It is obvious that the estimation algorithm for both angles has small value of errors.

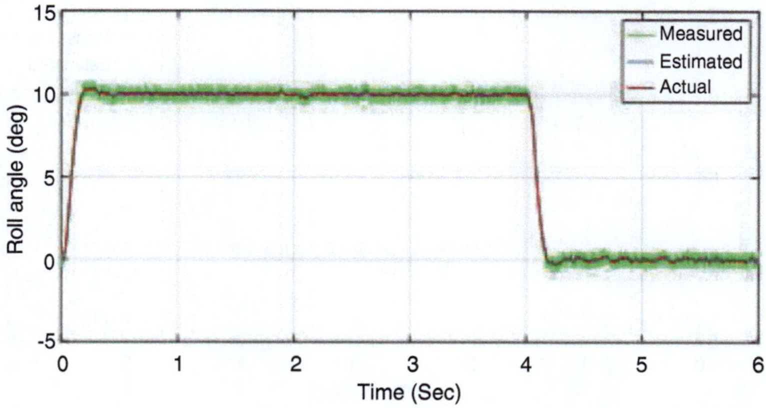

Fig. 16.3 Roll angle response at near-hover

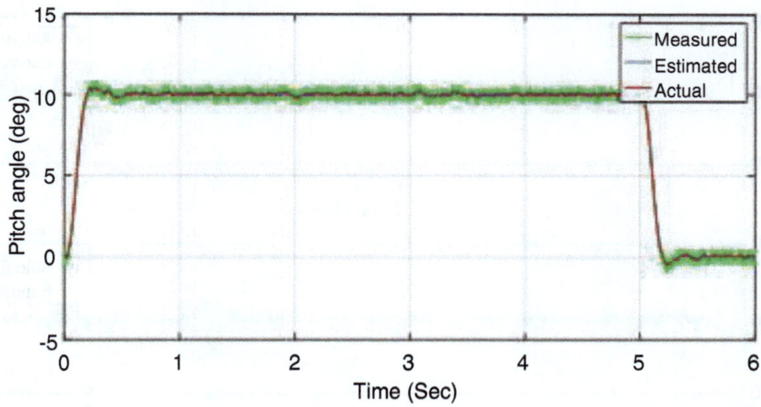

Fig. 16.4 Pitch angle response at near-hover

16.5 Experimental Validation

The proposed estimation algorithm is experimentally validated by conducting an actual near-hover flight. In real flight scenarios, the estimation error cannot be accurately calculated because the truth values of the helicopter states are not available. Nonetheless, in a near-hover experiment, the real values of the helicopter states are expected to have approximately known values. A Mazari autopilot was used to conduct the outdoor near-hover flight of the Maxi Joker 3. As shown in Fig. 16.10, the Mazari autopilot uses the MPC555 32 bits microcontroller hardware to sample the measurements of the MIDG sensor at 50 Hz. The specifications of measurements of MIDG are illustrated in Table 16.1.

The attitude response of the Maxi Joker 3 helicopter during the actual near-hover is illustrated in Fig. 16.11. The helicopter was controlled to perform short rolling and pitching to maintain the near hover dynamics.

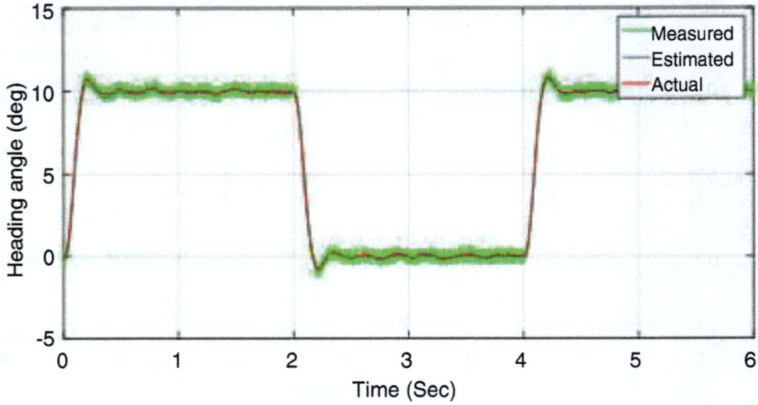

Fig. 16.5 Heading angle response at near-hover

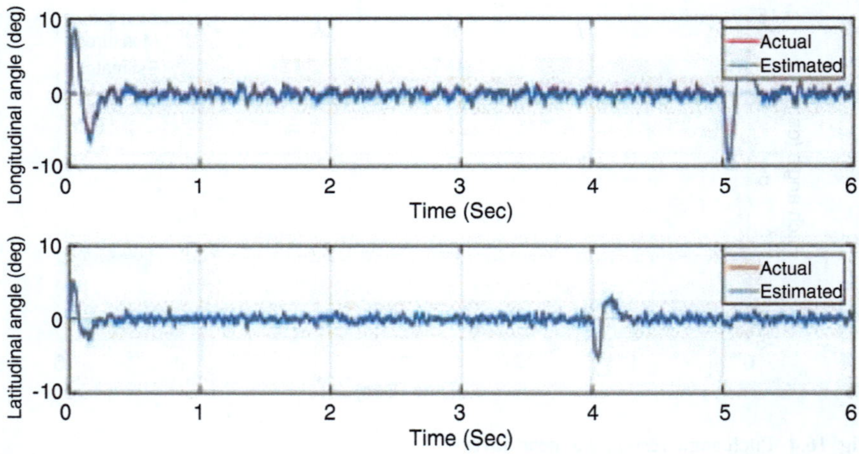

Fig. 16.6 Flapping angles estimation

It is expected that the flapping angles are tilting with small degrees while near-hover flight is being performed. Figures 16.12 and 16.13 display the performance of the flapping angles estimation algorithm at near-hover. Although the flapping states are unobservable, it can be seen that the estimated values coincide with the expected values at near-hover.

16.6 Chapter Summary

Accurate information of the small-scale helicopter's flapping states is needed for characterizing the main rotor dynamics. However, these angles cannot be directly measured by attaching sensors to the small rotor head to measure the main rotor flapping motion. Therefore, an estimation algorithm has to be used to estimate

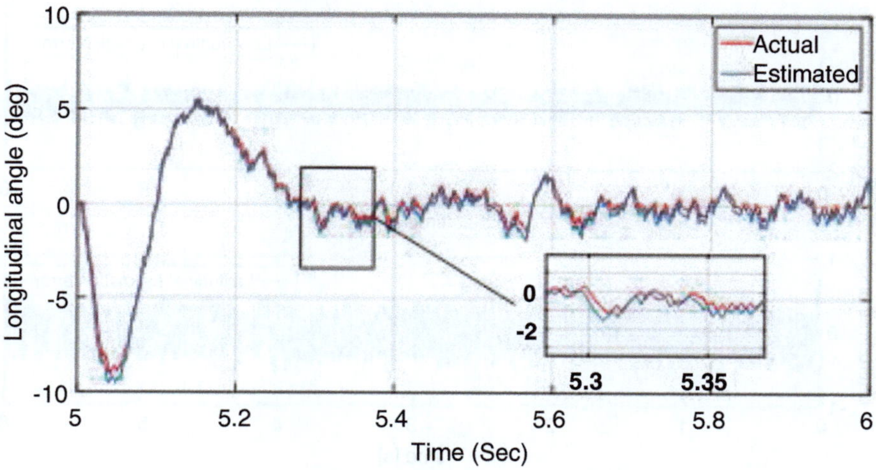

Fig. 16.7 Flapping longitudinal angle estimation at near hover-flight

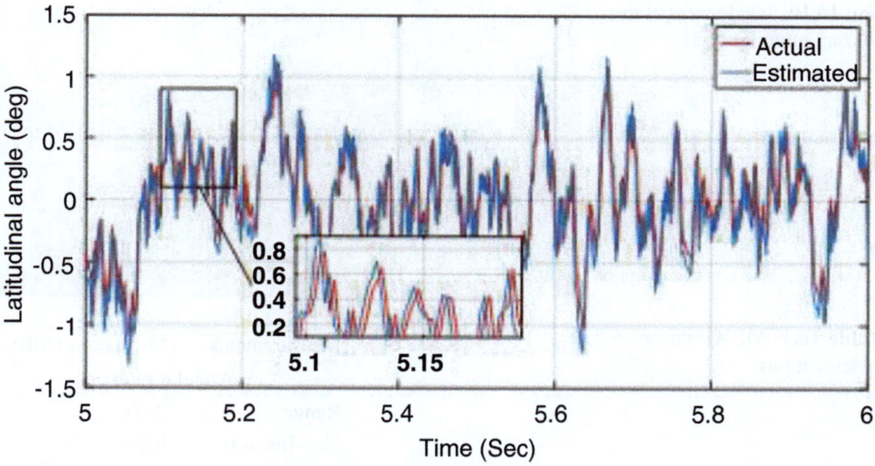

Fig. 16.8 Flapping latitudinal angle estimation at near-hover flight

these states. The existing research efforts in the field of helicopter's modeling and control has focused on reducing the vehicle model to exclude the flapping dynamics or replace them with algebraic expressions and numerical approximations. This chapter presents a model-based estimation algorithm for the unobservable flapping angles while a near-hover flight is being carried out. A Kalman state estimation algorithm is designed to provide accurate flapping estimates for the Maxi Joker 3. The proposed estimation technique was used to estimate the flapping angles while simulated and real near-hover flights are carried out. The results prove the accurate flapping estimation with small RMSEE values at near-hover conditions. Finally, despite the essentiality of the flapping states in describing the rotor orientation,

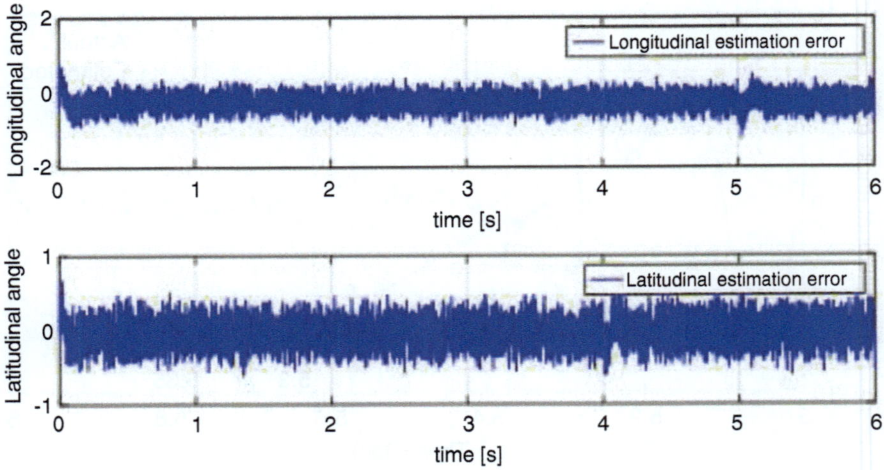

Fig. 16.9 Flapping angles estimation errors at near-hover

Fig. 16.10 Inside view of the Mazari autopilot box

Table 16.1 MIDG sensor specifications

Measurements	Mean error (ME)
Angular rates	
Range	$\pm 300°/s$
Non-linearity	0.1% of FS
Noise density	$0.05°/s/\sqrt{\text{Hz}}$
3 dB Bandwidth	20 Hz
Acceleration	
Range	± 6 g
Non-linearity	0.3% of FS
Noise density	$150\,\mu g/\sqrt{Hz}$
3 dB bandwidth	20 Hz
Attitude accuracy	
$0.4°(1\sigma)$	
Position accuracy	
2 m CEP, WAAS/EGNOS	

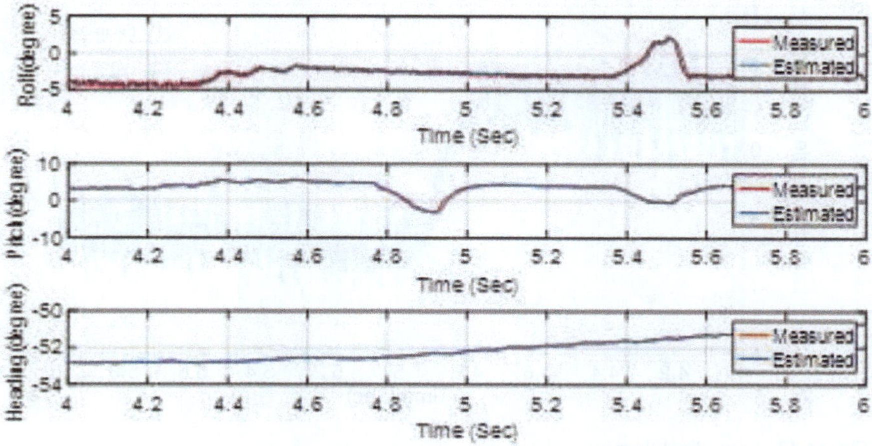

Fig. 16.11 Maxi Joker 3 attitude response at near-hover

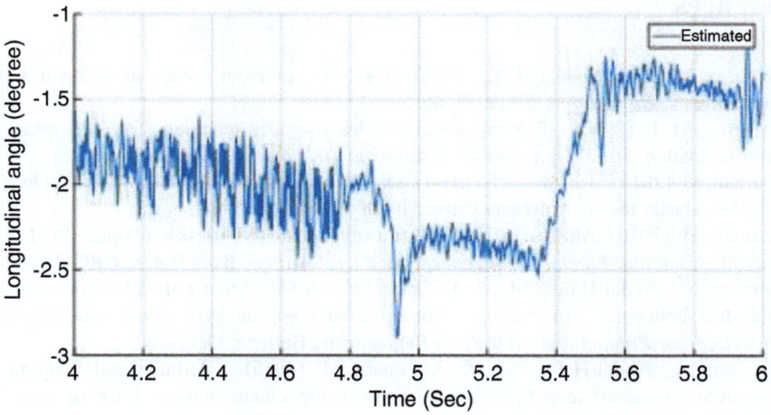

Fig. 16.12 Longitudinal flapping estimation

estimating them in a stochastic framework has not been yet studied sufficiently for small-scale platforms. As a future work of this research, intelligent state estimation methods will be implemented to improve the accuracy of the flapping angles state estimation. Another future research study could be in estimating the main and tail rotors forces based on the estimated flapping angles. This would help in designing a fault tolerant control design of the tail rotor at near-hover conditions.

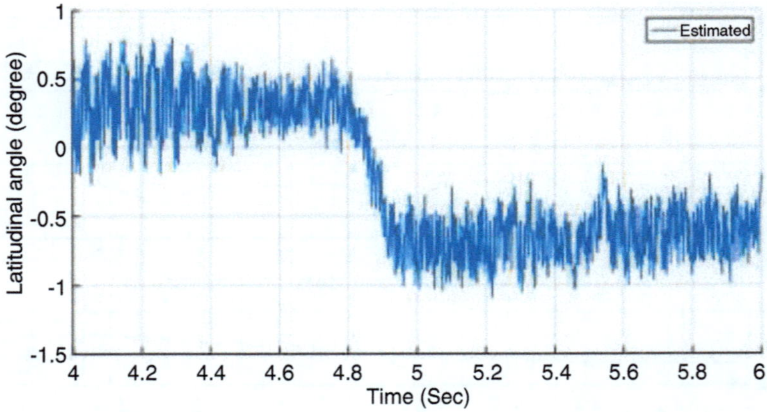

Fig. 16.13 Latitudinal flapping estimation

References

Abdel-Hafez, M. F., & Speyer, J. L. (2008). GPS measurement noise estimation in non ideal environments. *Navigation* , *55*(1), 55–66.

Abdel-Hafez, M. F. (2010). The autocovariance least-squares technique for GPS measurement noise estimation. *IEEE Transactions on Vehicular Technology, 59*(2), 574–588.

Al-Sharman, M. (2015). *Auto takeoff and precision landing using integrated GPS/INS/Optical flow solution*. Master thesis, American University of Sharjah, Sharjah.

Al-Sharman, M. (2016). Attitude estimation for a small-scale flybarless helicopter. In *Multisensor attitude estimation: Fundamental concepts and applications*. Boca Raton: CRC Press.

Al-Sharman, M., Abdel-Hafez, M., & Al-Omari, M. (2014). State estimation for a small scale flybar-less helicopter. In *2nd International Conference on System-Integrated Intelligence: Challenges for Product and Production Engineering*, Bremen, Germany.

Al-Sharman, M., Abdel-Hafez, M., & Al-Omari, M. (2015). Attitude and flapping angles estimation for a small-scale flybarless helicopter using a Kalman filter. *IEEE sensors Journal, 15*(4), 2114–2122.

Al-Sharman, M. K., Emran, B. J. M., Jaradat, A., Najjaran, H., Al-Husari, R., & Zweiri, Y. (2018a). Precision landing using an adaptive fuzzy multi-sensor data fusion architecture. *Applied soft computing, 69*, 149–164.

Al-Sharman, M., Al-Jarrah, M., & Abdel-Hafez, M. (2018b). Auto takeoff and precision terminal-phase landing using an experimental optical flow model for GPS/INS enhancement. *ASCE-ASME Journal of Risk and Uncertainty in Engineering Systems, Part B: Mechanical Engineering, 5*(1), pp. 011001-1–011001-17.

Allred, C. J., Churchill, D., & Buckner, G. D. (2017). Real-time estimation of helicopter rotor blade kinematics through measurement of rotation induced acceleration. *Mechanical Systems and Signal Processing, 91*, 183–197.

Alshoubaki, A., Al-Sharman, M., & Al-Jarrah, M. A. (2015). Flybarless helicopter autopilot rapid prototyping using hardware in the loop simulation. In *10th International Symposium on Mechatronics and its Applications (ISMA)*, Sharjah, UAE.

Chen, R. T. (1987). *Flap-lag equations of motion of rigid, articulated rotor blade with three hinge sequences*. Washington, DC: NASA.

Chen, R. T. N. (1990). A survey of nonuniform inflow models for rotorcraft flight dynamics and control applications. *VERTICA, 2*(14), 147–184.

Gavrilets, V. (2003). *Autonomous aerobatic maneuvering of miniature.* Ph.D. thesis, Massacheutes Institute of Technology.
Gavrilets, V., Mettler, B., & Feron, E. (2001). *Nonlinear model for a smallsize acrobatic helicopter.* In *Proceedings of AIAA Guidance, Navigation, and Control Conference*, Montreal, QC, Canada, 2001, Paper AIAA-2001-4333.
Gavrilets, V., Martinos, I., Mettler, B., & Feron, E. (2002). Control logic for automated aerobatic flight of miniature helicopter. In *AIAA Guidance, Navigation and Control Conference*, Monterey, CA, USA 2002.
Hald, U., Hesselbaek, M., Holmgaard, J. T., Jensen, C. S., Jakobsen, S. L., & Siegumfeldt, M. *Autonomous helicopter-modelling and control.* Aalborg: Aalborg University.
Johnson, W. (1981). Comparison of calculated and measured helicopter rotor lateral flapping angles. *Journal of the American Helicopter Society, 26*(2), 46–50.
Leishman, J. G. (2000). *Principles of helicopter aerodynamics.* Cambridge: Cambridge University Press.
Liu, H., Lu, G., & Zhong, Y. (2013). Robust LQR attitude control of a 3-DOF laboratory helicopter for aggressive maneuvers. *IEEE Transactions on Industrial Electronics, 60*(10), 4627–4636.
Marantos, P., Bechlioulis, C. P., & Kyriakopoulos, K. J. (2017). Robust trajectory tracking control for small-scale unmanned helicopters with model Uncertainties. *IEEE Transactions on Control Systems Technology, 25*(6), 2010–2021.
Padfield, G. D. (1996). *Helicopter flight dynamics: The theory and application of flying qualities and simulation modeling* (AIAA education series). AIAA, Washington, DC.
Perdomo, O., & Wei, F.-S. (2017). On the flapping motion of a helicopter blade. *Applied Mathematical Modelling, 46*, 299–311.
Taamallah, S. (2011). Small-scale helicopter blade flap-lag equations of motion for a flybarless pitch-lag-flap main rotor. In *AIAA Modeling and Simulation Technologies Conference*, Portland.
Zappa, E., Trainelli, L., Liu, R., Rolando, A., Cordisco, P., Vigoni, E., & Redaelli, M. (2016). Real time contactless sensor for helicopter blade angle measurement. In *2016 IEEE Metrology for Aerospace (MetroAeroSpace)*, Florence, Italy.
Zhu, B., & Zuo, Z. (2017). Approximate analysis for main rotor flapping dynamics of a model-scaled helicopter with Bell-Hiller stabilizing bar in hovering and vertical. *Nonlinear Dynamics, 85*(3), 1705–1717.

MIX
Papier aus verantwortungsvollen Quellen
Paper from responsible sources
FSC® C105338

If you have any concerns about our products,
you can contact us on
ProductSafety@springernature.com

In case Publisher is established outside the EU,
the EU authorized representative is:
Springer Nature Customer Service Center GmbH
Europaplatz 3, 69115 Heidelberg, Germany

Printed by Libri Plureos GmbH
in Hamburg, Germany